SCIENCE AND SOCIETY IN INDIA

DEPARTMENT OF HISTORY, UNIVERSITY OF CALCUTTA MONOGRAPH 29

SCIENCE AND SOCIETY IN INDIA
1750–2000

Edited by

ARUN BANDOPADHYAY

MANOHAR
2010

First published 2010

ISBN 978-81-7304-854-8

Published by

Ajay Kumar Jain *for*
Manohar Publishers & Distributors
4753/23 Ansari Road, Daryaganj
New Delhi 110 002

Typeset at

Digigrafics
New Delhi 110 049

Printed at

Salasar Imaging Systems
Delhi 110 035

To
Professor B.B. Chaudhuri,
my revered teacher

Contents

Preface

Academic developments, like many other things in real life, go on intermittently with their more dramatic and sluggish turns. Way back in 2005, in response to overtures of two of my colleagues, I organized a national seminar at the Department of History, University of Calcutta, on Science and Society in Modern India. The seminar, attended by a considerable number of historians and scientists from all over the country, was extremely successful: proposals were made to make the study and research on the history of science in India a more meaningful part of the university curricula in history and social sciences. I am extremely grateful to a number of distinguished academicians for making the inaugural session lively and scholarly. Professor Bharati Ray chaired the session, while Professor Asis Kumar Banerjee, then Vice-Chancellor, University of Calcutta, inaugurated it and Professor Suranjan Das, then Pro-Vice-Chancellor for Academic Affairs and later Vice-Chancellor, University of Calcutta, spoke as a Guest of Honour. The keynote address was delivered by Professor Deepak Kumar of Jawaharlal Nehru University, who called for breaking the barriers of the mainstream and peripheries of history in the studies of what he stated as HISTEM. The seminar went on for two days and ended in a great euphoria for the future of history of science in India. Suddenly, the contributors became members of a tiny club eager to do a lot of things for the social history of science.

Subsequent things, however, shaped slowly. The immediate likely offshoot, as we planned, was an edited volume. The revised drafts of papers came slowly and not always steadily. One paper was invited after the seminar. The editor was easily distracted with a medley of works. The net result was that the publication was delayed. Two of my younger colleagues who originally gave me the idea of organizing the seminar continued to press for the publication. They are Dr Madhumita Majumdar and Dr Raj

Sekhar Basu and I am obliged to them. My other colleagues who encouraged me at different stages of preparation include Professor Bhaskar Chakravarti, Professor H.S. Vasudevan, Professor Suparna Gooptu, Professor Nirban Basu, Professor Hasi Banerjee, Dr Sanjukta Dasgupta and Ms Shireen Maswood and of course the late Professor Basudeb Chattopadhyay. I am especially grateful to Mr Ramesh Jain of Manohar for taking a keen interest in the publication of the volume. My obligations to my wife Uma and son Manasij are deep and abiding but they are, as usual, more personally reserved.

There are sceptics for the entire venture. Is it going to be a history of science or a social history of science in India? Whatever it may be, what is its place in and relations with the mainstream of history? Are the boundaries of the so-called mainstream and peripheries of history permanent as far as the history of science is concerned? Sceptics have given us more occasions of and opportunities for thought. They are our great friends in this sense.

The publication is the delayed result of our initial enthusiasm as well as a response to the critical appreciation from many quarters for the studies in history of science in India. The otherwise mixed background of this edited volume should not be forgotten.

Kolkata ARUN BANDOPADHYAY
7 August 2009

CHAPTER 1

Introduction

ARUN BANDOPADHYAY

Certainly I am convinced that the history of science is of greatest importance. I even think that one does not know a science completely as long as one does not know its history. But such a study must be considered as entirely separate from the dogmatic study of science, without which the history would be unintelligible.

AUGUSTE COMTE[1]

We are trying to graft the science of the West on to an Eastern stem. . . . We have raised entire sections of the community from torpor to life, and have lifted India on to a higher moral plane. . . . In proportion as we teach the masses, so we shall make their lot happier, and in proportion as they are happier so they will become more useful members of the body politic.

LORD GEORGE CURZON[2]

The history of science such as I understand and try to practice it aims at grasping the evolution of this thought in the movement of its creativity itself. Towards this end, it is essential to place the (scientific) works in their intellectual and spiritual milieu, and interpret them with respect to the mental habits, preferences and aversions of their authors.

ALEXANDRE KOYRE[3]

. . . from the anthropological approach advocated by Koyre, the history of science became an autonomous discipline in relation to the philosophy of science and was animated by a critical approach sensitive to the behaviours and types of thought of scientific communities as well as to the problems linked to institutions.

PIETRO REDONDI ON KOYRE[4]

The objective component of scientific knowledge and its technological products must be recognized as very large; indeed,

> its power to create concepts and machines that can be shared by any large linguistic community lies near the base of its influence on social structures and practices of the modern world.
>
> SCOTT L. MONTGOMERY[5]

During the last two hundred years the study of the history of science has undergone critical changes before its tilt towards the social dimensions of the making of science. The epigraphs I quote indicate certain moments of these changes. Indeed, a problem remains. The development of the history of science in the last fifty years has been marked by a proliferation of methods and perspectives rather than by the emergence of a consensus. Helge Kragh has identified at least six grounds of justification for the study of history of science but is fully satisfied with none.[6] It is often stated, especially by the scientist historians of science, that the subject, when properly studied, may have a beneficial influence on science itself. A slightly different version of the argument is the assertion that the history of science should function as an analytical instrument for the critical evaluation of methods and concepts that appear in modern science. But none of these arguments give credence to the view that the history of science has a direct relevance to science. The relationship is no better than of a sporadic influence, like the one done by literature or religion.

Second, it has also been argued that the history of science has three separate aims: (a) it provides materials for a critical self-examination of science, (b) it bridges the gap between science and the humanities, and (c) it helps us to know the genesis of scientific theories. It is stated that the rise in interest in the history of science in the West was caused largely by the science-based technology of war. It is true that the history of science can be used, and ought to be used, to analyse the interplay between science, technology, and society. But empirical evidence that science always results in technology is weak; it is also not true that science and technology invariably contribute to people's welfare and safety.

Third, the history of science is stated to function as a background to related fields such as the philosophy and sociology of science. The links between the philosophy of science and the history of science have become steadily stronger in recent years, but it is a complex relationship, far from fully settled. The same is true about the relationship of these fields with the theory of science. Some scholars prefer to regard history of science as one element in a composite research programme, *the science of science*, which includes all studies relating to science in an interdisciplinary way. However, G. Krober has modified this argument by giving the history of science a more independent standing: 'It would be completely erroneous to talk of a simple incorporation of history of science writing under science of science or to imagine that it should be relegated to being a servant of science of science.'[7]

Fourth, the history of science may play a significant didactic role in demonstrating the true nature of scientific knowledge. Here the use of history is justified by its capacity to present the sciences in a 'softer' way to make them more attractive to a larger audience. The history of science can contribute towards a less dogmatic conception of science and scientific methods, thus acting as an antidote to orthodoxy and uncritical enthusiasm for science. But this argument cannot be stretched too far. For the history of science can also be used to support dogmas and strengthen scientific authority. In short, generally speaking, the question of didactic importance is problematic.

Fifth, the history of science is often stated as to reflect the humanist placement of science. G. Sarton reminded us of the connections and shared roots of science and the humanities in the 1930s and 1940s,[8] and C.P. Snow wanted to bridge the schism between two cultures in the 1950s and 1960s.[9] G.T. Clark went so far as to declare that 'the history of *science* is, in fact and in deed, the *new* humanism for our contemporary, irreversibly technological, and the present moment, beleaguered culture'.[10] However, such arguments should not be used to silence the critics of contemporary science. Ironically enough, it has even been stated that the more relevant question is why the greater part of science today is no longer regarded as an expression of humanist endeavour.

Finally, there is a view that the history of science does not require justification. As an important factor in general social, economic, and cultural development, science attracts historical attention in the same way as, for example, do religion and economic conditions. The history of science has no particular aim apart from that of revealing the past. As I.B. Cohen predicted in 1961, 'Not far off is the time when historians of science will be so numerous that they may produce scholarly works which need satisfy only the members of their own profession, the only requirement being that of high standards.'[11] In the process of the development of this history, occasions arise where the scientist studies nature and the historian studies the scientist.

Like the proverbial fable of the four blind men and the elephant, each of these justifications forms a significant part of the aims of the history of science, but none of them binds the discipline as a whole. Historians try to cope with infinity, integrating their subjects with the subjects and methods of other related disciplines. New perspectives, stimulated by social and economic history in particular, have been encompassed in the history of science. Nations, firms, political agencies, institutes, individuals, and scientific societies are being studied by a growing number of historians. Indeed, whatever the object of study, history of science is regarded by its practitioners increasingly as a field of history rather than of science. The other lesson from the history of science is that the positivist belief in a value-free, culturally independent science is largely a myth.

Given this general background, the Department of History, University of Calcutta, made a special effort to assemble a group of scholars at a National Seminar in 2005 to talk about 250 years of the development of 'science' at the Indian context, covering the late pre-colonial, colonial, and post-colonial periods. The contributions revolve around a few thematic contexts: the east-west encounter, the diffusion theory, the colonial impact, the nationalist and the post-colonial response, and globalization. Three specific periodic contexts may be identified around them: from the 1750s to the 1830s, from the 1840s to the 1940s and from the 1950s to 2000. This volume is a selective, revised, and enlarged version of the original presentations.

The east-west encounter is a long-drawn-out story, the different meanings of which can be deciphered in late pre-colonial, colonial and post-colonial era. Dhruv Raina has elsewhere identified two cardinal features of the encounter: 'how the sciences of the Orient were represented in the discourse produced by the Occident, but more importantly how Indian scientists and historians of science engaged with sciences of India'.[12] Over the short term the picture is one of the domination of political and intellectual power, as during the late pre-colonial era or the high noon of the colonial era. But over the long term, a more interesting picture had emerged. As Raina says,

> From the Orientalist period, through the nationalist period, to the period of Nehruvian socialism in India, the forces of decolonization sought to challenge a hegemonic conception of the history of science, such that history of sciences of the non-West would not be subordinate texts to the mainstream discourse of the history of Western science.[13]

A critique of 'the idea of science as a cultural universal' came into being only after 1970.[14]

The diffusion theory of science, indirectly connected with the theory of a civilizing mission, has been in vogue for a long time. Its connotations are manifold. The initiative was located in the West, with a chequered response—with all its potentialities, hesitation, partial fulfilment of promises and innate obstacles—in the East. Science is projected as the flag-bearer of the civilizing mission of the West, dissipating superstition and installing new technological systems over the old, primitive ones. Britain's changing role in the international economy led the colonial state in India to be innovative in founding technical institutions and colleges in the late nineteenth century. The colonial encounter, through a sort of diffusion of Western models and practices, contributed to the development of technical education in India.

> The several perspectives emerging argue, despite differences, that the standard tale of assimilation of modern science as a Western cultural import was inadequate and missed out the multifarious nature of exchange between modern science and so-called traditional knowledge forms.[15]

The criticism of the diffusionist model brings us closer to three arguments regarding the colonial impact that demand our immediate attention. The first is the close relationship of scientific and technological projects (like the railways) to nineteenth-century imperialism. Second, there was the impact of colonial rule on indigenous scientific knowledge and institutions. Third, the role of British and Indian scientists in the creation of scientific knowledge and institutions of science and engineering is to be critically examined. As Joseph Needham pointed out, scientific innovations are intricately enmeshed in socio-economic formulations.[16] Extending the argument in the case of colonial India, Ian Inkster has argued that though the overall financial impact of British rule is difficult to estimate, there is no denying the fact that (a) general British economic policy retarded structural change in the Indian economy, (b) there seems to have been no overwhelming socio-cultural obstacle operating within Indian society, (c) the wider scientific-technical environmental impact of the British in India was at best neutral, if not leading to retardation, and (d) growth, wherever it was possible in India, was enclavist.[17] Zaheer Baber has even pointed out that though the Indian scientists did not respond uniformly to Western science and technology, there was a group urging heavy industrialization which was dominant, and its views resonated in the Indian National Congress.[18]

The colonial impact is often matched by the nationalist response in the case of India. In constructing a national identity, a tradition is invented through choosing from the past, which then becomes instrumental in creating the required identity. Romila Thapar thinks that historical reconstruction is central as images are drawn from the past. She adds, 'If the history of science is tied to nationalist ideology, and it may be nationalism of any kind, it cannot be treated as universal. Categories such as Hindu science and Islamic science disallow a universality to science.'[19] However, the nationalist discourse in India is far more complicated than is presented by Thapar. There was latent and explicit racism, there were colonial reservations regarding the abilities of the Indians as scientists and, over and above all, there was competition from various quarters. In fact, J.C. Bose and Ramanujan sought to impress upon the people how traditional cognitive orders creatively enriched the practice of contemporary science, thus abandoning the epistemic privilege hitherto accorded to modern/Western science.[20] Aware of the pitfalls of 'universal science', Thapar adds 'What we are now experiencing is something which claims to be beyond nationalism and calls itself for trans-nationalism—a synonym for globalization, enriching us all to the siren of market.'[21]

In the historiography of science in India the post-colonial subframe is the most recent and is internally diverse. For the first time, we get an all-out criticism of colonial science. Concomitant with the beginning of the so-called decline of socialism and the end of scientism, the social history of science has now reached a critical stage. The epistemic version of science as transcendent and therefore culturally universal was severely challenged. There has been a sort of de-privileging of scientific activity in the wake of the environmental crisis. The 'progress' of science and its relation with society is critically placed in the wake of globalization, from within and without.

II

Our main focus in this volume is the relation and interaction between science and society in India, over a period of 250 years: from the early colonial, high colonial, and post-colonial perspectives. Most of the papers included in this volume relate to science and technology, supplemented by one on medicine and medico-botanical knowledge, two on public health, and one on the environment.

The volume opens with Deepak Kumar's masterly survey of the present state of the history of science, technology, medicine and environment, or what has been cryptically called HISTEM. Kumar finds constantly expanding frontiers of each of these areas of knowledge over time. He views the intellectual development in terms of an ever-increasing battle between what has been called mainstream and periphery in history. He is in favour of the accommodation of these new frontiers of knowledge, despite the tension caused in the discipline. He observes that today's periphery is tomorrow's mainstream.

While talking about the links between trade and medical networks in the eighteenth-century Coromandel coast, Pratik Chakrabarti also dwells on the connection between botany and imperialism. The networks of piety, profit, and knowledge that had developed in eighteenth-century India to study and preserve the 'Other' culture and to promote Christianity did serve to foster Western science in various ways. Chakrabarti shows that these networks demonstrate unique collaborations between spiritual, intellectual, and mercantile interests in the cause of science. His paper has attempted to assess the role of such informal peripheral networks in advancing scientific knowledge. At the same time, while such peripheral networks did gather information, verify facts, and facilitate experiments, they did not ensure scientificity of knowledge. Thus, Chakrabarti thinks, to focus solely on networks in the study of the formation of scientific knowledge and of peripheral creativity is inadequate. The increasing reliance of European practitioners on contemporary pharmacological explanations 'eroded' the status of cinchona as a 'specific' against intermittent fever. Its efficacy in treating both internally and externally manifest diseases (gangrene, ulcers, small pox) allowed cinchona to become a central therapeutic agent for both physicians and surgeons. It has been suggested that the perception of Peruvian bark as a general tonic eventually 'opened the floodgates' and made it a universal remedy. This led to Samuel Hahnemann's well-known 'principle of similarity', the centre-piece of homoeopathy.

S. Irfan Habib's contribution is concerned with an early Indian–European encounter through the eyes of Mirza Abu Talib (1752–1805). He was one of the first few converts from geo-centrism to helio-centrism. Abu Talib noted in his European sojourn that the British had a passion for mechanical devices, the French for mathematics. The existence of pumps was noted in place of Indian wells. In late eighteenth-century Europe, industrialism was based not on technical education but on empiricism. Abu Talib was impressed with the common people but disgusted with French peasant women. His accounts reveal three things: (a) the curiosity of a non-Western intellectual about the progress of science, (b) tensions and complexities of colonial politics, and (c) an encounter which enriched the search for, and the content of, knowledge.

Sujata Mukherjee writes about medicine and women's health in nineteenth-century India. There seemed to be an awareness among Indian men that a vigorous female domain of healing actually existed, which pre-dated British rule. These traditions came under attack by people influenced by the British civilizing discourse on science and progress. Western medicine in fact contributed to, and appropriated to itself, the cultural arrogance of post-1857 colonial rule. Disdain for indigenous medicine and faith in the superiority of Western medicine were buttressed by the bacteriological discoveries of Pasteur and others. As feared by nationalists like Bal Gangadhar Tilak, racial discrimination meant that medical education in colonial India developed on a tiered basis. Supervision remained in the hands of English physicians and Indian women were deemed fit for only subordinate posts. While it is true that the available choices of medical treatment for women widened with the promotion of Western curative medicine, it had very limited appeal. Most Indian women continued to depend on non-Western practices of healing and these were deprived of state patronage and discredited in the civilizing agenda of imperial rule in the post-mutiny era. Efforts to promote Western health care for Indian women was specifically made by the Dufferin Fund, and other organizations followed suit in subsequent years. These provided a new role for white women in India by promoting a new ideology of love and sympathy between the rulers and the ruled and at the same time contributed to the reproduction and maintenance of empire by utilizing resources that white women were uniquely qualified to provide in the sphere of health care for Indian women.

Arabinda Samanta's paper tries to focus on two things. First, it tries to delineate the different perceptions of people about the plague that ravaged eastern India in the closing years of the nineteenth and early decades of the twentieth centuries. Second, it explores the multiple sites where the rhetoric of Western medical intervention was contested by a popular construction of the disease. It views the disease and its remedial measures, the prophylactics, as functions of power and knowledge informed by a relationship between the ruler and the ruled in a colonial society. Perusing contemporary literary tracts and vernacular periodicals of the time and place, Samanta finds in retrospect that the incidence of plague was popularly viewed as a visitation of fate, to which one must submit with patience, but without an effort to do what was humanly possible to mitigate the calamity. People were distrustful of the methods which Western science had pointed as the most efficacious for the protection of public health and extirpation of epidemics. Added to this, both Hindus and Mohammedans viewed with the greatest dislike any intrusion into their homes, especially interference with the privacy of their women. In the case of eastern India, the riots, strikes, and disturbances were the actions of marginalized social groups of both communities, especially the municipal coolies, scavengers, sweepers, carters, butchers, durwans, menial servants, railways coolies, and mill workers.

Mridula Ramanna writes about the Indian responses to epidemics in the Bombay Presidency in the first two decades of the twentieth century. She begins her account with the measures taken by British officials to cope with epidemics and the Indian responses to these. Water supply schemes were recognized as a priority but the practical solution was often shelved. Village sanitation was an area of special concern. Although there were considerable smallpox deaths, vaccination was restricted among particular religious communities. Epidemics were widespread among the pilgrims. While there is no denying that the official figures of mortality may be unreliable, and many more fatalities may have occurred, during epidemics there was a perceptible change in the twentieth century in the official view of the health policy to be promoted. S. Harcourt Butler, in his presidential remarks at the Bombay Sanitary Conference in 1911, specified that the basis of all sanitary knowledge must be knowledge of the people, the conditions under which they live, their ways of life, their social customs, their habits and surroundings, and even their financial means. The *Memorandum on Indian Sanitary Policy* (1914) reiterated this view. For sanitary reform to be successful, was required 'a knowledge of the people' and a recognition of the diversity of local conditions; there were numerous castes, communities, and creeds, a variety of climates, temperatures and even levels of sub-soil water. After the ravages of plague and the discoveries of the results of medical research (which, the resolution contended, had made some real impression on the educated communities in India), the demand for sanitary surroundings and preventive measures had been made. As for Indian responses, their ambivalence is perceptible in the resistance to epidemic combating efforts and in the support extended to relief measures. The contemporary press expressed these mixed attitudes through their columns. During these years when national consciousness was spreading, the opinions of the nationalists were important. Hospitals organized on caste and community lines, made the concept of hospitalization acceptable. However, the fact that these institutions catered to their own groups would have meant that pollution prescriptions were maintained. There was also a notable change in pilgrim behaviour, showing a willingness to comply with regulations, attributable to greater awareness. Last, it was the promotion by Indian doctors of anti-epidemic measures such as better hygiene and information about the rat as a plague carrier, which made their fellow Indians adopt them. Trained in Western medicine, their conviction in the system was total and both private practitioners and doctors in municipal or government employment were responsible for whatever success was achieved in reducing casualties.

Rohan D'Souza talks about the hydraulic transition in the Bengal Delta in the first half of the twentieth century with special reference to a running debate between Engineer Willcocks and Governor Casey regarding the issue of irrigation and flood control. D'Souza says that this divide over interpreting and understanding the colonial responses to the delta's hydrology, throughout the nineteenth century, was indicative of general discomfiture with the gradual and sometimes abrupt hydraulic transition that was being then effected in the Bengal Presidency. Natural deltaic river regimes increasingly were made to give way to an altogether new hydraulic grid that was being assembled and shaped in the image of colonial authority in the twentieth century. Thus the hydraulic differences between Willcocks and Governor Casey sprang from a century-old debate about flood control, drainage, and irrigation. Radhakamal Mukherjee's work on the changing face of Bengal, published in 1938, is indicative of other views that D'Souza does not consider in this paper.[22]

Arun Bandopadhyay focuses on the technology and personnel employed in defence production with special reference to the Gun and Shell Factory, Cossipore, in the crucial decade of the 1890s. He shows that a new demand for expert personnel was generated in Cossipore in the 1890s when emphasis was placed on personnel with training in science and research and workshop experience in factory requirements. The knowledge bases of local expert personnel were not inferior by any standard, though in recruitments the primacy of Englishmen remained throughout the British rule. As regards technology, it was a double-edged affair in Cossipore as far as ordnance production and policy was concerned. The highly competitive world of the technology of warfare forced a partial modification of defence policy in India in the late nineteenth century. British science and technology had been weighed towards the military ever since the late nineteenth century, retarding industrial but encouraging military research and development in the process. The limited prospects of technology transfer in defence production in Cossipore should be seen from these perspectives. It is true that the potential experiments were temporary and success limited; the deployment of expert personnel from England was slow, sometimes half-hearted, and often not adequately remunerative. Nonetheless, the historical significance of the attempts of technological transfer should not be underestimated. The more important limitation was the virtual absence of a policy of recruitment of Indians as technical personnel at the higher levels of defence establishment in India, thus minimizing and virtually negating the diffusion effects. At the same time, the availability of qualified Englishmen was thwarted by many factors including their mobility, demand in Europe, and unattractive financial prospects in India.

Madhumita Majumdar concentrates on the communicating styles of J.C. Bose in the mature stages of his science career. Symbolized in the foundation of Vigyan Mandir in 1917, these styles indicate the colonial context of science practice in India even in the early twentieth century. Majumdar thinks that Bose's active seeking of a public may have in the final analysis been looked upon with regret but in the constraining circumstances of his career (in early twentieth-century Bengal) his attempts are worthy of commendation. The production of scientific knowledge in poorly equipped laboratories and in the face of uncertain financial resources may have been a self-defeating enterprise had Bose not actively courted an audience in Vigyan Mandir. The effort to look beyond the confines of the laboratory and create a public sphere of science had of necessity to be part of the scientist's brief if he had to sustain his pursuit in the midst of adversity. Majumdar argues that the singularity of Bose's style of communication (especially with regard to his research on the life of plants) have to be seen and appreciated in this context rather than through an explanatory framework that chooses to regard it as a sign of the scientist's eccentricity or as an expression of some abiding Eastern disposition that failed to read the conventions of the West.

Raj Sekhar Basu explores the engineering world of Visvesvaraya, especially his irrigation experiments. M. Visvesvaraya was perhaps the most important personality to influence the builders of irrigation projects in late colonial and early post-colonial India. Both engineer and economist, Visvesvaraya tried to draw the attention of the people from purely political wrongs to the fundamental economic wrongs. Colonial irrigation schemes aiming to utilize the flow of the Godavari and the Kistna rivers largely influenced Visvesvaraya's ideas on irrigation engineering, water supply and drainage and found the fullest expression in the construction of the Krishnarajasagar reservoir in Mysore. The Western civil engineering concepts guiding the Godavari system had led Visvesvaraya to believe that scouring sluices and regulators could be used to facilitate the flow of water to the main irrigation canals which made their way into the interior. But there was an obvious limitation. Basu thinks that environmental and other concerns, which recently have been regarded 'as disagreeable external impositions' hardly played a role in Visvesvaraya's project planning. Environmental issues were ignored and the preference seems to have been placed on revenue maximization and on the engineering aspects of dam construction. Visvesvaraya perhaps belonged to a school that held the view that water planning had to be left to engineers. There was very little idea that water planning needed to involve a multidisciplinary body comprising of agricultural experts, peasant organizations, environmental scientists, economists, sociologists, and lawyers.

Santanu Chacraverti builds the story of the construction of the *Shubhankari* tradition—certain computational verses once popular in Bengal—in colonial and post-colonial era. Though Bankimchandra categorically called them a starkly inferior kind of computation, Bangiya Sahitya Parishad and Ramendrasundar

Trivedi took serious note of them. There is no doubt about that the *Shubhankari* was popular in the nineteenth century. There were scholars such as Mahendranath Mukhopadhyay and Dineschandra Sen who spoke in favour of the system, the former for the excellence of the traditional mathematical system and the latter for its local and vernacular roots. Sukumar Sen went to the extent of calling this tradition (belonging to both Bengal and Assam), as going back to the fifteenth century. The emergence of ethno-science in academic discussion made the *Shubhankari* more relevant to certain analysts. However, Chacraverti suggests that there was a lack of interest in the vernacular system of computation in the post-colonial period. The other important fact is that the *Shubhankari* was differently looked by mathematicians and historians, and their approaches rarely converged.

The last two essays of Raina and Abrol deal with the post-colonial era. Dhruv Raina talks about the significant role of Needham and UNESCO in the development of the project of history of science in the wake of decolonization. Based on Needham–Nehru–Rahman–Chattopadhyay papers, the article makes certain interesting arguments. It begins with the similarities and differences of the cross-cultural studies of science. There are two modes of scientific internationalism, through the usual non-government and inter-government channels, but several ideological positions can be taken: traditional, leftist, liberal and pragmatist. Scientific causation has been seen from internal and external standpoints, and there have been remarkable changes in the historiography, especially in the intervening cold war period, when there was a marked decline in the ethos of social responsibility. There was a threefold manifestation of the historiographic revolution: (a) the integrity of European science, (b) the nature of the relation between European and non-European science, and (c) an interest in the past and the globalization of scientific research going hand in hand. Raina reviews the collaboration between Needham and his Indian colleagues in this context. This collaboration should be read as a history of the attempts to find an institutional home for the history of science in post-colonial India. Raina identifies three components of this collaboration. First, there was a legitimization of science as cultural activity. Second, it was taken as a collaborative venture of the humanist legacy of science after the Second World War. Finally, international organizations played a central role in this venture.

The last paper by Dinesh Abrol is concerned with the history of the Council of Scientific and Industrial Research (CSIR) in the post-colonial era. The paper takes a look at the patterns of changes made in the CSIR system of laboratories, in order to understand the dynamics of development in research organization over a period of more than sixty years. In the first part he talks about Science, Technology and Innovation (STI) planning during 1950–80. During this period the government was still planning for national development. In the Nehruvian period, the priorities had been basic raw materials and capital goods industry as well as consumer goods and small-scale industry. The recommendation of the 1948 Committee with members of big business was overturned because of protests from the Indian scientific community. During 1947–55, science funding was made through a relatively unstructured policy process. The Science Policy Resolution of 1958, already influenced by the linear model of innovation followed by advanced countries has been stated to have been planned too tightly, providing funds to help Indian industry to undertake technological development. During 1967–80, state intervention was designed to link science and technology with the productive system. However, it was in the last two decades of the twentieth century that CSIR emerged as a self-financing Research and Development (R&D) body, and there also emerged a global R&D platform in a competitive world. Abrol analyses reforms undertaken from time to time in respect mandates, priorities, and governance. It assesses the patterns of establishment of knowledge interests in the process of reforming the organization of research. In 2001, the country was induced to open up national laboratories to foreign companies. In the New Economic Policy (NEP), more funds were allotted for the transfer of knowledge than for the creation of technology. Abrol believes that the emerging need of technology detouring is unlikely to take care of the interests of business groups in the coming years.

NOTES

1. S. Alndreski, ed., *The Essential Comte*, London, 1974 quoted in Helge Kragh, *Introduction to Historiography of Science*, New York, 1987, p. 12.
2. Quoted from Curzon Papers cited by Zaheer Baber, 'Science, Technology and Colonial Power', in S. Irfan Habib and Dhruv Raina, eds., *Social History of Science in Colonial India*, Delhi, 2007, p. 103.
3. Alexandre Koyre in his lecture at the College de France in 1951, quoted by Pietro Redondi and P.V. Pillay, eds., *The History of Science: The French Debate*, Hyderabad, 1989, p. 22.
4. Ibid.
5. Scott L. Montgomery, *Science in Transition: Movement of Knowledge through Cultures and Time*, Chicago, 2000, quoted by Habib and Raina, op. cit., p. xxxv.
6. Helge Kragh, *Introduction to Historiography of Science*, pp. 32–40.
7. Quoted by G. Krobe's, 'Wissenschaftswissenschaft und Wissenschaftsgeschichte', ibid, p. 36.
8. G. Sarton, *An Introduction to the History of Science,* Baltimore, 1927–48, *passim.*
9. C.P. Snow, *The Two Cultures and a Second Look*, Cambridge, 1966.
10. J.T. Clark, 'The Science of History and the History of Science', in H.D. Roller, ed., *Perspectives in the History of Science and Technology*, Norma: Okhlahama, 1971, quoted by Kragh, 1987, p. 37.
11. I.B. Cohen, 'History of Science as an Academic Discipline', in A.C. Crombie, ed., *Scientific Change*, London, 1961, quoted in ibid., p. 38.
12. See Dhruv Raina, *Images and Contexts: The Historiography of Science and Modernity of India*, Delhi, 2003, p. 2.
13. Ibid.
14. S. Irfan Habib and Dhruv Raina, eds., *Social History of Science in Colonial India*, Delhi, 2007, p. xiv.
15. Ibid., p. xvi.
16. Dhruv Raina, 'Introduction', in S. Irfan Habib and Dhruv Raina, eds., *Situating the History of Science: Dialogue with Joseph Needham*, Delhi, 1999, pp. 1–15.
17. Ian Inkster, 'Science, Technology and Imperialism in India', in Habib and Raina, eds., *Social History of Science*, pp. 196–226. See also Sabyasachi Bhattacharya, 'Cultural Constraints on Technological Innovation and Economic Development: Some Case Studies', in *Indian Economic and Social History Review,* vol. III, no. 1, 1966, pp. 240–67.
18. Zaheer Baber, 'Science, Technology and Colonial Power', p. 151.

19. Romila Thapar, 'History of Science and the *Oikoumene*', in Habib and Raina, eds., *Situating the History of Science,* p. 27.
20. 'Introduction', in Habib and Raina, eds., *Social History of Science*, p. xxx.
21. Romila Thapar in Habib and Raina, *Situating the History of Science*, p. 27.
22. Radha Kamal Mukherjee, *The Changing Face of Bengal: A Study in Riverine Economy*, Calcutta, 1938; 2nd edn., Kolkata, 2009. See specially the new introduction by Arun Bandopadhyay in the second edition of the book, as published by the University of Calcutta.

CHAPTER 2

HISTEM in South Asia: An Overview

DEEPAK KUMAR

Take the cylinder out of my kidneys,
The connecting rod out of my brain,
The camshaft from under my backbone,
And assemble the engine again.

So went a song sung by airmen during the Second World War.[1] Every generation has a right to reassemble the 'engine'. This is the relevance and beauty of historical investigation. History is always contested intellectual territory, and no historiography can ever be neutral. Every society, indeed every individual, has a right to reflect on it. We need not come to a single or consolidated conclusion: rather, the more the merrier. And then there is no guarantee that the 'new' will work better than the earlier. What appears as a new explanation may actually be old wine in a new bottle. Nevertheless, the variations in arguments or thrust deserve no less appreciation. Certain portrayals or arguments, based on a somewhat imaginative use of sources, may have a longer validity and wider acceptance. But this does not mean that those which go against the grain be consigned to dustbin. New evidence or a fresh look may induce a rethinking. And the cycle goes on. In the process, theories often melt and very often historians appear better than their theories.[2]

* This paper, originally presented at the Seminar on 'History of Science and Society in Modern India' at the University of Calcutta in March 2005, builds upon my earlier presentations at Calicut, Chandigarh and Sambalpur.

Another issue that calls for attention is the concept of 'mainstream' history. What constitutes the 'mainstream'? Does it really exist? Where do the 'tributaries' stand and when do they become mainstream? Agreed that a study of material culture is mainstream history, what would this study be without a history of tools and techniques? Again, what would the history of ideas be without a study of scientific ideas? Some scholars, perhaps due to a misplaced academic ego, think that what they do is mainstream history.[3] So, in the context of modern Indian history, those writing on the national movement or on Partition tend to treat their themes as 'mainstream' research. Changes, however, did occur from political accounts to social explanations, peasant studies, then 'subaltern' studies, post-modernism, and now, probably, the history of science, technology, environment and medicine (HISTEM). Scholars working in hitherto unexplored areas are sometimes themselves keen on 'mainstreaming' their themes. A book on the history of medicine, for example, claims, 'Mainstreaming takes a historical sub-specialty, like the history of medicine, lifts it out of the confining limits of a disciplinary channel, and refloats it in broader historical currents.'[4]

I consider this a meaningless exercise. There is no mainstream, instead there are numerous streams that collectively produce the ocean that is history. Such an approach not only admits both macro and micro studies as equally relevant but also allows other disciplines a significant role and thereby ensures interdisciplinary interaction.

The most significant feature of HISTEM lies in its necessarily interdisciplinary nature. It borrows and benefits from different disciplines. Philosophy, parts of physical and biological science, and social sciences such as sociology, anthropology, and political economy provide valuable insights. But any dependence on or overuse of any one of these may interfere with the simplicity of a historical narrative and may even make a fine historical construction jargon-ridden. Conceptual analyses need not always result in epithets, labels, or jargon. Concepts and empirical study should and do often mesh together, and HISTEM provides ample opportunities for this interplay. This agenda must necessarily be not only interdisciplinary, but also comparative in both content and analysis. Comparisons are relatively easy in terms of centre and periphery relationships, but these are not sufficient. The major question is whether a periphery can alter the terms of the centre.[5] The traffic is complex and we need to undertake comparative studies, for example, in regard to different colonial experiences in different parts of the globe. As Michel Paty cautions, only by 'differential' studies, whose subject is restricted but which are accurate and varied concerning different but in some respects comparable situations, can legitimate comparative statements be made.[6] The incorporation of these comparative facts and data may provide a correction to 'mainstream' histories of sciences. I am consciously using the term 'histories', as there need not be one history, there can be many. Technology, for example, can be seen from the perspective of social history, economic history, and even cultural history. So is the case with medical history. HISTEM, like any other historical project, thus involves a study of several cross-sections representing events and ideas that are inter-connected and which exemplify the cause and effect relationship.

THE DISCURSIVE TERRAIN

Every society, however simple it might appear from the benefit of hindsight, has a certain amount of scientific rationality and endeavour. Science is intrinsic to human nature. So has been man's fascination for tools as *homo faber*. South Asian society has always nurtured a thinking civilization. It never lived an isolated existence and never displayed xenophobic tendencies. The techno-scientific tradition in South Asia has largely been a synthetic tradition, continuously evolving as a result of each politico-cultural interaction with the outside world and social change within the region.[7] In pre-modern times South Asia was known for its contribution to astronomy, medicine, and mathematics.[8] In ancient India, medical men, in spite of their scripture orientation, insisted on the supreme importance of the direct observation of natural phenomena and on the technique of a rational processing of the empirical data. The *Charaka Samhita*, an ancient medical text, says, 'To one who understands, knowledge of nature and love of humanity are not two things but one.'[9] Nothing could illustrate better the links between science and society. In relatively recent times, the relevance of science to society was not lost, either, on colonizer or colonized. In its very first issue *The Calcutta Review* wrote:

> The history of science is itself a science and one of the most interesting and important of them all. To trace the stream of discovery from its lofty well-head, to follow its various windings, mark its frequent disappearances, its rapids and its stagnancies, is a work at once of the greatest interest, the greatest importance, and the greatest difficulty. The interest of investigation is derived from our very nature and constitution as members of the great human brotherhood.[10]

Fascinated by the intricacies of the man–nature relationship, the founder of the Asiatic Society (1784) had set as its goal, 'Man and Nature; whatever is performed by the one, or produced by the other'. William Jones could not have been more apt and precise. Taken together, man and nature form the basis for the HISTEM or STEM in history.

The discursive terrain of HISTEM was never flat. It had its ups and downs, its ruptures and dichotomies. The ancient Greeks had talked of the 'desire to know', which later gave way to a 'desire for power'. Ancient Indians had hailed knowledge as 'a liberating force', Bacon popularized it as the 'power to control'. For long, scholars debated the distinctions between *theoria* and *praxis,* between *episteme* and *techne.* Historians of science have wrangled over internalism and externalism. Many have played 'historiographer royal' to science, transforming an individual genius into an icon. In this sense, the history of science helped an *ex post facto* legitimation and pandered to the scientists' *amour propre*. 'We are liable to optical illusions if we only have eyes for the mountain-peaks', warns Butterfield.[11] On another occasion he wrote, 'Sir Isaac Newton is the starting-point of a new age not merely for us but also for the Indians and the Chinese.' However, in recent years there has been an almost universal realization of HISTEM within 'the wider spheres of thought, culture and society'.[12] Men of science are no longer the sole 'creators and consumers' of this new discipline. It now has a wider appeal. And in this sense the history of science belongs to the mainstream social and cultural debates in history.

In recent years, a good deal of attention has been paid to the place and role of techno-scientific change and developments in the colonial process. New questions have been asked and certain explanations attempted. For example, what shape does 'modern' and 'universal' science take in a colony? To what extent were scientific discourses used to achieve political or economic goals? How were the indigenous scientific traditions perceived and how did the indigenous people react or respond to the introduction of 'new' science? Exact sciences like physics or astronomy may appear cognitively 'insular' in a colony, but several other branches of scientific knowledge were not. How 'colonial' do they become in a colonial setting? These questions lead to a new research agenda, which includes the generation of national currents and traditions in science, the interaction between 'traditional' sciences and modern sciences, and the diffusion, confrontation, and integration of sciences.

TECHNOLOGY AND HISTORY

It is universally agreed that techniques are part of, rather than the basis of, what Braudel calls material civilization. Any account of *homo faber* would be a history of technology. A social history of technology, to quote Cowan, assumes a mutual relationship between society and technology; it also assumes that changes in the one can, and have, induced changes in the other.[13] In many ways technology is central to human history; everywhere it shapes and is shaped by political, cultural, social, and economic changes.[14] Such being the case, can the history of technology be segregated from the rest of history?

Technology was earlier defined as a technical artefact and science as knowledge. Is technology science's other? Probably not. Both are historical variables. Science in part is knowledge about technology, and technology can be embodied knowledge. So why distinguish the two? They are two sides of the same coin, enmeshed in a 'symbiotic relationship'. Techno-scientific developments can be presented as a non-determined, multidirectional flux that involves constant negotiation and renegotiation among the groups and between the forces shaping history. This approach is flexible and eclectic, and wards off the fears of technological determinism.[15]

To some scholars, however, technology forms an entirely separate area with its own internal evolution and dynamics. Others would make a sharp distinction between pure and applied sciences. This I believe goes against the spirit of interdisciplinarity that constitutes the core of HISTEM.

During the last two decades, work on the social construction of technology has virtually revolutionized our understanding of the technology–history relationship.[16] This is fundamentally a sociological approach to technology that analyses artefacts in the context of society. It focuses on those social groups that play a role in the development of a technological artefact. Based on the notion of 'interpretive flexibility', it argues that radically different meanings of an artefact can be identified for different social groups.[17] This has interesting ramifications for gender studies as well. Women, for example, pump water and occasionally operate field machinery, whereas it is men who fix a leaky pump, and oil, grease, or even redesign a machine. Technical competence thus helps to define them as masculine (in opposition to a non-technically competent femininity) and thereby reinforce the patriarchal system.

During the eighteenth century, technology was understood in terms of 'knowledge of the arts'. It was knowledge produced and transmitted by craftsmen. In the next century, it came to be described as 'the application of the sciences to the useful arts'. Now it involved both knowledge and machines. This was an era of romanticism as well. The romantics were worried about what technology was doing to skill and craftsmanship. There appeared a reasoned critique of machines and industrialization. Romantics like Thoreau believed that machines were destroying not only nature but also human artistry, creativity and sensitivity. Later the Luddites were to sharpen these arguments. Yet there were many who fancied the new Juggernaut. In the 1880s an American poet sang:

> Singing the great achievements of the present,
> Singing the strong light works of engineers,
> Our modern wonders (the antique ponderous Seven outvied)
> In the Old World the east of the Suez Canal,
> The New by its mighty railroad spann'd,
> The seas inlaid with eloquent wires.[18]

It is generally agreed that technology—whether as tool or form of knowledge—is not value free; it always manifests socio-political qualities and drives the economy. In industrialized high-energy societies, it may become 'messy' and 'complex', but in colonial conditions it naturally acquired its contours from the colonial power, both commercial and administrative. What impact did colonization have on technological systems and the capabilities of the colonizers on the one hand, and the colonized on the other? How did colonialism determine the transfer mechanisms? Did it mean geographical relocation of technologies, or could it encourage an acculturation as well? Did British colonization accelerate India's 'fall' or did the British presence, however inadvertently, prepare what became in Indian hands a vital basis for a later economic, scientific, and technological 'take off'? Technologies are significant not only as 'tools' but also as forms of knowledge (sometimes referred to loosely as 'technical education' in colonial records). How was this knowledge generated, used, and transferred, and to whom? The 'token' industrialization that did take place in certain sectors (textiles and later steel) had no 'multiplier effect' on the industrialization of the colonial economy as a whole. Morris D. Morris has argued that the total structure of poverty and high factor input costs (but low raw labour costs) combined to discourage investments in new technology and thus prevented rapid economic growth.[19] But this does not explain what went into the making of that 'structure of poverty', and what sustained it. Was that structure a continuation or legacy of pre-colonial times? Was it the result of some cultural proclivity to remain poor or was it the inevitable consequence of the imperial system? The answer could be a combination of all these and much more. Aware of the darker side of Western industrialism, many in India wished to gain its benefits without the loss of indigenous culture. How can we have the best of both worlds? The dilemma continues even at the turn of the millennium.

In the Indian context it may be useful to recall the views of Mahatma Gandhi. Amidst the growing demands for self-rule, democracy, industrialization, and development, Gandhi had emerged as an extraordinary dissenter. He condemned the West for precisely those virtues in which it took pride: modernization and industrialization. Gandhi seldom used the terms 'science' and 'technology'. His concern was with civilization and mechanization, and on these topics he talked and wrote profusely. He considered machinery the 'chief symbol of modern civilization'. 'It represents a great sin,' he wrote. 'It is machinery (and Manchester) that has impoverished India. . . . Machinery is like a great snake-hole which may contain from one to a hundred snakes. . . . I cannot recall a single good point in connection with machinery. Books can be written to demonstrate its evils.'[20] Yet many times Gandhi would say he was not opposed to machinery *per se*:

> How can I be when I know that even this body is a most delicate piece of machinery? The spinning wheel itself is a machine. What I object to is the craze for machinery . . . today machinery helps a few to ride on the backs of millions. The impetus behind it all is not philanthropy to save labours, but greed. It is against this constitution of things that I am fighting with all my might.[21]

Gandhi, with all good intentions, probably wanted to force the Ganga back to Gangotri. This was simply impossible. The sharpest criticism of Gandhian views came from Meghnad Saha (1893–1955), a pioneer astrophysicist. Saha wanted India to choose the 'cold logic of technology' over the vague utopia of Gandhian economics. While he appreciated Gandhi's 'genuine sympathy with the victims of an aggressive and selfish industrialism', Saha firmly refuted the claim that better and happier conditions of life could be created by 'discarding modern scientific technique and reverting to the spinning wheel, the loincloth and the bullock cart'.[22]

At a global level, the two world wars not only rescued *Prometheus* but established him as the presiding deity of modern times. Leading intellectuals from different disciplinary backgrounds like Karl Popper (philosopher), C.P. Snow (novelist), David Landes (historian), and W.W. Rostow (economist) saw a connection between the rationality of science, technology, and industrialization, on the one hand, and the values of democracy on the other.[23] Relevant and reasoned criticism from critics such as Lewis Mumford, Jean Meynaud, and (in recent years) Asis Nandy have no doubt illuminated the debate but these in no way have discounted the significance of historical investigation in this area.

HISTORY OF MEDICINE

The history of medicine is no less significant a field than history of science and technology. Medicine has enormous social implications and also equally important political and cultural dimensions. Earlier, Indologists studied the Indian medical tradition as part of philosophical and cultural enquiry.[24] Now scholars of modern Indian history too have taken to it in a big way. Several works have appeared on medicine in politics and the politics of medicine.[25] At a micro-level, anthropologists and sociologists have contributed a great deal. Comparative and prosopographical studies of medical men can be equally instructive.[26]

Unlike the Cartesian world-view, in India nature and culture, subject and object, were never seen in an adversarial mould.[27] Pluralism was recognized, but at the core there remained a quest for synthesis and a holistic understanding. This trend is manifest especially in the Indian medical system. It is this trend that takes *yukti-vyapashraya bheshaja* (rational medicine) into the mirth and mire of metaphysics.[28] How does this take place? From magico-religious beginnings to rationalistic therapeutics and then back to decadence is a fascinating journey. Or is it possible to think of another trajectory? What were the factors responsible for the ups and downs? Was it because of the priestly class and the varnashrama hierarchy? The priests would naturally have been interested in supernaturalism and the mystification of nature. Why did the medical men bow to them?[29] We must explore the options open to them. Linear, hagiographical accounts and philosophical explanations do not suffice. We need critical sociological enquiries, as shown by Debiprasad Chattopadhyay.[30]

Apart from the above-mentioned concerns of a general and societal nature, one may also ask certain specific questions, as Roy Porter does in the context of late medieval England.[31] We need to know how healing was practised and who practised it, how disease was perceived. Medical anthropologists have looked into magico-religious rituals and shamanism. Can these be contextualized historically? Who were the grass root healers, and how did professionalism emerge? What were the contours of medical pluralism? How can one chart the interaction between the great and little traditions in terms of folk medicine and ethno-history?[32] Disease histories are many but we may need to look at them from the patients' eyes.[33] How did the sick evaluate doctors? How did the many distinct and competing practitioners relate to each other? This question was put in sharp focus when modern medicine entered new lands riding the colonial wave. Colonialism required human bodies to travel from one place to another and this influenced the relations between the bodies and the pathogens. The colonizing bodies were naturally anxious about their fragility either in the face of larger natural and social environments or in relation to other bodies (indigenous or foreign) that constituted an implicit threat.[34] What were the concerns for sanitation and the public health? And how 'public' was public health?[35] Colonial expansion strengthened the alliance between science and the state and the concept of state science/ medicine emerged. Questions arise about its function and its impact. Is there anything specifically colonial about colonial medicine?

Like technology, Western medical discourse functioned in several ways: as an instrument of control, which would swing between coercion and persuasion as the exigencies demanded, and as a site for interaction and often resistance. This discourse was mediated not only by consideration of political economy but also by several other factors. Polity, biology, ecology, the circumstances of material life, and new knowledge all interacted and produced this discourse. The emergence of tropical medicine at the turn of the last century may need to be seen in this light. It may be argued that tropical medicine itself was a cultural construct, 'the scientific step child of colonial domination and control'.[36] In a now burgeoning literature, terms like tropical medicine, imperial medicine, and colonial medicine have often been used interchangeably, but they have specific connotations. Tropical medicine and imperial medicine emphasize the tropics and the empire as units of analysis while colonial medicine stresses the colony, and each attracts different sets of questions. In tropical medicine what ought to be the determining factor—

climate, race, geography or all taken together? What was carried over from the old medicine of tropical civilizations into the new tropical medicine? What attempts were made outside Europe to reconcile the older discourse of body humours and environmental miasmas with the new language of microbes and germs? Interestingly enough, a medical historian described colonialism as 'literally a health hazard'.[37] But to dismiss the colonial doctors reductively as the handmaidens of colonialism or capitalism would also be to ignore a more complex, and more interesting, reality.[38] Doctors had little choice but to assume multiple roles. Still one can ask, what role did the 'peripherals' play? Could a synergetic relationship between the core and the periphery develop? These questions assume special significance when viewed against the four centuries of European's struggles in the 'torrid zones' and the transition from early explorer, traveller, and trader to conqueror and ultimate arbiter of the trampled tropics. Earlier the 'tropical discourse' was viewed through its pioneers; now issues and dichotomies have been given primacy. However, these still abound in metropolitan theorizations and do not include the study of indigenous (non-settler) societies through their own literature and practitioners.

Certain specific diseases have received special attention. Some are explained in terms of topography, some on grounds of race, and others in terms of microbes. In the old colonies all these explanations had almost the same result—let the 'natives' fend for themselves. This was a theoretical justification for opposition to consistent sanitation projects and fundamental medical research. Successive sanitary commissioners were to rely more on statistics than research. They lacked the will, if not the resources, to take effective action. Research, however, was not totally ignored. The work of Koch, Ross, Donovan, and Haffkine in India show the importance of fieldwork in India, the agenda coming from Europe. Compare this with nineteenth-century Brazil, an independent settler colony.[39] The works of Bahian Tropicalista School of Medicine during 1860–90 were not imitative but innovative. It made full use of Western clinical methods, tools, and statistics but framed its own agenda and worked without official funding. For the first time in Brazil, Otto Wucherer isolated hookworm parasites in 1865 and later isolated the embryonic filaria. Pereira worked on beri-beri, while Silva Lima started a medical journal. All of them worked to inculcate autonomy and self-esteem in their profession. Living in a tropical climate, they argued, made them merely 'different' from the Europeans but not better or worse. Could their contemporaries in India make such claims? Though to consider their works 'national' or 'Brazilianized medicine' may be a little over-enthusiastic, the Tropicalists deserve notice and recognition.

Nevertheless, medicine did emerge as a nationalist issue in India. Even at the height of colonial power, voices were raised against the dominant medical discourse. Indigenous practitioners vehemently denied that their system was unscientific or irrational, yet they did not see anything wrong in learning and benefiting from the new knowledge. Their emphasis was on reforming the system by adopting 'scientific' methods, not on changing the fundamentals of the system. A critical anti-colonial spirit permeated the indigenous response. To quote a verse from a hakeem in 1910:

Kuch-ilaj Aya na Kuch Charagiri Ayee:
Tibb-e-Unan ke Munh Doctory Ayee:
Band Sheeshe Mein Vilayat se Pari Ayee:
Lal-Peeli Hui, Gusse mein Bhari Ayee:
Chaman-e-Tibb se Guldasta Uda Kar Layee:
Nayee Tarkeeb se Bandish Saja Kar Layee.

(Knows no method of treatment, but Doctory dared to challenge Unani. In a closed bottle a fairy has come full of anger from foreign lands. The bouquet stolen from the garden of Unani Tibb has been rearranged in a new fashion.)[40]

There were several areas in which the Western and indigenous systems could collaborate but did not. The former put emphasis on the cause of the disease, the latter on *nidana* (treatment). Microbes and microscopes constituted the new medical spectacle,[41] whereas *vaidyas* put emphasis on the power of resistance in the human body. The Westerners were forced to take cognisance of indigenous drugs and the *vaidyas* took to anatomy, ready delivery of medicine, quick relief, and so on (as in case with P.S. Varier and Hakim Ajmal Khan). But the comparison ends there. As a recent critique argues, 'they were inclined to borrow but could not create a dialogue between the two epistemics'.[42] Borrowed knowledge seldom develops into organic knowledge.

ENVIRONMENTAL PERSPECTIVES

In recent years environmental history has aquired a new thrust and relevance.[43] There is a need to integrate environmental perspectives with the existing discourses on history of science, technology and medicine. Both stand to gain in this mutual give and take. How can nature as an area of scientific research help historians understand the dynamics of the past? The sciences of nature have given us bodies of data on all possible dimensions of our material existence. They have given models and new paradigms. What place should these have in historical thinking and historiography? How do the environmental ideas emerge? Are they socially constructed, or do they emerge from the top as instruments of authority and explanation? Contest and defiance have played no less significant a role, the Chipko and Narmada movements being recent examples.

Environmental ideas have undergone numerous changes. The 'world history' perspective is rewarding, but we can also 'zoom in' when needed to explain certain special circumstances. One such special circumstance that comes to mind is the history of imperial expansion, arguably the most potent historical force in modern times. What impact did this expansion have on knowledge and use of natural sciences? How did the latter shape the 'pattern' of imperial expansion? One need not be obsessed with Europe while looking into these questions. The Eurocentric account is probably at its best while dealing with the Enlightenment project and the subsequent imperial 'burden'. The empire brought to Europe the shock of the world's 'diversity', especially so in terms of flora and natural resources. It held immense possibilities for commerce and power. One can easily justify the empire as an instrument of development. But here the key question is: 'whose development'? Was it a largely one-way traffic? The evidence of commercial considerations are so abundant that no imageries or search for the Eden hypothesis can give an alternative explanation or turn history 'upside down'.[44] As for the communities and environments destroyed in the midst of European expansion, a Cambridge scholar asks, 'Should we care? Were the pyramids and so much human grandeur not built on slavery and despotism?'[45] He is not alone; some are convinced of 'constructive imperialism'. However, many more recognize that there did exist a dark side of both topicality and hegemonism and its history is 'almost as long and as complex as the Edenic'. The debate continues.[46]

We need to move away from state centric approaches and look to more complex engagements, which could be grouped under broad themes such as 'nature', 'culture', and 'science'.[47] Another significant point that makes the environmental perspective relevant is that the human response to nature must necessarily be collective (to quote Radhakamal Mukherjee, 'synecological'). It cannot be individual-oriented 'autoecology'. In any case, while political economy creates and revels in barriers, environmental concerns unite us. This is a silver lining and adds to the relevance of HISTEM.

POSTSCRIPT

Notwithstanding the arguments about 'ruptures', 'ambivalence', etc., one cannot lose sight of the fact that colonialism on the one hand, and modern thought, science and technology on the other form two aspects of one phenomenon, two branches stemmed from one root, two sides of the same coin, or two inseparable characteristics of Western bourgeois civilization.[48] This may be criticized as too nationalistic. Yet within this critique of colonial power relations, it is possible to locate instances of 'colonial nomadism' or 'intellectual migrancy', for which Warwick Anderson pleads so passionately.[49] Recently he has alluded to new configurations under the rubric post-colonial 'techno-science'.[50] Sites of hegemony are also sites of refutation, negotiation, and even exchange. Multi-sited interdisciplinary studies would definitely be more illuminating. Social constructions, historical anthropology, and cultural insights can bring forth new dimensions of HISTEM and enrich our understanding as nothing else has done so far.[51] Help from scientists should be equally welcome. How can historians tap the scientific literature without becoming scientists themselves? They can do so by consulting their colleagues in sciences. The study of human welfare (be it in terms of technology or health) requires a joint effort on the part of scientists and historians; neither side has an advantage in method or truth over the other.[52] In fact, both stand to gain. As Raina enquires, what did the Indian scientists and science historians inherit from the occidental discourse about the orient and where did they depart from the former?[53] Another significant aspect is that STM has always been much more *plural* than most people appreciate; there have always been lots of different ways of knowing and of making.[54] But this pluralism has to be both critical and contextual. HISTEM recognizes this and testifies to the complementary co-existence of the natural and social sciences. Apart from its research significance, it can be a useful teaching-aid. It deserves a place in our curricula; long ago in 1918 an official report had asked for it,[55] and the need remains.

In the past decade or so it has become fashionable to talk of knowledge-society, knowledge-economy and the new century as the century of knowledge, as if the previous centuries or societies were not (or less) concerned with knowledge. *Homo sapiens* have always gathered and valued knowledge! It is the pace and pressures that are greater. Here one may pause and ask, knowledge for what purpose and for whom? For a plausible answer, permit me to return to Haldane:

> When I look at history, I see it as man's attempt to solve the practical problems of living. The men who did most to solve it were not those who thought about it, or talked about it or impressed their contemporaries, but those who silently and efficiently got on with their work.[56]

NOTES

1. Quoted by J.B.S. Haldane, *On Being the Right Size,* New York, 1985, pp. 83–4.
2. Charles Gillespie quoted by Peter Dear, 'Cultural History of Science: An Overview with Reflections', *Science, Technology and Human Values,* vol. 20, no. 2, 1995, pp. 150–70.

3. Some recent titles read 'situating' social history or the history of science. We now want to 'situate' history and 'fix' people; mere study is not enough!
4. Mary Lindemann, *Medicine and Society in Early Modern Europe*, Cambridge, 1999, p. 1.
5. Pratik Chakrabarti, *Western Science in Modern India: Metropolitan Methods, Colonial Practices*, Delhi, 2004, pp. 1–26.
6. Michel Paty, 'Comparative History of Modern Science and the Concept of Dependency', *Science, Technology and Society*, vol. 4, no. 2, 1999, pp. 171–204.
7. A. Rahman, ed., *Science and Technology in Indian Culture: A Historical Perspective*, Delhi, 1984, pp. 22–3.
8. Centuries ago Said-al-Andalusi (1029–79) in his *Tabaqat al Uman* (probably the first work on history of science in any language) referred to India as the first nation which cultivated the sciences. M.S. Khan, 'Qadi Said-al-Andalusi's Account of Ancient Indian Sciences and Culture', *Journal of the Pakistan Historical Society*, vol. XLI, 1997, pp. 1–31.
9. D. Chattopadhyay, *Science and Society in Ancient India*, Calcutta, 1977, p. 7.
10. *The Calcutta Review*, vol. I, 1844, p. 247.
11. H. Butterfield, 'The History of Science and the Study of History', *Harvard Library Bulletin*, vol. 13, 1959, pp. 329–47.
12. R. Porter, 'What is History of Science?' in J. Gardiner, ed., *What is History Today*, London, 1988, pp. 69–71.
13. Ruth S. Cowan, *A Social History of American Technology*, New York, 1997, p. 3.
14. Michael T. Allen and G. Hecht, eds., *Technologies of Power*, Cambridge, 2001, p. 3.
15. Refuting technological determinism, Melvin Kranzberg says, 'All history is relevant, but the history of technology is the most relevant. . . . Technology is a very human activity, so is the history of technology', Melvin Kranzberg, 'Technology and History: Kranzberg's Laws', *Technology and Culture*, vol. 27, no. 3, 1986, pp. 544–60.
16. For details, W.E. Bijker, T.P. Hughes, and T.J. Pinch, eds., *The Social Construction of Technological Systems: New Directions in the Sociology and History of Technology*, Cambridge, Mass, 1987. This work was inspired by Peter Bergger and Thomas Luckman, *The Social Construction of Reality: A Treatise in the Sociology of Knowledge*, New York, 1966.
17. For example, high-wheel bicycles were 'macho machines' for young men of means and nerve, but for older people and women they had the radically different meaning of 'unsafe machines', Trevor Pinch,

'The Social Construction of Technology: A Review', in Robert Fox, ed., *Technological Change,* Amsterdam, 1996, pp. 17–36.

18. Quoted by Cowan, op. cit., p. 215.
19. Morris D. Morris, 'Some Factors Affecting the Rate of Economic Growth and Technical Change in India after 1800', in Peter Gaeffke and D.A. Utz, eds., *Science and Technology in South Asia*, Philadelphia, 1985, pp. 125–9.
20. A.J. Parel, ed., *Gandhi: Hind Swaraj and Other Writings*, Cambridge, 1997, pp. 109–10.
21. *Young India,* 13 November 1924, pp. 377–8.
22. M.N. Saha, 'Editorial', *Science and Culture*, vol. 1, 1935, pp. 3–4.
23. Michael T. Allen and G. Hecht, eds., *Technologies of Power*, p. 6.
24. Bhagvat Singjee, *Aryan Medical Science, 1895,* rpt. Delhi, 1993; U.C. Dutt and G. King. *The Materia Medica of the Hindus,* Calcutta, 1922; Shiv Sharma, *The System of Ayurveda,* 1929, rpt. Delhi, 1993; J. Filliozat, *La Doctrine Classique de la medicine indienne,* tr. Dev Raj Chanana, Delhi, 1964; P.V. Sharma, *History of Medicine in India*, New Delhi: 1992; M.S. Valiathan, *The Legacy of Caraka,* Delhi, 2002.
25. In the Indian context the important ones are: Poonam Bala, *Imperialism and Medicine in Bengal: A Socio-Historical Perspective,* Delhi, 1991; David Arnold, *Colonising the Body: State Medicine and Epidemic Disease in Nineteenth Century India*, Delhi, 1993; Mark Harrison, *Public Health in British India, 1854-1914,* Cambridge, 1994; Anil Kumar, *Medicine and the Raj,* Delhi, 1998; Biswamoy Pati and Mark Harrison, eds., *Health, Medicine and Empire: Perspectives on Colonial India,* Delhi, 2001; Deepak Kumar, ed., *Disease and Medicine in India,* Delhi, 2002.
26. Bruno Latour, *The Pasteurisation of France,* Massachusetts, 1988; Deepak Kumar, 'Colony under a Microscope: The Medical Works of W.H. Haffkine', *Science, Technology and Society*, vol. IV, no. 2, 1999, pp. 239–71.
27. Ole Brunn, and A. Kalland, eds., *Asian Perceptions of Nature: A Critical Approach*, Copenhagen, 1995.
28. R.C. Majumdar's enthusiastic account of Ayurveda is an example. See D.M. Bose et al., eds., *A Concise History of Science in India,* Delhi, 1971, pp. 213–62.
29. This question is relevant for our own time as well. If a particular political dispensation imposes astro-medicine, should our medical institutions accept it?
30. D. Chattopadhyay, op. cit.
31. Roy Porter, *Disease, Medicine and Society in England 1550–1860*, Cambridge, 1993, p. 2.

32. D.V. Hart, *Bmisayan Filipino and Malayan Humoral Pathologies: Folk Medicine and Ethno-History in South East Asia,* Ithaca, Data Paper No. 76, n.d.
33. Elaborating on the history of tuberculosis a new work has called for a new kind of history of disease that proceeds 'from the perspective of the patient' rather than the doctor. For details, Shiela Rothman, *Living in the Shadow of Death: Tuberculosis and the Social Experience of Illness in American History*, New York, 1994.
34. Alan Bewell, *Romanticism and Colonial Disease*, Baltimore, 1999, p. 24.
35. C. Hamlin, *Public Health and Social Justice in the Age of Chadwick, Britain 1800–1854*, Cambridge, 1998, pp. 1–14.
36. Lenore Manderson, *Sickness and the State: Health and Illness in Colonial Malaya 1870-1940,* Cambridge, 1996, pp. 10–14.
37. Donald Denoon, *Public Health in Papua New Guinea,* Cambridge, 1989, p. 52.
38. Heather Bell, *Frontiers of Medicine in the Anglo-Egyptian Sudan, 1899-1940*, Oxford, 1999, pp. 10, 52.
39. Julyan G. Peard, 'Tropical Medicine in Nineteenth Century Brazil', in David Arnold, ed., *Warm Climates and Western Medicine*, Amsterdam, 1996, pp. 108–32.
40. Jogi Dakani, quoted in Neshat Qaiser, 'Colonial Politics of Medicine and Popular Unani Resistance', *Indian Horizons*, April–June 2000, pp. 29–42.
41. W. Anderson, 'Laboratory Medicine as Colonial Discourse', *Critical Inquiry*, vol. 18, 1992, pp. 506–29.
42. K.N. Panikkar, 'Indigenous Medicine and Cultural Hegemony: A Study of the Revitalization Movement in Keralam', *Studies in History*, vol. VIII, no. 2, 1992, pp. 283–307.
43. The pioneers are Ramachandra Guha, Richard Grove, Madhav Gadgil, and Mahesh Rangrajan. For bibliography, S. Sangwan, *Science and Environment in the Age of Imperialism: A Handbook of Source Materials*, Delhi: Report to INSA, 2001.
44. Abhijit Mukhopadhyay, 'National Science in Colonial Context: The Calcutta Botanic Garden and the Agri-Horticultural Society of India, 1787-1870', unpublished Ph.D. thesis, Jadavpur University, 1996, p. 350.
45. Richard Drayton, *Nature's Government: Science, Imperial Britain and the 'Improvement of the World'*, London: Yale University Press, 2000, p. 272.
46. David Arnold, *The Problem of Nature: Environment, Culture and European Expansion*, Oxford, 1996, p. 140.
47. Mahesh Rangrajan quoted by Rohan D'Souza, 'Disparate Narratives:

Environmental History Writing is Beginning to Look Beyond Colonial Forestry', *Down to Earth*, 31 January 2003, pp. 46–8.

48. Abdul-Hadi Hairi, *The Early Encounters of the Iranian Thinkers with the Two-Sides Civilization of Western Bourgeoisie*, Tehran, 2002.
49. Warwick Anderson, 'Where is the Postcolonial History of Medicine', *Bulletin of the History of Medicine,* vol. 72, 1998, pp. 522–30.
50. Warwick Anderson, 'Postcolonial Technoscience', *Social Studies of Science,* vol. 32, nos. 5–6, 2002, pp. 643–58.
51. Robert Fox, ed., *Technological Change,* Amsterdam: Harwood, 1996; Bryan S. Turner, *Medical Power and Social Knowledge,* London, 1987; C.E. Rosenberg and Janet Golden, eds., *Framing Disease: Studies in Cultural History,* New Brunswick, 1992.
52. Bruce S. Fetter, 'History and Health Science: Medical Advances Across the Disciplines', *Journal of Interdisciplinary History*, vol. XXXII, no. 3, 2002, pp. 423–42.
53. Dhruv Raina, *Images and Contexts: Essays on Historiography of Science in India*, Delhi, 2003, p. 2.
54. J.V. Pickstone, *Ways of Knowing: A New History of Science, Technology and Medicine*, Manchester, 2000, p. 25.
55. H. Sharp, *Report of the Committee to Enquire into the Position of Natural Science in the Education System of Great Britain*, Delhi, 1918, pp. 25–6.
56. Haldane, op. cit., p. 84.

CHAPTER 3

Networks of Medicine: Trade and Medico-Botanical Knowledge on the Eighteenth-century Coromandel Coast

PRATIK CHAKRABARTI

In recent times the links between trade and medical networks have been explored as have been those between botany and imperialism. In the specific case of India, both trade and medico-botany formed important parts of European engagement from the seventeenth century. It has been suggested that missionary engagement with plants and souls of the colonies was a mission in 'improvement'.[1] On the Coromandel Coast of south-eastern India were settled missionaries who studied the local flora and medical texts. They had interesting links with the European trading companies and their surgeons as well as with local practitioners and experts. Scholars have suggested that the project of natural history (the collection, improvement, cultivation and classification of species) had a strong evangelical connotation. Natural history representations articulated the notions of spiritual development and sanctification.

Along with the project of spiritual improvement of subject populations, the material improvement of the metropolis and the garden has been well documented by Lucile H. Brockway, *Science and Colonial Expansion: The Role of the British Royal Botanic Gardens*. Focusing on the cases of cinchona, rubber, and sisal, Brockway shows that despite internal rivalries, the industrialized and imperialist nations of Europe shared a common interest in 'extracting and mobilizing the energy of the world for its own purposes'.[2]

More recently, Richard Drayton has located the 'garden' at the genesis of the modern world.[3] He traces the idea of improvement as the cornerstone of the modern world. But the problem remains of how to locate a 'garden' at the centre of such a complex experience.

At one level, placing such great trust in material 'improvement' as the metaphor of the modern world does put a major stake on the 'imperial vision' in the history of the modern world. Drayton's predominant focus on Banks and the Hookers can be cited for this. This tends to overshadow the peripheral negotiations and networking that were simultaneously taking place. These negotiations were not only about improvement; they were also about appreciation, incorporation, and preservation. The process of collection of knowledge and material, as this paper will show, was far more complex.

In the project of improvement, the argument of Christian missionary interest in gardening and improving of souls is an important one; Joseph Banks himself had an evangelical motivation in 'developing' plants in the colonies and at Kew. The logic of 'improvement' sits very well with such missionary interest in botany, agriculture, and gardening in the periphery. But that perhaps was not the only motivation. In exploring local flora there was an encounter with the exotic; we find several Jesuits and Pietists, for example, involved in the study of indigenous culture, practice, language, texts, and flora from a very early time on the Malabar and the Coromandel coasts, where the logic of 'improvement' does not appear to have been central. Rather, it reflected the duality of the colonial experience; an engagement with the 'modern', the Linnaean botanical study and identification of specie (with its logic of improvement), as well as the 'ancient', the study of the classical texts, exotic knowledge systems, and the search for a different world. For that we have to look at the diverse negotiations between missionaries, botanists, and nations taking place at different sites.

In the specific case of India, both trade and medico-botany formed important parts of European engagement from the seventeenth century. This paper studies various aspects of medical botany on the Coromandel coast in the eighteenth century. The focus would be on an interesting world where various ideas were being debated and assimilated; the journey of therapeutics through the forests, texts, and bazaars of the East to the laboratories, apothecary shops, and dispensaries of the West involved missionaries, colonial officials, doctors, and local practitioners. The distinct motivations of the various participants in these networks will be analysed.

The Malabar and the Coromandel coastlines had become crucial both for spices, and medicinal herbs from the seventeenth century. Although by the eighteenth-century spices had ceased to be the crucial commodity of the eastern trade, the resonance of such interest could still be found in the European search for medicinal herbs and their local uses. Tranquebar on the Coromandel coast was itself an interesting site, with a heterogeneous population of Hindus, Muslims, and Catholics of varying ethnicity including Tamil-speaking people, Arabs, and the Portuguese.

THE COMPANY BOTANISTS

The medicinal properties of the flora of the East had long attracted the attention of the Europeans. The Malabar region especially had continuously attracted attention for its medicinal foliage and expertise. Whitelaw Ainslie, East India Company surgeon, while compiling the Indian Materia Medica elaborated a sense of the exotic and mentioned how vital the Travancore and the Malabar vegetation could be for medicinal reasons:

> The greater part of the articles mentioned in this volume are parts of plants which grow in India; and are to be met with in the jungles, amongst the woods of Malabar, and mountains of the lower tracts of the peninsula, and more especially, in Travancore, that country so beautiful, so rich, I may say, in vegetable productions. . . .[4]

He added:

> I cannot help expressing a regret that, in that singular country, fitted by climate and soil for the cultivation, perhaps, of every vegetable produce which any part of the torrid zone affords, it has never been attempted to rear those articles of the Materia Medica for which the

world is now solely indebted to America. Travancore, and Malabar also, no doubt, possess (notwithstanding the great exertions of Rheede) many curios and useful plants which have not yet come within the reach of scientific investigation.[5]

The Coromandel region retained its twin attraction for medicine and spices. William Roxburgh, East India Company botanist posted at Samulcottah north of Madras, combined research on the medical botany of this region with experiments on growing cinnamon, clove, tobacco, indigo, and teak, about which there was continuous correspondence between him and Ross.[6]

A succession of European physicians, particularly in south India, had taken the first significant steps in this field in India. Among the earliest was Samuel Browne, a Company official who was studying the plants, spices, and herbs of the region.[7] Browne's account, much like that of others, was not just a description of the medicinal plants, roots, barks, and the different decoctions prepared with those by the local people, but also a detailed description of the local cultural practices linked with their medicinal uses. Along with this, Petiver was involved in comparing them with similar accounts from the West Indies, Guinea, and East Asia,[8] as well as other similar accounts on India, in the process drawing a comprehensive treatise on medical botany. What finally was published in the *Philosophical Transactions of the Royal Society* was a rich account that also included the Portuguese connections in bringing the aromatic and medicinal plants to these parts.[9]

Dr. Johann Gerhard Koenig, a Baltic German, was a student of Linnaeus in 1757, and came to India as a surgeon in 1768. Engaged solely in the study of flora of the Madras region, in 1774 he became a naturalist to the Nawab of Arcot, and made trips to the surrounding hills and also to Ceylon.[10] His detailed illustrations and botanical descriptions of Indian plants 'useful in medicine and the arts', never saw the light of day, his successor Dr. Patrick Russell turning to fish and snakes became an authority on snake pills.

Dr. James Anderson, horticulturist and surgeon at the Madras Presidency who arrived in Madras in 1769, started botanical research in 1771. He established a Nopalry (to cultivate the nopal plant) in Saidapet. By 1791, this had been developed into a botanical garden, the first in India, and flourished till 1800 when William Roxburgh was transferred to the Calcutta Botanical Gardens.

Anderson had developed his private botanical gardens in the 111 acres around his house in Nungambakkam. The gardens survived until at least 1828, long after his death. Anderson also established the Madras Agri-horticultural Society. Having found that the root of the plant *Asclepias vomitoria* was used by European and Indian troops with great success in dysentery, (epidemic in the camp), Anderson tried to apply it, but, Koenig's note says, 'finding the practice of the black doctors much more successful than his own, was not ashamed to take instruction from them, which he pursued with good success; and collecting a quantity of the plant which they pointed out to him, he sent a large package of the roots to Madras'.[11] The plant was considered an important substitute for the Ipecacuanha root, which was scarce in India.

The most important figure was William Roxburgh, who studied medicine at Edinburgh University. Beginning as surgeon's mate on the Company's ships, in 1776 he accepted the post of assistant surgeon at the Madras General Hospital. Roxburgh's work, benefiting from Koenig's compendium, led to the publication of the monumental *The Native Plants of the Coast of Coromandel* in 1798. By then he had moved from Madras to Samulcottah, for Lord Wellesley had summoned him to Calcutta to administer that city's now-famed botanical gardens. In a few years, the gardens of Madras were to become as neglected as they are today.[12]

An important aspect of the research in medicine and botany in the eighteenth century was its predominantly international character, both in Europe and in the colonies. Medical students moved from British universities to Leiden, where they studied with Albinus and Boerhaave, or they went to Paris or Orange, and later to Göttingen to learn from Haller. Dutch and other European students came to Glasgow and Edinburgh. Such travel was a notable stimulus to thought, and doctors began to explore the association between environment and disease.[13] This internationalism was reflected in the Indian subcontinent as well. One aspect of Roxburgh's broad medico-botanical networks is his links with the missionaries and the Moravians in the Danish settlement of Tranquebar, as well as with the Orientalists.

Their engagement eventually provided a very interesting characteristic to the entire project. Harold J. Cook has explored the significant links between medicine and natural history from the sixteenth century. That was a period which witnessed a shift from philosophical reasoning to investigations of nature. Medicinal studies in many leading European universities included a curriculum on studies of nature, probably the only institutional training of that kind in seventeenth-century Europe. The message was that a physician needed knowledge of nature, an understanding of plants and waters, oils, spirits, and salts, as well as of diseases, medicines, and climate.[14] Because of the natural history education the physicians had received in Europe, they became prominent in studying the soil, rocks, climate and particularly plants of the European lands. Thus by the mid-seventeenth-century natural history had become the 'keystone of both the physicians' *ars* (the skill or method of treating disease, the uses of plants, and minerals) and their *scientia* (the knowledge of health and disease)'.[15] In India, as we shall see, Company physicians played an important role in discovering the medical virtues of certain plants.

William Jones was a prominent figure in this culture of gardens and libraries, collecting both medicinal plants as well as texts. In his inaugural address to the Asiatic Society of Bengal in 1784, he stated, 'You will investigate whatever is rare in the stupendous fabric of nature . . . you will examine their methods in . . . medicine. To this you will add researches into their agriculture. . . .'[16] Among the English surgeons on the Coromandel coast, Whitelaw Ainslie carried out the Orientalist tradition of reading Tamil texts. He looked as much in the market as into Tamil texts such as the work of Agashtier on the practice of medicine, *Agashtier Vytia Anyouroo*.[17]

This same coastal region was the site for missionary settlement and activities over some time, which included Pietists, Lutherans, and Moravians who had played an important role in the study of plants and cures. Father Mathew of the Order of St. Joseph's Carmelite, for example, helped Henry Van Rheede, the Dutch governor, in his compilation of the twelve-volume *Hortus Indicus Malabaricus*, from his own observations and with the advice and assistance of local experts.[18]

THE MISSIONARIES

The Danish East India Company earned a monopoly for trade between Denmark and Asia for a period of twelve years in 1616. Over a period it maintained trade contacts with the Portuguese in Ceylon, Nagapattam. During 1621, the Danes began their occupation of Tranquebar. From 1670 to 1729 the second Danish East India Company resumed trade with Tranquebar. The Danes acted as neutral third parties in the intra-Asiatic or 'country trade', trading pepper and cloves, and taking part in the regional trading structure on a par with local merchants. This was of crucial importance to the continued viability of the Danish colony. Furber termed the Danish activities, particularly in country trade, as 'clandestine' because they opened means and ways for many European company servants to engage in private trade and remit profits home.[19] With much less political ambition than either the French or the English, the Danes built-up an intricate country-trade network in South Asia from Tranquebar on the Coromandel coast. This Danish settlement and its country trade played a crucial role in supplying medicines to English surgeons. During war with France, William Roxburgh at Samulcottah sent specimens to Banks in England through 'neutral' Danish ships.[20]

The Danish settlement brought the first Protestant mission, the Danish-Halle Mission, to Tranquebar (1706–1845). The 'neutral' position of the Danes had allowed them, particularly the Danish Lutheran missionaries, to develop informal local contacts. In 1706, Bartholomäus Ziegenbalg (1683–1719) and Heinrich Plütschau (1678–1747), German Lutheran missionaries of the Pietist tradition at the University of Halle, landed at Tranquebar. They were the first Germans to take up the Danish king's orders. Hence all these missionaries, though Germans by birth, came to be called 'Royal Danish Missionaries'. But the Danish East India Company was not too keen to assist them, and often posed hindrance to their activities.

These missionaries had to contest with other missionaries like the Moravians in the region, Danish officials at home, religious sceptics, and a constant paucity of funds.[21] They found solace in interacting with their English neighbours, the naturalists and surgeons. Their common interest in the local natural history and local culture served to create the bond. The Danish missionary Revd. John once mentioned to William Roxburgh, 'Indeed I would wish to have or make myself wings like Dædalus to fly to you to see your large tract of country and to have a share in the superintendence of it. I would forget my chief employment but teach the inhabitants at the same time how to plant and rear Coconut. . .'.[22] Ziegenbalg and Grundler wrote too letters to the English; they considered the English settlements in the East Indies an ideal area for the spread of Christianity as Tranquebar was too small 'to afford us an open Door to preach among the Neighbouring Heathens'.[23] The contacts were friendly and open and the Danes were often invited by the British, who did not have any missionary representative from home in these parts.

The missionaries collected plants and grew them in their own gardens from quite early on. Particular mention has to be made of the Danish settlements in southern India, in Tranquebar.[24] They studied and documented not only the religious life but also the plant and animal life of south India. Among them were the linguist Benjamin Schultze (1689–1760), the Hebrew scholar Christopher T. Walther (1699–1741), the diplomat and royal priest Christian Frederick Schwartz (1726–98), the Bible translator Johann Philipp Fabricius (1711–91), the educator and natural history enthusiast Christopher Samuel John (1747–1813), and the lexicographer John Peter Rottler (1749-1836). The missionaries had friendly and open contact with their English neighbours in Madras.[25]

Other missionaries on the coast were the Moravians, who had arrived rather unexpectedly in 1760 and settled in Tranquebar, an international Protestant evangelical group. They played a significant role in the internationalism of medical botany in the eighteenth century, and learnt Portuguese and Tamil. It seems that the Moravians established a 'rival market' to that of the Germans. In two years the number of the Moravians increased; they soon became more popular among the Tamils and Europeans.[26] The two important figures in the Moravian Mission in Tranquebar were Johann Gottfried Klein and Benjamin Heyne.

From the middle of the eighteenth century the Moravian church too was marked by a certain 'Pan-internationalism'.[27] The mission had begun in colonies of the Danish crown. The first, launched in 1732, was among slaves in the West Indies and it was followed a year later by efforts in Greenland. Guidelines for Moravians were directed towards these missions in West Indies and Greenland[28] but they also settled in Nancavery, one of the Nicobar islands, until 1788.[29] The Moravians followed a distinct lifestyle with a focus on financially supporting themselves in these distant lands. As Spangenberg pointed out, the brethren and sisters maintained a common housekeeping, and wherever practicable they tried to earn their livelihood by the 'work of their hands'.[30] Often the Moravians bought and cultivated land for their subsistence.[31]

Although the Danish missionaries in Tranquebar were initially hostile to these Moravians, within a very short period the latter seemed to have become popular and they certainly helped the Danes to broaden their missionary network. Oluf Madrup, a Danish missionary, grudgingly admitted, 'I cannot describe how the Moravians have insinuated themselves in so short a time into the goodwill of the Danes, French and even Hindus by their voluntary humility and angel-like conduct as they consider it'; he added, ' . . . if they were as pure in their doctrine and teaching as their life is outwardly to the eyes of the world, there would not be a sect or race to equal them in the whole of Asia. . . .' Madrup continued, 'The Natives therefore call them the "Saints" or the "Nyanigol" which means "the wise men", that being the name which they gave here to some of their own holy men.'[32] In Ceylon and Nicobar, however, severe health casualties restricted Moravian influence. By 1803, the Moravians had almost disappeared from Tranquebar, the last of them having left for Europe after selling their land and garden.

It is true that a major part of the engagement of the Danish missionaries in India was with Western medicine. Until the end of the eighteenth century, medicines prepared in the pharmacy of the Francke Foundations were sent to India. Schwartz introduced vaccination against chicken pox. Similarly, the Moravians were trained both in medicine and natural history as a part of their missionary preparation.[33] In this globalization of plants and therapeutics the Moravians seemed to be unwilling participants, as their relationship with modern science and knowledge was rather ambivalent.[34] Their own world-view was militantly anti-rationalist. They wanted only to live by their understanding of God's word, rejecting certain forms of modern life, particularly the increasingly prominent material and secular aspects of its social existence.[35] The Moravian settlements in these parts sought new ways of life, a fresh repository, and an alternative morality.

The missionaries had among them people of varied interest. The early missionaries, Ziegenbalg, Plütschau, and Grundler, had as part of their religious duties built-up important linkages with 'locals' including the Sidhha practitioners. They had also read various texts. Ziegenbalg and the other German missionaries maintained diaries and sent them home periodically. These diaries, together with many station registers, letters, and private papers focused on Tamil life—festivals and temples, arts and crafts, music and dance, legends and fables, rituals and religious practices, ceremonies, customs and manners, diseases prevalent among them and the medicines used.[36] Ziegenbalg initiated conferences and exchanges of letters with Hindu priests. The reports of these conferences are compiled in the *Genealogy of the Malabarian Gods*, which was sent to Europe for publication. In a letter to Europe about local medical practices and how they would amaze the European physicians back home, Plütschau wrote:

> *Ars Medica*, or the Art of Physick, is held by the *Malabarians* in great vogue and Esteem, and exalted above all other Arts and Sciences. Those that do professedly study it, take the greatest Pains imaginable to search into the Secrets of Nature. There are extant many useful Composures on the *Art* of *Physick*, some whereof I have read with

Pleasure and Satisfaction, and often wished to be at Leisure to translate but one into *High Dutch* for your Perusal. I do not question [that] our Physicians in *Europe* would wonder at the Performance of our *Malabar* Doctors here.[37]

His comments regarding the efficacy of European medicine in Indian climates were interesting, 'Our black Physicians here do give out that Medicines prepared in *Europe*, and transported to *India*, have not that Power and Efficacy here, which perhaps they may have in those Places where they have been prepared.'[38]

John Peter Rottler, a Lutheran missionary with the Danish settlement in Tranquebar, published descriptions of the new plant species of his own collection in Europe; they appeared in Nova Acta Acad. Nat. Curiosorum at Berlin along with a *Dictionary of Tamil and English Languages* published in 1834. Rottler was responsible for the garden at the settlement.

The Hospital Board on receiving Patrick Russell's Coromandel specimens from London in 1788, requested the government for an apartment for the specimens. They added that some 'enclosed ground that might serve as a Botanical Garden for supplying new and fresh specimens should be appreciated and an intelligent Botanist appointed to the immediate care'. They recommended Rottler as he, they mentioned, 'has made great proficiency in the Knowledge of Botany, in which way it will serve as a deposit of Botanical discoveries and excite the exertions of such as may be qualified to make additions'.[39] The government, however, showed not much interest and insisted on appointing someone from the English military establishment.[40] Eventually one Ponton from the Guntoor Circar was appointed for the work on Russell's chest.[41]

Rottler was acknowledged for his contribution by Ainslie in his *Materia Medica* which was dedicated to him, 'To Reverend Doctor J.P. Rottler, the following pages are inscribed, as a token of respect for his scientific celebrity, of Esteem for his Private Character, and in Grateful Acknowledgement of the kind and liberal aid which has been received from him, by his sincere friend, Whitelaw Ainslie.'[42] However, the most important medical compilation was that by Johann Ernst Gruendler. He never left India, spending his life studying Tamil medical texts. To study

Tamil culture and habits closely he left Tranquebar and settled in Poreiar, a nearby village. There he started eating and dressing like the indigenous people and acquired from the local people, many medical palm-leaf documents on diseases, medicines, and herbs. This comprised his *Malabar Medicus* (Malabar Physician).[43]

Another Danish missionary, Christopher John, with whom Roxburgh had regular contact, prepared a list of several plants and their medical uses from this text for Roxburgh, with names written in the local script.[44] Roxburgh had persuaded John to study the local vegetation and send him specimens and notes. This corresponded with John's own interest in gardening and local flora, and they formed a close bond. John inspired Roxburgh in his pursuits in remote Samulcottah, 'Go on in your zeal in studying the beauties and wonders in the works of our Glorious God and in promoting this excellent knowledge for the benefit of your fellow creatures.'[45] This relationship unfolded in many terms, in medicine, natural history and friendship. John often referred to Roxburgh as 'My best and most dear friend.'[46] John once wrote to Roxburgh, 'The days when I get a letter from you revive my spirits again, I want you therefore as friend, as Doctor and as Natural Philosopher.'[47] About Father John, Roxburgh wrote to Charles Taylor, Secretary to the Society for the Encouragement of Arts, Adelphi, 'I must tell you that John is the Revd. Dr. John, a most worthy Dane, at Tranquebar near Madras, an able friend and zealous philosopher.'[48] He also sent breadfruit from Nicobar to Roxburgh. John had also appointed a 'black Natural Philosopher' to collect plants, shells (of particular interest to John), seeds, insects, fishes, in Nicobar, which he shared with Roxburgh.[49] This assistant was particularly instructed to send poisonous roots from Nicobar to Roxburgh.[50] John also offered to get cinnamon seeds from his 'Dutch friends' in Ceylon.[51]

Roxburgh was particularly insistent on medicinal plants, about which he often wrote to John, and the missionary was happy to oblige with his collections in the localities as well as where his missionary friends were posted.[52] Father John was thus the crucial link for Roxburgh in collecting medical plants from Ceylon and the Nicobar islands. He talked of his missionary links (his 'indolent friends' whom he likened to 'the ancient inhabitants of Creta') in Ceylon, who could help Roxburgh acquire plants.[53] In Ceylon, the Danish Lutheran Engelhard collected for Roxburgh cinnamon plants.[54] John's Dutch friends also obliged for the same.[55] In 1793 John wrote to Roxburgh of a species of *Pandanus* found in the Nicobar, about whose local medical and nutritive uses the Moravian Hensel had informed him. He told Roxburgh that collecting plants was not easy in the Nicobar because of the high casualties, but added, '. . . I shall do all in my power to get you some' (A tree or shrub of the genus so called, belonging to the family Pandanaceæ, native to Malaysia, tropical Africa, and Australia, this plant is distinguished by forked trunks with thick aerial roots, long, narrow, prickly leaves arranged in spiral tufts, and large, sometimes edible fruits resembling a pineapple.[56]) He was particularly interested in spreading Roxburgh's fame as a botanist, particularly in the countries of his and Moravian origins. As early as 1789 he mentioned, 'Your great zeal and success in Botanicks affords me the highest pleasure and will afford the same to the learned Botanists in Denmark and Germany to whom I publicly will mention it together with the useful observations you frequently make in your letters.'[57] He also offered to get books on botany for Roxburgh from Europe, through his contacts.[58] John also shared his own research on snakes and snake poison with Roxburgh.[59]

Revd John was also urging the need for informal peripheral networks of collections. He added:

> Those who cultivate this study in Europe of whom very few can travel into foreign countries must depend upon their friends who have better opportunities for gratifying their desires by sending them their observation and collections. Amongst these Missionaries and Mission servants, and those who correspond with them, their native pupils and itinerant catechists have often been of great service on their journey by procuring curiosities from distant countries and islands.[60]

He mentioned that the Danish mission in Tranquebar always maintained through its connections a garden and nursery where various local and exotic plants were cultivated.[61]

William Roxburgh sent three papers on the hunting of the

wild hogs, on the pearl fishery, and the Description of Royal Tube that he received from John at Tranquebar for publication in the *Transactions* of the Society.[62] Benjamin Heyne was in regular contact with the Asiatic Society. He forwarded to Codrington Edmund Carrington a draft for 4 gold *mohurs* to cover his admission fee and also for quarterly subscription for 1797.[63]

Christian Friedrich Schwartz who arrived in Tranquebar in 1750 as a Lutheran missionary also worked in Trichinopoly and Tanjore, and learnt Tamil, Telugu, Sanskrit, and Persian. His linkages with the English were particularly strong. During the Anglo–French rivalries of 1764, he was requested by the British Commandant Major Preston to attend those wounded in the battle.[64] Although originally part of the Danish mission, from 1767 Schwartz was sponsored by the Anglican 'Society for the Promotion of Christian Knowledge' of London and in 1768 the English East India Company appointed him chaplain to the garrison of Trichinopoly.[65]

Among the many English doctors Schwartz befriended was Paul Jodrell who went to India as Physician to the Nawab of Arcot in 1787. Jodrell lived in the Nawab's palace, and had his own garden where he reared plants. Schwartz was a regular visitor, although Jodrell who had recently arrived from England did not approve of his missionary skills. In his diary Jodrell wrote, 'In the afternoon I went to Church and heard a Sermon from Mr. Swartz the Missionary who preached extempore in vile language and with a disgusting manner.' He scoffed at the Lutheran's fascination for the English, 'He (Schwartz) considered himself an "English missionary" and signed himself "Swartz".'[66] Incidentally the English records always refer to him as 'Swartz'.

The Moravian connection in this collection of medical plants commenced when Schwartz wrote to Father John asking for Roxburgh for help in settling Moravians in the British territories where he hoped to gain from their knowledge of agriculture in these parts. This not being possible, he transferred the Tranquebar Moravian Benjamin Heynes to Samulcottah.[67] In August 1793 Heyne was allowed by the Moravian order to join the British territories.[68] Along with improving plantation activities in Samulcottah, like introducing potato and breadfruit cultivation,

Heyne wrote the *Tracts, Historical and Statistical, on India and Sumatra 1814,* an important compilation regarding plants, natural resources, and medical preparations of the area, including the medical text, *Kalpastanum.*[69]

These textual and botanical interests were part of their larger engagement with the study of Indian languages, particularly Tamil, and culture. The Tranquebar missionaries were eager to learn Tamil and study its rich and ancient literature, mainly to get closer to the people among whom they had to work. As Ziegenbalg arrived at Tranquebar he found that the predominant idea among missionaries that Malabarians were 'barbarians', strange. Ziegenbalg conveyed his impressions to Francke, the pastor in Halle, who was instrumental in sending him to Tranquebar and requested him to have the letters published so as to remove this stereotype. He instead found that the local people discussed 'the same philosophical subjects as the Savants of Europe, and that they had a regular written law, wherein all theological subjects were treated of and demonstrated'. Francke, however, said that 'the missionaries were sent out to exterminate heathenism in India, not to spread heathen nonsense all over Europe'.[70]

This influenced the study of medicine as well. Benjamin Heyne, considering most of Indian practitioners to be quacks, argued that their knowledge was not altogether false:

> The medical works of the Hindoos are neither to be regarded as miraculous productions of wisdom, nor as repositories of nonsense. Their practical principles, as far as I can judge, are very similar to our own; and even their theories may be reconciled with ours, if we make allowance for their ignorance of anatomy, and the imperfections of their physiological speculation.[71]

However, sceptics in Europe like Francke were not always ready to acknowledge such studies or the missionary medical contribution. As pursuers of modern science, missionaries had little standing in Denmark. In 1791 when the Danish mission surgeon Martini died, the missionaries decided to appoint Klein, who had just reached the shores and had attended medical lectures in Copenhagen, as the new surgeon. They wrote to the physicist, Professor Christian Gottlieb Kratzenstein of the Medical

College in Copenhagen, for his opinion. Kratzenstein was dismissive of the whole plan and of the possibilities of scientific pursuits:

> I do not know Klein, but I think it would be advisable to grant the doctor to the Missionaries who is proposed by themselves; for the situation of a medical man in Tranquebar is not an enviable one; it is something like a penitential parish here. Any one who wishes to obtain such a place must either be a zealous Pietist desirous of gaining a higher place in Elysium, or an enthusiastic Linnéanist, like König (who had held this position before) who after all became blind and died.[72]

The French scholar Sonnerat, Commissioner of the Marine, Naturalist to the King in his 'Voyage to the East Indies' (Sonnerat was sent to India at the behest of Louis XVI) commented that Indians were mostly pretenders to some knowledge of medicine with no understanding of anatomy; the physicians were mostly comprised of individuals who have been washermen, weavers, or blacksmiths a few months before. . . .'[73] And Hamilton Buchanan found the Sanskrit texts extremely confusing, if not untrustworthy, for the identification of local medicinal plants.[74]

The medicinal botany projects were attempts at 'enrichment', both for the heathens as well as the modern sceptics. The Pietists and the Moravians settled on the Coromandel Coast, like the Orientalists, were dissenters and their work was shaped by a rejection of certain forms of modern life, particularly the increasingly prominent material and secular aspects of social existence. These pursuits were shaped by a search for a new way of life, a new world-view, a fresh repository of experience, and an alternative morality.

One strain of imperial historiography highlighting the dynamic, creative, and regenerative aspects of the colonial encounters, particularly of the eighteenth century, in an attempt to situate these local histories as important sites of global understanding. C.A. Bayly has suggested that it is important to see these zones not as sites of degeneration, but of vitality. This has been the suggestion for history of science for the colony as well: to highlight the growing creativity and vitality of colonial knowledge systems, resembling the vibrant material culture emerging in these regions out of interactions, encounters in the eighteenth century. Richard Grove looks at this local-global position, however, in another direction and reveals the problems engendered in such approaches. In a search for the roots of present global ecological concern, his focus is on ecology, highlighting the 'global' emergence of ecological thought, and the significance of the local in this apparently metropolitan culture. This is an example of work focusing on interactions and peripheral creativity and thus accretion with the social and political developments influencing the local sites.

I suggest that eighteenth-century history tends to suffer from an exoticization of networks and collections as hybrid exchanges. There is a tendency to see in these exchanges an agenda that blunted the violent edge of colonial modernity. The journey to the empire and the eighteenth century from what appears to be a monolithic Western science is sometimes marked by a search for the myriad roots of that science. To that extent this is sometimes an ecumenical journey. Yet networks were not always idyllic nodal points of exchanges. Kapil Raj has sounded a very important word of caution. In his work on colonial encounters in the forging of new knowledge, he says that Indians and Europeans participated equally in a dialogue process in an idyllic 'commonwealth of letters'. On the contrary, the kinds of knowledge discussed in Kapil Raj's essay could only be constructed and sustained within a strong framework of formalized institutions with their imperatives of teamwork and a stratified division of labour.[75]

Londa Schiebinger has provided an interesting example of transfer of knowledge. She shows that the 'peacock flower', brewed by slave women into an abortifacient, to ensure that they would bear no children into oppression, never flowed into Europe while other plants and herbs were being shipped frequently. Sometimes networks did not translate to scientific knowledge.

I now discuss the bark *Swietenia soymida* discovered by Roxburgh and tested and promoted through these local and international botanist-missionary networks. It will help to demonstrate how networks were not always conducive to the free and liberal flow of knowledge and information.

INTRIGUES OF THE SWIETENIA BARK

Confusion reigns regarding the botanical and ethnological roots of the *Swietenia soymida* tree. Roxburgh claims to have collected the species from the forests of the Rajahmundry region (present Andhra Pradesh) and that he 'accidentally' came across its medical virtues. He does not mention how exactly that accident happened or if local practitioners traditionally used the bark for medicinal purposes. But there is some evidence that it was indeed traditionally used as a medicine, as a febrifuge although not for malaria. Forbes had mentioned that soymida was 'reckoned a good stomachic and powerful febrifuge' and that the bark of *Cedrela toona*, a congener, as well as that of *Chikrassee* was used as an astringent.[76] In John A. Parrotta's *Healing Plants of Peninsular India*, Soymida is 'a popular drug in traditional Indian medicine, and is credited with antiperiodic properties. It is used in the treatment of diarrhoea, dysentery, and intermittent fevers, including malarial fever, and as a bitter tonic to treat general debility.'[77] It is also mentioned by V.V. Sivarajan and Indira Balachandran that the bark is traditionally used as a stimulant and in diarrhoea and dysentery.[78]

Roxburgh discovered the root and its virtues in 1791 and published his treatise on it in 1793, which was in fact written to the Court of Directors to promote his bark.[79] The report is an interesting text; it includes the detailed chemical researches that he undertook to establish the qualities of the bark. He experimented with chalybeate, limewater, vitriolic acid, vinegar, mild or caustic vegetable alkali, magnesia, and found that the antiseptic powers of this bark kept pace with its bitter and astringent virtues.[80] Roxburgh's conclusions were that the active parts of the bark of *Swietenia febrifuga* were more soluble than those of the Peruvian bark, particularly in watery menstruums; that it contained a much larger proportion of the active (bitter and astringent) powers than the Peruvian bark; and finally that the watery preparation of the bark remains good much longer than similar preparations of the Peruvian bark, or similar preparations of it.[81]

Andreas-Holger Maehle has suggested that a shift took place in eighteenth-century ideas of 'specifics' when a remedy was no longer defined through a single target, such as intermittent fever, but through its uniquely powerful and not fully understood mode of action, which was beneficial in a number of diseases.[82] Roxburgh promoted the bark as useful not only for malarial fevers, but also for gangrene and other putrid maladies. He conferred with William Cullen of the Edinburgh University, who described the recurrence of the paroxysms in terms of the recurrence of the atony (or the lack of tone) in the extremities of the arterial system and the use of simple bitters to obviate this atony.[83] Roxburgh concluded that 'this New Bark, from the successful experience I have had with it in intermittent fevers. . . . I have every reason to imagine it will prove equal, if not superior, to the Peruvian bark, for every purpose where that medicine is used.'[84] Significantly, in the same paper Roxburgh also mentioned the several plants and roots 'perfectly unknown to European botanists' that the poor natives of the mountainous parts of the coastline consumed for nutritional value.[85]

What is important to us is that Roxburgh's conclusions were drawn from an interesting network of trials and experiments carried out on the bark in the Madras Presidency. He had sent the bark soon after discovering it in 1791 to his trusted friend Revd. John to try out at the local hospitals and to the mission surgeons in Tranquebar.[86] In July 1792 John also wrote to him about the trials on the Swietenia bark.

A month later Father John reported the success of Swietenia, 'Your Bark Swietenia Febrifuga gets here the highest reputation by Dr. Klein and Dr. Folly and is applied in the Hospital here.'[87] In the same letter he asked for seeds of the Swietenia tree. He added that they had plans of sending it to Copenhagen as well:

> Dr. Folly writes to me just a letter mentioning his best Respects to you and his wishes to get more of this valuable bark and that he will send to you and Prof. Jade at Copenhagen . . . the cases in which he has applied it. As we have sent last February, only small quantities to Copenhagen, Berlin . . . and Trasburgh which will not suffice to make experiments we wish to send now more and especially to Prof. Jade at Copenhagen.[88]

In November of the same year, John mentioned that eighty samples of the bark had been sent to Winslow and Jade at

Copenhagen, 'both of high reputation'.[89] Soon, for its interesting networks, Tranquebar rather than Madras or Samulcottah became the site of success for the Swietenia. Results were coming in regularly, as John wrote:

> . . . Mr. Johnson thanks you for his perfect recovery (from the) long luring fever by your bark and Capt. Wuhldorff was likewise restored to health by it. In the hospital it has not had always this effect as Dr. Folly says. Our Doctor Klein has cured Mr. Johnson by it, but is now with Dear Rottler who was near his death by a nervous fever, since two months in Madras and Vellore. . . . Amongst the many Febricitands this season Dr. Folly was the most dangerous but escaped death and is now recovering. He thanks you very much for his part of the bark. . . .

Andrew Ross, the company merchant in Madras, helped in the trials elsewhere in the Presidency. He wrote to Roxburgh that Dr. Duffin at the hospital in Madras had found it useful in the limited use that he had put it to. He added, 'I shall obtain a Declaration from him, to be sent home in support of it. . . .'[90] In 1792 Ross wrote, '. . . Dr. Duffin & his assistants at the Hospital made some trials of your Bark, but not enough to give Certificates—as yet.'[91]

Trials of the bark had thus started within the Madras medical establishment. Roxburgh was keen about the results of the experiments at the hospitals. In March 1793, Ross mentioned that he had spoken about the bark, to Chancellor Oakley who suggested that Roxburgh write about it to the Madras Hospital Board.[92] Roxburgh sent the bark to James Anderson in Madras in February 1793.[93]

> Two years having elapsed since I accidentally discovered that the Bark of a new species of Swietenia was possessed of the sensible qualities of Peruvian Bark in a very concentrated state, this led me to ascertain its virtues in the Cure of such Diseases as the Peruvian Bark is employed to remove, and I am now happy to say that it has been very successfully administered, and not only by myself but by many others in various parts of India for the last two years.

On 5 March 1793, Valentine Conolly, Secretary to the Hospital Board, at Fort St. George, wrote to the head surgeons in the provinces, Raine, Gahagan, Binny and Mein:

The Hospital Board have received from Mr. Roxburgh a quantity of the Bark of a new species of Swietenia which from several experiments and observations he has made seems to possess the qualities of Peruvian Bark in a more concentrated state, and for which it is proposed as a Substitute. In order that its Virtues may be more fully ascertained the Apothecary has been directed to transmit to you a specimen of it and the Hospital Board request you will take every favourable opportunity of exhibiting it in your Hospital and report to them the result of your trials.[94]

Gahagan soon replied to Conolly from Vellore promising to try the Bark.[95] Nicol Mein wrote to Conolly on 26 March 1793 from Trichinopoly saying that he had only one patient suffering from remittent fevers but that he was recovering, and added that the Bark seemed to contain a higher degree of astringency than the Peruvian one.[96] George Binny wrote from Masulipatam to James Anderson and the Hospital Board, saying that he had heard of the Bark from Roxburgh and had tried it in some cases. 'I believe it will turn out a very useful Medicine, but I wish to speak of it with caution, till by greater experience, its effects are more clearly ascertained.'[97] On 4 June 1793, Nicol Mein wrote to Valentine Conolly, Trichinopoly, regarding his success with Swietenia,

. . . there have been several cases of Intermittent fever in the General Hospital now in which it proved a successfull [*sic*] cure when the Cortex Peruvian had been tried and failed. . . . This account of the efficacy of the Cortex Swietenia having been immediately under my own eyes and the cure of the Intermittent fevers corroborates Dr. Roxburgh's observations and I hope will prove satisfactory to the Physician General and the Members of the Hospital Board.[98]

Ross kept Roxburgh informed about the progress of the trials, and that the doctors in Madras had promised to promote the use of the Bark. 'Dr. Berry has read your dissertation upon it & will also support it.'[99]

In 1793 there seems to have been a turn of events. Apparently impatient of the time taken in confirmation from Madras hospitals and probably suspecting Anderson's role in that, Roxburgh seems to have suggested that Ross offered a share of the advantages that might arise from the new Bark to Dr. Berry, to gain his support. Ross wrote back that such a step '. . . would amount to an indirect acknowledgement that its (Swietenia's) intrinsick [*sic*] quality, was not good enough . . . it would be a derogation to your character, to take any indirect course, of bringing it to any further test, which may be required, of public enquiry. . .'.[100] He continued, 'I have no reason to doubt that the means which Dr. Anderson is pursuing (which I find he has himself just mentioned to you) of ascertaining the true character of the Bark—is not only in due form, but fair in intention.'[101] A few months later Ross wrote to Roxburgh with more news:

of the four hospitals to which applications had been made—an answer has been received from Mr. Meine of Trichinopoly alone and herewith I send you a copy of it—why the other three had not given an opinion he cannot account for—tho' surely it is very improper—nor the less that Capt. Alex Reid (who is here on a visit from the unwholesome country of Burrah-mal) told me that he has applied for the Bark both at Vellore and here & and that they have none to give him; . . .[102]

Ross then wrote to John Forbes (Sr.) in Bombay about the Swietenia bark, asking him to try it out in the hospitals there. He stressed that this new bark was as good as if not better than the Peruvian bark,

[It] . . . is possessed of the same virtues as are found in the Bark of Peru & even in a higher degree in which opinion he (Roxburgh) is confirmed by repeated trials & and which have been made of it by himself with the best success & have been confirmed by experiments made by some of the most eminent men in . . . both at Bengal & on this coast . . . an article of so much utility to the health of mankind should be more generally used than it is & therefore make no difficulty in conveying that sample . . . to you that you may put it into the hands of Physical Gentlemen at your place to give them the opportunity (we cannot but be grateful to them) of making proper trials of an article. . . .[103]

Ross elaborated that the 'great advantage' in commerce that such a discovery would have. It would be 'of the highest utility' to India especially because of 'its greatest freshness over that of Peru', as 'great advantage' would accrue to the public 'by its being the produce of the Company's own estate instead of the large imbursement of its value which is made to another nation. . . .'[104] He also mentioned that Roxburgh intended to test it as far as Constantinople, where malaria was reported to be rampant. He imagined that this would be a way for it to get known among the European physicians. In the next letter which went with the same despatch, Ross added that he was sending a larger quantity so that the trials could be made not only in the hospitals of the Presidency, but at each of the principal hospitals at the subordinates and added that that was the method adopted in Madras.[105]

If verification in the colony was a problem, recognition from the metropolis was a further complicated issue. In 1791 Roxburgh complained to the Madras government about the neglect of the present Court of Directors in

> . . . promoting and promulgating useful discoveries made by their own official Servants, and in their own territories, however, that, I find from experience, cannot make me neglect my usual pursuits and there are many ways of making known to the community anything that I may be able to discover that promises general benefit, while that can be done, their neglect shall give me but little uneasiness, your assistance I conceive to be of more real use, and your letters now before me that I may rely on having it, in the fullest extent of the word.[106]

His complaint was particularly against Alexander Dalrymple (FRS and Hydrographer to the Company and the Royal Navy). He mentioned that Dalrymple had published materials from his collections without consulting him. He added that he had sent to Dalrymple a paper on Swietenia together with a letter addressed to the Court of Directors, as well as a copy of the same to the Madras government, to be delivered to the court by them separately.[107] Roxburgh's concerns about Dalrymple had grounds. By this time he had established his position in London. He was a close friend of Joseph Banks and a regular member of the Royal Society, dining weekly under Banks' chairmanship. He became involved in the geographical aspects of many questions of science and government.[108]

This was also precisely the time that Roxburgh was trying to utilize several networks for the international recognition of his bark. In 1791 he sent Joseph Banks a package of the Swietenia bark, which he had analysed and employed medicinally. He also asked Banks to pass some on to his friend Dr. Patrick Russell.[109]

Roxburgh was eager that his discovery of Swietenia should attain popularity in England and that the paper be published in the *Transactions of the Royal Society.*

Around the same time Roxburgh was continuing his communications with James Edward Smith (1759–1828), scientist, collector, and founder of the Linnaean Society in London, about his new bark. In 1792 he sent him the 'fever bark' through William Farquhar (1774–1839) and wrote, 'I have hitherto found this bark infallible in the cure of the intermittents . . . and gangrenous ulcers. I refer you to Dr. Russell for a reading of my paper on it.'[110] In March 1793 he wrote again to Smith about the success of the Swietenia bark, 'It gives me particular pleasure to observe to you that the new Swietenia bark continues to support to the utmost of my wishes, the character I first gave it, seldom or never failing to perform speedy cures in all kinds of Remitting and Intermittent fevers, even after the Peruvian bark had faild [*sic*] . . .'. He added that the Circars were particularly suited for the experiments, '. . . our intermitting fevers throughout the Circars have been both frequent and severe this season, so that we have had abundance of cases to try it in, you would receive some of it from Mr. Farquhar the purser of the Atlanta.'[111] Roxburgh continued to write to Smith about the bark, particularly expressing his hope of success in Europe, from Calcutta after his appointment as the Superintendent of the Garden there. 'The Swietenia Bark continues to support its character well, I hope it proves of equal benefit in Europe. It has answered so well with the Batavia Fever, that the physician general there applied for and got 1000 pounds of it a few weeks ago.'[112]

A few years later, in 1796, Banks wrote to Roxburgh. 'The plants you have sent home by the Royal Admiral arrived in the finest order possible.' He wrote candidly and rather critically about the physicians in England at their lack of enthusiasm for 'new' knowledge, 'I rejoice to hear that your Swietenia Bark gains credit in India. Here it's scarce possible for a new medicine to gain a place in the Dispensary. Physicians here who gain more in a week probably than they deserve in a year care not how few medicines their Materia Medica is composed of, the fewer their number the less expense of thought in making use of them.'[113] Thus the scientists in the metropolis did not always warmly welcome networks of new knowledge and information.

Banks's pessimism was justified as the final outcome of the trials of the Swietenia bark in England demonstrates. Roxburgh's bark never gained acceptance in Western medicine. Behind the many chemical experiments and clinical 'trials', personal ambitions and commercial pressures were barely hidden. There was also a bias for indigenous or locally developed astringents. Johannes Fawssett in Edinburgh (who wrote the *Dissertatio medica inauguralis de febribus intermittentibus medendis*, Edinburgh 1798) advocated in its place a combination of the astringent bistorta and the bitter calamus aromaticus. He claimed this had been successful in 'almost a hundred cases' of intermittent fever in Horncastle Hospital, Lincoln and even better than the yellow bark.[114] In all this play of personal interests and searches in London and elsewhere in Europe Roxburgh's bark had little chance.

The Edinburgh physician Andrew Duncan wrote his dissertation on the Swietenia based on the samples sent by Roxburgh. He admitted that he never saw the bark in its original form and depended on Roxburgh's description of it.[115] In his *Edinburgh New Dispensatory* he wrote that that it was found in a condition where they could not be distinguished from the 'kind of the shops'.[116]

Trials of the anti-malarial specific took intriguing turns in the years to come. The same Duncan had also made the important discovery of isolating the principle *cinchonine* from cinchona, as related in Nicholson's *Journal.*[117] Cinchonin was an alkaloid, $C_{19}H_{22}N_2O$, which Duncan derived from the bark of various cinchona trees, to be used as an antimalarial agent. But there is a prehistory to the discovery. The metropolis at this time was faced with a great variation of similar species and substitutes of Peruvian bark, and often with very confusing botanical and ethnological roots. There were strong efforts made around 1800 to identify the 'active principle' of Peruvian bark. Thus in his *Dispensatory*, Duncan, following his chemical orientations, expressed a desire to identify the essential cinchona: 'the preparation of an extract of cinchona, which may contain its active principles in a concentrated form, is a desirable object'.[118]

Later in the same year, 1803, he announced the discovery of the *cinchonine.*[119] Duncan was motivated by the French chemist and assistant of Lavoisier, M. Seguin's 'Abstract of a Memoir on the Febrifuge Principle of Cinchona', published in an earlier number of Nicholson's *Journal.*[120]

Cinchonine seemed to hold some clues to the essential principle of many confusing anti-malarial astringents, including Ipecacuan, Colomba, and Angustura.[121] Accordingly, he suggested determining the 'cinchonin' content of barks to assess their quality. In *Swietenia febrifuga,* Duncan did not find any cinchonin in 1803.[122] The French chemists Pelletier and Caventou in 1820 finally discovered that the febrifuge principle was actually present in two alkaloids, called *quinine* and *cinchonine*, with the same virtues, but much more powerful in *quinine.*[123] Writing in 1829 on the discoveries on Pelletier and Caventou, Duncan commented that their discovery of *quinia* and the salts formed by acids with both bases, 'completed our knowledge of the *essential* principles of cinchona bark' (emphasis added).[124] He was careful to point out that his claim as the 'first discoverer of the essential principle of cinchona bark is now fully admitted by the French chemists' although 'it was long erroneously ascribed to Gomes'.[125] In this debate on priorities and essentials, involving personal stakes, Swietenia was never tried for quinine once the latter was discovered and never became an antidote for malarial fever.

CONCLUSION

The networks of piety, profit and knowledge that had developed in eighteenth-century India for the study of the 'Other' Eastern culture and flora and for the promotion of Christianity did serve to foster Western science in various ways. However, the processes of collection of knowledge had diverse motivations. These networks demonstrate unique collaborations between spiritual, intellectual, and mercantile interests in the cause of science. I have attempted to assess the role of the informal peripheral networks in the advancement of scientific knowledge. While such peripheral networks did gather myriad information, verify facts, facilitate experiments, they did not ensure scientificity. Thus a focus on networks alone is thus inadequate.

Andreas-Holger Maehle has demonstrated that European practitioners' increasing reliance on contemporary pharmacological explanations 'eroded' cinchona's status as a 'specific' against intermittent fever, leading instead to its recommended use against a broader array of diseases including gangrene, ulcers, and smallpox. Its efficacy in treating both internally and externally manifest diseases allowed cinchona to become a central therapeutic agent for physicians and surgeons. Maehle suggests that the perception of Peruvian bark as a general tonic eventually 'opened the floodgates' and made it a universal remedy. This led to Samuel Hahnemann's 'principle of similarity', the centerpiece of homoeopathy.[126]

But this is as much a story of the disappearance of the alternatives as it is of the emergence of cinchona and quinine as the panacea. The search for a universal remedy was not limited to cinchona. The case of swietenia suggests that it was tried with other barks as well, but only cinchona could become the universal remedy. Not that soymida was the only bark or remedy to be discarded, but it is illustrative of those which used networks, discussed by all the leading men of science of that time and yet dropped out from ultimate factual verification for various social and cultural factors. It was one of the many competing remedies that struggled to get recognition in Europe.

NOTES

1. Sujit Sivasundaram, 'Natural History Spiritualized: Civilizing Islanders, Cultivating Breadfruit, and Collecting Souls', *History of Science*, vol. 39, no. 126, 2001, pp. 417–43.
2. Brockway, *Science and Colonial Expansion: The Role of the British Royal Botanic Gardens*, New York, 1979, p. 8.
3. Richard Drayton, *Nature's Government: Science, Imperial Britain, and the 'Improvement' of the World*, New Haven, 2000.
4. Whitelaw Ainslie, *Materia Indica, Or, Some Account of Those Articles which are Employed by the Hindoos, and Other Eastern Nations, in their Medicine, Arts, and Agricultural; Comprising also Formulae, with Practical Observation, Names of Diseases in Various Eastern Languages, and a Copious List of Oriental Books Immediately Connected with General Science*, n.d., p. xxxvii.

5. Ibid.
6. 'Letter from Andrew Ross to John Forbes, 11 December 1789', Madras, Mss ROX, 1793, Botany Library, Natural History Museum.
7. Sam Browne and James Petiver, 'An Account of Part of a Collection of Curious Plants and Drugs, Lately Given to the Royal Geographical Society by the East India Company', *Philosophical Transactions*, vol. 22, pp. 579–94.
8. *Transactions*, vol. 22, 1700–1, pp. 849, 856. Also *Transactions*, vol. 23, 1702–3, p. 1456 (Dr. Solan's History of Jamaica).
9. *Transactions*, vol. 23, 1702–3, pp. 1252–3.
10. Ray Desmond, *The European Discovery of the Indian Flora*, London, 1992, p. 39.
11. John Fleming, 'A Catalogue of the Indian Medicinal Plants and Drugs, with Their Names in the Hindustani and Sanscrit Languages', *Asiatic Researches*, vol. XI, 1812, p. 158.
12. *The Hindu*, 11 June 2001.
13. A.M. Lysaght, *Joseph Banks in Newfoundland and Labrador, 1776: His Diary, Manuscripts and Collections*, London, 1971, p. 34.
14. Harold J. Cook, 'Physicians and Natural History', in N. Jardine, J.A. Secord, and E.C. Spary, eds., *Cultures of Natural History*, Cambridge, 1996, pp. 92–3.
15. Ibid., p. 103.
16. It was partly the influence of this call for knowledge on Lord Wellesley, the Governor-General, that led to his encouraging his brother Arthur, later the Duke of Wellington, to launch in south India the three great surveys that led to the founding of the Zoological, Botanical, Anthropological and Archaeological Surveys of India and the Survey of India. The route surveys that led to the first group of institutions were undertaken in 1800 by Francis Buchanan, whose findings were recorded in *A Journey from Madras Through the Countries of Mysore, Canara and Malabar*, London, 1807.
17. Whitelaw Ainslie, *Materia Medica of Hindoostan, and Artisan's and Agriculturalist's Nomenclature*, Madras, 1813, p. i.
18. *Hortus Indicus Malabaricus*, Amsterdam, 1678.
19. Furber, *John Company at Work*, Harvard, 1948/1951, pp. 110–20.
20. Roxburgh to Banks, Samulcottah, 24 November 1782, Banks Correspondence, BL, vol. 1, 1765–84, p. 181.
21. C.S. Mohanavelu, *German Tamilology: German Contribution to Tamil Language, Literature and Culture during the Period 1706-1945*, Madras: South India Saiva Siddhanta Works Pub. Society, 1993, pp. 35–7.
22. Letter from Christopher John to William Roxburgh, 17 August 1793, Tranquebar, Mss ROX, 1793, Botany Library, Natural History Museum.

23. *Thirty Four Conferences between the Danish Missionaries and the Malabarian Bramans (or Heathen Priests) in the East Indies, Concerning the Truth of the Christian Religion: Together with some Letters written by the Heathens to the said Missionaries*, Translated out of High Dutch by Mr. J. Thomas Philipps, London, MDCCXIX, pp. 343–52.
24. Denmark established a trading station at Tranquebar in 1616. Almost one hundred years later two Germans, Ziegenbalg and Heinrich Plutschau, arrived in Tranquebar as missionaries sponsored by King Frederick IV.
25. Johan Ferdinand Fenger, *History of the Tranquebar Mission: Worked out from Original Papers*, published in Danish and tr. P.H. Emil Francke, Tranquebar, 1863.
26. Mohanavelu, *German Tamilology*, pp. 12–27.
27. J.C.S. Mason, *The Moravian Church and the Missionary Awakening in England 1760-1800*, Suffolk and New York, 2001, p. 187.
28. August Gottlieb Spangenberg, *An Account of the Manner in Which the Protestant Church of the Unitas Fratum, or United Brethren, Preach the Gospel, and Carry on Their Mission Among the Heathen, Translated from German*, London, 1788, p. 39.
29. Ibid., p. 53.
30. Ibid., p. 58.
31. Ibid., pp. 83–4.
32. Ibid., p. 268.
33. James Hutton, *History of the Moravian Church, Book II—The Revival Under Zinzendorf*, Chapter 6.
34. Jon J. Sensbach, 'Globalization and Its Discontents: Religious Radicals Confront the Modern Age', Symposium, 'German Moravians in the Atlantic World', April 2002, in celebration of the 250th Anniversary of the Wachovia Settlement, Department of History, Wake Forest University, Winston-Salem, North Carolina.
35. Ibid.
36. Daniel Jeyaraj, *Bartholomaus Ziegenbals*, Delhi, 2005.
37. 'Extract of another Letter, relating to some diseases incident to the Malabarians: Likewise of some remedies they commonly use against them', *An Account of the Religion, and Government, learning, and Oeconomy, &c of the Malabarians: Sent by the Danish Missionaries to their Correspondents in Europe, Translated from the High-Dutch*, London, 1717, pp. 61–2.
38. Ibid., pp. 62–3.
39. The Hospital Board 10 October 1788, Surgeon's General's Records, vol. 3, from 1 January 1788 to 22 December 1788, G. no. 12440, Tamil Nadu Archives (hereafter TNA), pp. 200–1.
40. 'Letter from the Military Board to the Hospital Board', Fort St. George, 14 October 1788, ibid., pp. 209–10.

41. Hospital Board 22 May 1789, Surgeon's General's Records, vol. 5, from 1 January 1789 to 25 December 1789, G. no. 12442, TNA, p. 17.
42. Whitelaw, Ainslie, *Materia Medica of Hindoostan, and Artisan's and Agriculturalist's Nomenclature*, Madras, 1813.
43. Mohanavelu, *German Tamilology*, pp. 43–5.
44. John to Roxburgh, 20 June 1791 Tranquebar, Mss Eur D 809, OIOC, British Library.
45. Letter from Father John to Roxburgh, 24 November 1789, Tranquebar, Mss Eur D 809, OIOC, British Library.
46. For example, letter from Father John to Roxburgh, 24 November 1789, Tranquebar, Mss Eur D 809, OIOC, British Library.
47. Letter from Father John to Roxburgh, 13 October 1790, Tranquebar, Mss Eur D 809, OIOC, British Library.
48. Letter from William Roxburgh to Dr. Charles Taylor, undated, Mss ROX, 1789, Botany Library, Natural History Museum.
49. Ibid. John added that this 'black Natural Philosopher' was the third one that he had sent there, the other two having died in these excursions.
50. Letter from Father John to Roxburgh, 29 September 1789,Tranquebar, Mss Eur D 809, OIOC, British Library.
51. Ibid.
52. In 1790 Father John sent a list of medicinal plants collected by him to Roxburgh with details of its local uses, as desired by the latter. See letter from Father John to Roxburgh, 13 October 1790, Tranquebar, Mss Eur D 809, OIOC, British Library. In 1791 in another letter he sent a list of plants and information of their medical uses, as collected from 'Malabar Physicians', with their names written in Malabar script. Letter from Father John to Roxburgh, 20 June 1791, Tranquebar, Mss Eur D 809, OIOC, British Library.
53. Letter from Christopher John to William Roxburgh, 17 August 1793, Tranquebar, Mss ROX, 1793, Botany Library, Natural History Museum.
54. Ibid.
55. Letter from Father John to Roxburgh, 29 September 1789, Tranquebar, Mss Eur D 809, OIOC, British Library.
56. Letter from Father John to Roxburgh, 7 April 1793,Tranquebar, Mss Eur D 809, OIOC, British Library.
57. Letter from Father John to Roxburgh, 24 November 1789, Tranquebar, Mss Eur D 809, OIOC, British Library.
58. Ibid.
59. Ibid.
60. Christopher Samuel John, *On Indian Civilization, or, Report of a Successful Experiment, Made . . . in Fifteen Tamul, and Five English*

Native Free-schools; with Proposals for Establishing a Native School Society, London, 1813, p. 38.

61. Ibid., p. 40.
62. AR/L-15 1797, 5 September (Illegible) (HLS), 5 pp. (Torn.)
63. Proceedings of the Asiatic Society of Bengal, AR/l-7 1797, 9 April, Samulcottah (HLS), 3 pp.
64. Letter from Schwartz to the Mission College, 10 July 1766, in J. Ferd. Fenger, History of the Tranquebar Mission.
65. W.J. Wilson, *History of the Madras Army*, Madras, 1882, vol. 1, p. 219.
66. Copy of a Journal of Sir Paul Jodrell (1746-1803), physician to the Nawab of Arcot 1787–1803, transcribed and annotated by R.S. Seale, a descendant of Jodrell's; also copies of letters, dated 1795–98, from Rev. Christian William Gericke (d. 1804) to Lady Jodrell, Mss Eur Photo Eur 136, European Manuscript, OIOC. 'The Journal of Sir Paul Jodrell, M.D., F.R.S. July 1st 1792–May 31st 1793', p. 74.
67. Benjamin Heyne, a naturalist and Moravian missionary, resided in India for over twenty years (he was at Tranquebar in 1792) under the patronage of the East India Company. He was superintendent of the Bangalore Garden from 1802 to 1808, and collected the plants that were published by A.W. Roth in *Novae Plantarum Species praesertim Indiae Orientalis Ex collectione Doct. Benj. Heynei*, Halberstadt, 1821. He wrote the *Tracts, Historical and Statistical, on India and Sumatra*, 1814.
68. Letter from C.S. John and the Moravian brothers to Ross, August, 1793, Tranquebar, Mss ROX, 1793, Botany Library, Natural History Museum.
69. Heyne, *Tracts*, pp. 125–48.
70. Quoted in S. Viswanathan, 'Tamilology and a German Quest: Material Relating to the Study of Ancient Tamil Literature and Culture by German Missionaries in the 18th Century Could Provide New Insights into the Tamil Past', *Frontline*, vol. 15, no. 09, 1998.
71. Heyne, *Tracts*, p. 124.
72. Fenger, *History of the Tranquebar Mission*, p. 297.
73. Ainslie, *Materia Indica*, p. xxxi.
74. Buchanan, *An Account of the District of Bihar and Patna*, vol. 2.
75. Kapil Raj, 'Colonial Encounters and the Forging of New Knowledge and National Identities: Great Britain and India, 1760-1850', *Osiris*, 15(2000), p. 134.
76. Forbes, *Illustration of the Botany and Other Branches of Natural History*, p. 143.
77. John A. Parrotta, *Healing Plants of Peninsular India*, Wallingford, New York, 2001, p. 501. He also mentions that the bark forms the Ayurvedic drug *mamsarohini*, used to promote blood coagulation,

and to treat oedema, wounds, dental diseases, uterine bleeding, diarrhoea, dysentery, haemorrhage, and malarial fever, p. 502.

78. *Ayurvedic Drugs and their Sources,* Delhi, 1994.
79. William Roxburgh, *A Botanical Description of a New Species of Swietenia (Mahogany) with Experiments and Observations on the Bark thereof, in order to determine and compare its powers with those of Peruvian Bark for which it is proposed as a substitute, Addressed to the Honourable Court of Directors of the United East India Company by their most obedient humble servant Wm Roxburgh*, London, 1793.
80. Ibid., p. 16.
81. Ibid.
82. Andreas-Holger Maehle, *Drugs on Trial: Experimental Pharmacology and Therapeutic Innovation in the Eighteenth Century. Vol. 53 of Clio Medica*, Wellcome Institute Series in the History of Medicine, Amsterdam, 1999, pp. 60–1.
83. Roxburgh, *A Botanical Description of a New Species of Swietenia,* p. 18.
84. Ibid.
85. Ibid., 24.
86. Letter from Father John to Roxburgh, 26 July 1792, Tranquebar, Mss Eur D 809, OIOC, British Library.
87. Letter from Father John to Roxburgh, 17 August 1792, Tranquebar, Mss Eur D 809, OIOC, British Library.
88. Ibid.
89. Letter from Father John to Roxburgh, 26 November 1792, Tranquebar, Mss Eur D 809, OIOC, British Library.
90. Letter from Andrew Ross to William Roxburgh, 6 October 1791, Madras, Mss ROX, 1793, Botany Library, Natural History Museum.
91. Letter from Andrew Ross to William Roxburgh, 17 July 1792, Madras, Mss ROX, 1793, Botany Library, Natural History Museum.
92. Letter from Andrew Ross to William Roxburgh, 5 January 1793, Madras, Mss ROX, 1793, Botany Library, Natural History Museum.
93. Roxburgh's letter dated 17 February 1793, from Samulcottah, to James Anderson and the Hospital Board about the Bark, Surgeon's General's Records, vol. 9, from 1 January 1793 to 30 November 1793, G. no. 12446, TNA, p. 24.
94. 'Copy of a letter addressed to Messrs. Raine, Gahagan, Binny, and Mein, Head Surgeons', Valentine Conolly, Secy. to the Hospital Board, Fort St. George, 5 March 1793, Surgeon's General's Records, vol. 9, from 1 January 1793 to 30 November 1793, G. no. 12446, TNA, p. 21.
95. Ibid., p. 22.
96. Ibid., p. 28

97. Ibid., p. 32.
98. Ibid., p. 50.
99. Letter from Andrew Ross to William Roxburgh, 26 February 1793, Madras, Mss ROX, 1793, Botany Library, Natural History Museum.
100. Letter from Andrew Ross to William Roxburgh, 21 March 1793, Madras, Mss ROX, 1793, Botany Library, Natural History Museum.
101. Ibid.
102. Letter from Andrew Ross to William Roxburgh, 28 September 1793, Madras, Mss ROX, 1793, Botany Library, Natural History Museum.
103. Letter from Andrew Ross to John Forbes, 2 November 1793, Madras, Mss ROX, 1793, Botany Library, Natural History Museum.
104. Ibid.
105. Letter from Andrew Ross to John Forbes, 6 November 1793, Madras, Mss ROX, 1793, Botany Library, Natural History Museum.
106. Letter from Roxburgh to the Madras Government, Samulcottah, 30 August 1791, Mss Eur F 18 (I) K148.
107. Ibid., p. 244.
108. Andrew S. Cook, 'Dalrymple, Alexander (1737-1808)', *Oxford Dictionary of National Biography*, Oxford University Press, 2004 [http://www.oxforddnb.com/view/article/7044, accessed 4 August 2005].
109. Roxburgh to Joseph Banks, Samulcottah, 10 December 1791, Banks Correspondence, vol. 1790–94), Add. Ms 33979, Manuscript Collection, British Library, pp. 116–17.
110. Letter from Roxburgh to J.E. Smith, 4 August 1792, Samulcottah, 8.182, J.E. Smith Manuscripts, Linnaean Society Library, London.
111. Letter from Roxburgh to J.E. Smith, 20 June 1793, Samulcottah, 8.184, J.E. Smith Manuscripts, Linnaean Society Library, London.
112. Letter from Roxburgh to J.E. Smith, 14 January 1794, Calcutta, 8.188, J.E. Smith Manuscripts, Linnaean Society Library, London.
113. Joseph Banks to Roxburgh, London, 29 May 1796, Banks Correspondence, vol. IV (1795–1801), Add. Ms 33980, Manuscript Collection, British Library, p. 65.
114. Maehle, *Drugs on Trial*, pp. 266–7.
115. Andrew Duncan, *Tentamen inaugurale, de Swietenia soymida. Quam, . . . pro gradu doctoris, eruditorum examini subjicit Andreas Duncan*, Edinburgh, 1794.
116. Duncan, *The Edinburgh New Dispensatory*, p. 356.
117. *Journal of Natural Philosophy, Chemistry and the Arts*, 2nd series, vol. 6, December 1803.
118. Duncan, *The Edinburgh New Dispensatory*.
119. 'Letter from Andrew Duncan, M.D. F.R.S.E. Containing Experiments and Observations on Cinchona, tending Particularly to Shew that it does not contain Gelatine, to Mr. Nicholson', *Journal of Natural*

Philosophy, Chemistry and the Arts, vol. VI, December 1803, p. 225.

120. Cit. Seguin, 'Abstract of a Memoir on the Febrifuge Principle of Cinchona', *Journal of Natural Philosophy, Chemistry and the Arts*, October 1803.
121. 'Letter from Andrew Duncan, M.D. F.R.S.E. Containing Experiments and Observations on Cinchona . . .' *Journal of Natural Philosophy, Chemistry and the Arts*, vol. VI, p. 228.
122. Duncan, *The Edinburgh New Dispensatory*, p. 356.
123. Markham, *Peruvian Bark*, p. 32.
124. Duncan, *Supplementary*, p. 186.
125. Ibid.
126. Maehle, *Drugs on Trial*, p. 64.

CHAPTER 4

Mirza Abu Talib and his European Sojourn: An Indian Savant's Encounter with Modernity and Science in the Early Nineteenth Century

S. IRFAN HABIB

There have been several accounts of travel from Europe to Asia ever since the sixteenth century by men who engaged with the non-Western world, reconstructing its culture, society, and knowledge practices. That India was an object of curiosity for the expanding European nations is well known. However, the reverse was possible only after the eighteenth century, when a few Indian travellers 'accompanied Englishmen back to their native land, in one or other capacity, as *munshis*, envoys, or gentlemen of leisure[1] (as with Mirza Abu Talib). It is possible that quite a few travellers were encouraged by the British to cross the seas to see for themselves 'the "wonders" of *wilayat* and the superiority of Western culture and technology, even if they equally contained disparaging remarks on food, manners or climate'. While no clear evidence is available we may assume that quite a few Muslim travellers were inspired by the Islamic idea that travel brings enlightenment (*agahi*). Such Indo-Persian travel accounts (*safar nama*), whether to the West or to Western Asia, 'elucidate some aspects of a history of mentalities that so far has been developed only rather superficially in South Asian historiography'.[2]

*A revised draft of the paper presented at the University of Calcutta seminar on Science and Society in Modern India, 7–8 March 2005.

The first Indo–Persian travel accounts of Europe begin to make an appearance in the late eighteenth century with a bureaucrat of Emperor Aurangzeb (1659–1707), Mutamad Khan Badakhshi, who went to Portugal to learn Latin and translate Clavius's *Eight Books of Gnomics* into Arabic.[3] As a matter of fact Portugal remained the reference point until direct contact was established with the West, while there existed only a rudimentary knowledge of Europe. The Emperor Akbar's attempts to establish a channel of communication with Europe remained largely unsuccessful. To begin with, the term *farang* or *farangistan* was used in Indo–Persian literature to denote Europe as a whole: *farang* meant an inhabited part of the world, 'a world of cities and settled agriculture' without any implications of a qualitative distinction between the East and the West.[4]

Mirza Abu Talib Khan was born in 1752 in Lucknow and died in 1805 in the Bundelkhand region and was buried in Lucknow. His father, Haji Mohammad Beg Khan, was of Turkish origin, was born in Isphahan, and sought employment with the Nawabs of Awadh. Abu Talib moved to Bengal in 1766 and acquired his early education in Murshidabad. In 1791, Abu Talib published a scholarly commentary entitled *Diwan-i-Hafiz*; this was followed by an anthology of ancient and modern Persian poetry entitled *Khulasat-ul-Afkar*. Before his trip to Europe, this literary figure wrote the *Lubb-us-Siyar*, a book on general history that discussed Europe and America. In 1799 he sailed for Europe with his friend Captain David Richardson. Abu Talib had been going through a troubled phase when his friend suggested a change of scene.

Though born into the north Indian Muslim social gentry, Abu Talib was known in British circles as the 'Persian Prince'.[5] He did nothing to downplay this image, since it opened a number of doors for him within the British aristocracy.[6] He even declared in his travelogue that 'I never assumed the title; but I was so much better known by it than by my own name, that I found it in vain to contend with my godfathers'.[7] Six fashionable artists painted his portrait, two of which were displayed at the Royal Academy exhibition of 1801.[8] Prominent British journals recognized him as a 'man of considerable experience and knowledge of the world', well known in the highest and most respectable circles during his residence in England'.

Between 1799 and 1803 Mirza Abu Talib Khan Isfahani spent most of his time in England and France. He was a keen observer of both nature and man and had the curiosity to learn and the desire to record whatever he saw.[9] Writes Humayun Kabir,

> It therefore occurred to him that if he were to write all the experiences of his journey through Europe, to describe the curiosities and the wonders which he saw and to give some account of the manners and customs of the various nations he visited, of which little is known to Asiatics, it would afford a gratifying banquet to his countrymen. He was also of the opinion that many of the customs, inventions, sciences and ordinances of Europe, the good effects of which are apparent in those countries, might with great advantage be imitated by Mohammedans.[10]

However, Abu Talib feared that his countrymen would throw away his exertions due to their indolence and want of energy. He goes on to point out categorically that

> the great and the rich, intoxicated with pride and luxury, and puffed up with vanity of their possessions, consider universal science as comprehended in the circle of their own scanty acquirements and limited knowledge; while the poor and common people, from the want of leisure . . . have not time to attend to their personal concerns, much less to form desires for the acquirement of information on new discoveries and inventions.[11]

Interestingly, Abu Talib makes no distinction between the rich and the poor and is convinced that the passion to acquire knowledge 'has been implanted by nature in every human breast as an honour and an ornament to the species'. Perhaps to justify his worst fears, the first Urdu translation and that too only of the first part of his book, appeared from Moradabad only in 1904.[12] Despite being a little despondent, Abu Talib assured his reader 'that by reading this account of the state of the arts and sciences in Europe, he will considerably add to the stock of his own knowledge'.[13]

In his observations he compared England with France and both of them with India. He did not write about science in great detail, as his major preoccupation was European society and culture, even though he had authored some tracts on astronomy and wrote about modern astronomy (*hiyat-i jadid*) after his return to India. In 1793, Abu Talib, before his trip to Europe, duly notes in the synopsis of his projected scheme of world history:

> Europe was witnessing the birth of the great philosophers (*failasuf-i'azam*) to the degree that even a great number of common people (*awamm al-nas*) had developed philosophical disposition. The Europeans were seeking to relate themselves with the Greek philosophers, as if the Greeks themselves had become part of Europe's soil.[14]

Abu Talib did not consider Greece a part of Europe although, in the emergent historiography in the Occident, Greece was the very fount of European science. What he did find rather perplexing was the endeavour to redefine the European intellectual tradition as directly descended from that of the ancient Greeks. In fact, Abu Talib is bewildered by the contemporaneous European project to appropriate the Greek legacy.[15] It is possible to trace the evolution of Abu Talib's historiography, truly incisive for its time, to his essays in the *Lubb-us-Tawarikh* and *Lubb-al-Siyar*.[16] Abu Talib is one of the first writers to take note in his scheme of world history, of Copernicus and Galileo for their contribution to scientific learning, of the discovery of America, and of the geography and history of Europe.[17] He provides brief accounts of Greek and Islamic philosophers as the past heritage of the learned world.

During his stay in London, the clubs and the professional societies immensely impressed Abu Talib. The societies were important as they provided space to scholars to discuss new ideas in their subjects openly, and to present inventions and discoveries to their colleagues. This was seen as one of the major factors in the flourishing of the sciences in Europe. These societies facilitated the institutionalization of knowledge. Abu Talib himself occasionally attended the meetings of the Royal Society and wrote that such meetings were held on weekends in the house of Sir Joseph Banks, the president of the Society.[18] Abu Talib wrote:

During these meetings British scholars and scientists (*ulama wa hunarmandan*) displayed their inventions and discoveries to their colleagues. They discussed their inventions (*ikhtira't*) with each other, sought each other's views and opinions in order to improve their inventions and exchanged ideas on abstruse matters.[19]

Abu Talib looked for factors that had led to the scientific and industrial progress of Britain and he identified among these the attitude of British people towards knowledge and their philosophers/scientists. They believed that 'man through continuous efforts had progressed from being a savage and cannibal to the high status of the philosopher Newton. It might happen that in times to come man achieves such a high degree of perfection that even Newton the philosopher might (in comparison to the future man) look like a savage.'[20] This statement reveals Abu Talib's ideal of progress where the established scholarship and icons could be surpassed with the expansion of knowledge.

Abu Talib found that European scholars, particularly the British, were deeply concerned about the common people and desired an effective material change in their lives through the invention of new devices in industry that would facilitate labour. The recognition of merit and scholarship by the state had a positive impact on the cultural milieu; it raised the social status of the scholar among his fellow citizens and among the community as a whole; scholars in turn enhanced the prestige and esteem of their nation.[21]

Abu Talib writes about a philosopher called Walker, who hired an old playhouse where he exhibited, every night during summer, an astronomical machine called the *Orrery*, by which all the revolutions of the planets and heavenly bodies were perfectly described. The machine 'exhibited at one view, all the revolutions of the Solar system, with such perspicuity as must convince the most prejudiced person of the superiority, nay, infallibility, of the Copernican system'.[22] One can say that Abu Talib was among the first few converts, particularly among Muslims, from geocentrism to heliocentrism, as we know that a modernist reformer like Syed Ahmad Khan was not quite prepared to accept the heliocentric theory till the latter half of the nineteenth century.[23]

Abu Talib did not manifest great interest in cosmology; he wrote a treatise after his return from Europe only at the insistence of his friend Farid al-Din Ahmad, who wanted to know about contemporary developments in astronomy in Europe. The metrical treatise he composed is known as *Miraj al-tauhid* (The Unity of Heavens) and was completed on in September 1806.[24] It carefully reports the latest advances in observational astronomy and specifically points out that apart from the seven known planets, another planet had been discovered in historical times, the Sitara jaj that is the Georgium Sidus.[25] He stresses at various places in the treatise the centrality of telescopes in the modern advancement of astronomical sciences and is also aware that the satellites of planets (Jupiter and Saturn) could be seen only with the aid of a telescope; with its help, 3,000 stars had been observed. He explains further that it was modern knowledge (*aqwal-i hukma-i jadid*) that he was presenting in the tract, regarding fixed stars, planets, satellites (*aqmar*), comets, spheres (*aflak*), the pure air, nature of colours, seas, mountains, motion of earth, reasons of solar and lunar eclipses, etc.[26]

We find evidence of Abu Talib's intellectual energy and alertness on almost every page of his *Travels*. The problem of refraction attracted his notice when he found that on approaching the Nicobar Islands, he could see the land with naked eyes, but when he wished to look at it minutely through a telescope, the land seemed to disappear. One of the naval officers explained to him that what he had first seen was not actually the islands of Nicobar but their reflection in the atmosphere. He removed Abu Talib's doubts through a simple experiment.[27] European scientists had further strengthened their ideas and proofs with the aid of modern instruments (*alat*). Theirs was a far more accurate, reasonable, rational, and superior system than any system previously upheld. Abu Talib was of the opinion that some people raised objections to the premises of the new astronomy because they preferred to cling to their old views on the universe, but their arguments were irrational and unsound.[28]

Abu Talib realized that the things he was observing in Europe were in various respects far more advanced than what was available in his own country. He was convinced that all he observed should be communicated to his countrymen, and provided a fairly comprehensive account of the industrialization process in Britain. Abu Talib intended to supplement his textual descriptions of British technical innovations with drawings of inventions that he knew to be unknown in India. He believed that visual representation would greatly facilitate his readers' understanding of his subject, since what he described was entirely new to his intended readers.[29] He pointed out the use of these 'wheels and the tools' (*charkh u alat*) which were the reason for the high quality of the arts and crafts (*sanat*) in Britain and the main reason for the low prices of various commodities. The British had a passion for mechanical devices and spent much time and money putting together machinery for any major work. The French, on the contrary, although good mathematicians, were content with manual labour, if any difficulty were to occur in erecting the machinery.[30]

Abu Talib was fascinated and awed by some of the devices he saw, like the hydraulic machine to supply water to the city of London. He gives a detailed account of the water lifting process from the Thames River. Besides this machine, there was generally 'in every square or large yard a machine called a *pump*, whence, by the slightest exertion of the arm, the water is easily forced: it is a very simple contrivance, and much preferable to our wells'. He felt that the English carry their passion for mechanics to such an extent that machinery is introduced into their kitchens, and a very complete engine is used even to roast a chicken.[31] He found the art of printing to be one of the most admirable inventions of Europe. It enabled the production of thousands of copies of any scientific, moral, or religious book. The English used this art to produce *newspapers* every morning, without which life would be irksome to them.[32]

Shipbuilding and the development of the various navigational techniques made a serious impact on our observer's mind and he noticed and reported such developments in detail. He was convinced that Europe's expansion overseas was the result of a continuous effort to improve seamanship and the navigational instruments that facilitated distant voyages. Abu Talib discerned a clear relationship between scientific knowledge and technical progress, as the principles derived from the new science of mechanics were applied in industry to facilitate labour. After observing the shipbuilding industry he was convinced that industrialism is not based on education but empiricism, because tasks like drilling could be performed even by an old woman or child.[33]

In Britain Abu Talib met British scholars such as Sir William Ousley, Charles Wilkins, and William Jones. Ousley was a Persian scholar with such a devout commitment to the study of Oriental languages and literature that he named his son after the Persian poet Hafiz. Abu Talib considered Wilkins the first European Sanskritist. He had known William Jones from his days in Calcutta and had been very critical of the latter's Persian Grammar which he considered inaccurate and erroneous. Abu Talib had his reservations about the scholarship of this first generation of Orientalists. The major drawback, according to him, was that the British were very thorough and methodical while studying their own society and culture, but were not so while looking at the non-European world.[34]

Abu Talib was always conscious of the economic disparity between the ruling classes and the masses, which he felt not only militated against material development, but also impeded cultural progress. The common people were too absorbed in the bare struggle for existence to have the energy or time for the pursuit of knowledge.[35] He was drawn towards this conclusion by his study of conditions in Oudh, but it was only after his travels in Europe that he was confirmed in his belief. Emphasizing the role of economic factors in the development of history, he wrote about the causes of the downfall of the *ancien regime* in France:

> Previous to the late Revolution, the French Government spent immense sums on public buildings, gardens and illuminations. The nobles lived in a superb style, while the lower classes were reduced to the most abject poverty. Their patience was finally exhausted and they readily joined the leaders of faction and drove their inconsiderate domineering masters out.[36]

Abu Talib's account is replete with concern for the common people and their plight under the feudal dispensation. His narrative can be said to adhere closely to a genre of the social history of intellectual and economic development. This history shared the Voltairean disdain for military campaigns and palace intrigues that had overwhelmed eighteenth-century Indo-Muslim historiography.[37] While recognizing that court intrigues influenced the course of events, he went on to suggest that they were themselves often products of deeper economic and political causes.

Abu Talib was highly impressed by the French countryside which he crossed while travelling from Calais to Paris. He observed that during the whole journey, the country was beautiful and highly cultivated with grain, vineyards, or orchards of delicious fruits; rivulets of clear water crossed the road in various places and there were constructed neat stone bridges. Another interesting observation he makes is that it was on this journey that he first observed oxen used in Europe to draw carriages.[38] He found the villages of France 'exceedingly mean, and do not at all resemble their towns'. He also found the female peasants very disgusting, both in their manners and their dress: the attire of the village girls in India in comparison with these was infinitely superior.[39]

Our traveller was impressed by a large number of hot and cold baths frequented by the inhabitants; particularly delightful were those constructed on boats and mooted in the river. This institution was well deserving of imitation.[40] He also noticed the facility to wash clothes available on the boats: the boilers and tubs were below, and the deck was used for drying. 'Owing to this simple invention, the clothes are much better washed and bleached here than in London, where the occupation is often practised either in a cellar or in a garret, amidst smoke and dust.'

Abu Talib found that Paris abounds in coffee-houses which were generally filthy; as many French smoked *segars* or *cheroots* in them at all hours of the day, they smelled shockingly of tobacco. He was also intrigued by the annoying presence of a large number of beggars at these places: 'they follow a gentleman into the room and sometimes even take hold of his hand, to move his compassion . . . they are however content with a trifle, and will sometimes be satisfied by a piece of bread.'[41] Abu Talib found that in some of the streets of Paris there were few lamps, 'which yield glimmering light . . . and as the shopkeepers do not light up their windows as in London, the city has then a very sombre appearance.'[42]

The place he was impressed most was the Louvre, where he

felt 'all persons of science and liberal education may find an inexhaustible fund of amusement and information'.[43] The Louvre is a very extensive and lofty building. Its lower story is filled with statues and heavy articles, about the perfections of which Abu Talib declared was not a competent judge. After viewing the Louvre he was convinced that what he saw at Dublin and London appeared like 'mere children's playthings'. He also expressed surprise that the Louvre as well as other such places in Paris are 'supported at the public expense, and the people are admitted gratis: this serves to amuse them, and renders Buonaparte's [*sic*] government popular'.

Abu Talib was particularly entertained in Paris by an invention called the *phantasmagoria*. He expressed inability to give the details of its principles but said that 'it can only be exhibited in a dark room, and is effected by transparent paintings and shadows'. He commented that by means of this science, one may introduce on a stage the exact semblance of any hero of antiquity, or the likeness of any absent friend; and this art was formerly used by necromancers to deceive people, by leading them to suppose they possessed supernatural powers.

Abu Talib visited the Public Library of Paris and wrote that it contained nearly a million books in various languages, and upon all subjects. He felt that it is the most liberal of the institutions he had come across 'as the people are permitted to enter gratis, and have not only permission to read there the whole day, but to make extracts, or even to copy any book in the collection'.

In Paris, Abu Talib got a chance to meet several French literati, some of whom possessed a considerable knowledge of the Oriental languages, especially M. Langley and M. De Sacy. Both would visit Abu Talib almost every evening, 'neither of them spoke Persian well; but as they had studied Arabic, and read a few poetical Persian books; they translated several of my odes into French verse with much facility'. Abu Talib found that the French scholars were quicker in comprehending Persian than the Englishmen and he had his own reason for that. He felt 'perhaps the French language approaches nearer to the Persian idiom than English or, that our poetical expressions are more congenial to their ideas than to those of the inhabitants of a colder climate'.[44] The climate, we must not forget, was often evoked to explain cultural differences in the eighteenth century. It was a common causal explanation evoked by the savants of the Enlightenment, and it is not certain whether Abu Talib had turned the argument on Europe, or whether it was an element of Indo-Iranian social theory.[45]

This early nineteenth-century traveller's account reveals to us the curiosity of the non-Western intellectual about the progress of history of science, in a period when the tensions and complexities of colonial politics had not yet come into play. Abu Talib Khan, situated within the pre-colonial Indo–Persian travel writing, is inspired by an ideal of knowledge wherein the encounter with the Other enriches the search for, and the content of, knowledge.

NOTES

1. Muzaffar Alam and Sanjay Subramanyam, 'Empiricism of the Heart: Close Encounters in an Eighteenth-century Indo-Persian Text', *Studies in History*, vol. 15, no. 2, 1999, p. 264.
2. Ibid., p. 262.
3. Gulfishan Khan, *Indian Muslim Perceptions of the West During the Eighteenth Century*, Karachi, 1998, p. 270.
4. Ibid., p. 369.
5. *Travels of Mirza Abu Taleb Khan*, (henceforth *Travels*), tr. Capt. Charles Stewart, in 1814, rpt 1972, Delhi, p. 111.
6. Michael H. Fisher, 'Representing "his" Women: Mirza Abu Talib Khan's 1801 "Vindication of the Liberties of Asiatic Women" ', *The Indian Economic and Social History Review*, vol. 37, no. 2, 2000, p. 216.
7. *Travels*, p. 111.
8. Fisher, op. cit., p. 216.
9. Humayun Kabir, *Mirza Abu Talib Khan*, Patna, 1961, p. 6.
10. *Travels*, pp. xiv–xv.
11. Ibid., p. xv.
12. Kabir, op. cit., p. 7.
13. *Travels*, p. xvi.
14. *Lubb al-siyar u jahan-numa*. Cited in Khan, op. cit., p. 265.
15. Dhruv Raina and S. Irfan Habib, 'The Voyages of Abu Talib and Le Gentil: A Preliminary Comparative Study of the Scientific Imagination

in 18th Century Indian and French Travelogues', presented at the *XXI International Congress of History of Science*, 8–14 July 2001.
16. Kabir, op. cit., p. 16.
17. Ibid., p. 15.
18. *Travels*, p. 86.
19. Khan, op. cit., p. 278.
20. Cited in Khan, op. cit., p. 279.
21. Ibid., p. 280.
22. Ibid., p. 106.
23. David Lelyveld, *Aligarh's First Generation: Muslim Solidarity in British India*, New Delhi, 1996, p. 111.
24. S.M. Razaullah Ansari, 'Modern Astronomy in Indo-Persian Sources', in E. Ihsanoglu, ed., *Transfer of Modern Science and Technology to the Islamic World*, Istanbul 1992, p. 127.
25. The Georgium Sidus was discovered by F.W. Herschel (1738–1822) in 1781, who named it after George III. Its discovery won for Herschel a fellowship and medal of the Royal Society. The King made him King's astronomer with a salary of two hundred pounds a year. The planet was later named Uranus by another scientist, J.F. Bode, and it is with this name that the planet came to be known. Khan, op. cit., p. 325.
26. Razaullah Ansari, op. cit., p. 128.
27. Kabir, op. cit., p. 7.
28. Khan, op. cit., p. 289.
29. *Travels*, p. xvi.
30. Ibid., p. 118.
31. Ibid., p. 120.
32. Ibid., p. 110.
33. Khan, op. cit., p. 307.
34. Ibid., p. 256.
35. Kabir, op. cit., p. 22.
36. Ibid., p. 23.
37. Ibid., p. 16.
38. *Travels*, p. 205.
39. Ibid., p. 206.
40. Ibid., pp. 206–7.
41. Ibid., p. 207.
42. Ibid., p. 209.
43. Ibid., p. 211.
44. Ibid., p. 219.
45. Raina and Habib, op. cit., p. 8.

CHAPTER 5

Imperialism, Medicine, and Women's Health in Nineteenth-century India

SUJATA MUKHERJEE

In nineteenth-century India, an institutional and administrative infrastructure necessary for the spread of Western medicine was gradually set up by the colonial state. Hospitals,[1] dispensaries,[2] asylums opened in British India formed the nucleus from which Western medicine sought to establish its superiority over indigenous forms of medicine.[3] Scholars have suggested that in the first half of the nineteenth century, in an essentially male-oriented and male-operated system of medicine, the primary areas of concern of the state were the army, the jails, and the hospitals—exclusive male domains.[4] Yet it might be argued that Western curative medicine had, since its initial stage of expansion, served a wider agenda of imperialism and gender—albeit in a limited way—through hospitals, and dispensaries, and went on to expand its sway and gradually establish a kind of cognitive control over a section of the Indian population. This contributed to the formulation of a civilizing agenda of imperial rule, linking together the issues of gender and health care.[5]

The aim of my paper is to investigate how far and in what ways the gradual institutionalization of Western medicine in colonial India affected Indian women and informed this civilizing agenda. This will involve an analysis of the growth of the hospital system and its relation with female patients, the medical intervention of the state in the shape of health regulations and sanitary measures, the role of the missionaries as well as of private philanthropists in providing medical care for women and finally, how all these processes affected women.

The earliest measure taken by the colonial state to ensure medical care of a section of India women sprang out of the perceived need to protect the military interests of Empire. This meant the creation of 'lock hospitals' in cantonment areas throughout India for the treatment of prostitutes suspected of suffering from contagious venereal diseases.

It was in the early eighteenth century that lock hospitals flourished in London in asylums formerly used for lepers. Patients were kept in confinement or locked up until they were cured. Elderly women who were superintendents in the *lal bazaar* or red light area (brothel) of the regimental *bazaar* were entrusted with ensuring that infected prostitutes were either expelled or sent to hospitals.

Before the end of the eighteenth century, the Governor-General-in-Council sanctioned the building of 'hospitals for the reception of diseased women' at Berhampur, Kanpur, Dimapur, and Fatehgarh. In 1807, also, hospitals for 'diseased publick women' were established in Agra. Already in Madras in 1805, the Medical Board, consisting of several senior medical officers, recommended the establishment of lock hospitals, the appointment of special police to control the prostitutes, and compulsory medical examinations in order to tackle the problem of venereal disease among soldiers. The lock hospital system continued to operate in the Madras Presidency until 1835. In the Bombay Presidency, lock hospitals were established only when the necessity was reported, and were closed down when no longer needed.

Apparently the dominating criterion in policy discussions regarding lock hospitals was the health of the British soldiers. It was admitted by the colonial authority that sexual contact was necessary for unmarried soldiers (and regarded as markers of their manliness) and to control the demasculinizing effect of infections that ensued. The authorities condoned sexual contact between European soldiers of lower rank and Indian prostitutes, rather than bearing the extra financial burden of providing greater salaries for soldiers to enable them to maintain a family in India. It was pointed out, 'for a young man who cannot marry and who cannot attain to the high moral standards required for the repression of physiological natural instincts, there are only two ways of satisfaction, viz., masturbation and mercenary love. The former, as is well known, leads to disorders of both body and mind; the latter, to the fearful dangers of venereal.'

Venereal disease among European soldiers, the result of sexual contact between Indian prostitutes and soldiers, was both 'troubling' and 'embarrassing' to the authorities. The debilitating effects of venereal disease on the soldiers' capacity to fight were regarded as fearful and pernicious. According to one report, the disease not only kept a large proportion of the Army constantly in Hospital, but the necessary treatment 'undermines the constitution, cripples the bodily powers on the march in the field, and predisposes to other diseases, especially rheumatism which affection swells very largely the annual number invalided'.[7]

The foundation of the lock hospitals was seen as the most acceptable device to check VD, but the authorities also felt that some amount of caution was needed to run the system, so that it did not arouse excess of ill feeling among the Indians. The Governor-in-Council once stated that the army officers (who were entrusted with the task of deciding which women should be subjected to restrictions) must understand that the system was to be applied only to those persons who may be 'liable from their general habits to occasion the diffusion of the fatal complaint which it is important to counteract, as from the nature of the local usages of India much confusion might be produced by any measure tending to create an interference with the establishments attached to many of the principal places of Hindu worship'.[8]

At the other end of the scale was the opinion that the system of lock hospitals was slack and needed rigorous improvement. Loopholes were pointed out such as dearth of efficient police to take 'all infected or suspicious women' to hospital. Those who were most wretched or starving came to the hospital, but the richer section of the prostitutes preferred affordable native medical aid to treatment in the hospital, where all castes were 'huddled together in a miserable choultry'. Sometimes it was thought that to reduce costs lock hospitals should be closed; but this aroused protests in the military circles. Lieutenant Colonel Gibbs wrote in 1810 that VD among the soldiers had increased steeply after the closure of lock hospitals at Bangalore. In fact, according to statistics, one man in seven was then suffering from VD. It was opined that the Bangalore hospital should be reopened and there should at least be one hospital for each European regiment station. Members of the Medical Board suggested that the system should be improved by organizing special police and proper buildings. The proposal for special police was approved by the government.

T.P. Weeks, surgeon to the 1st Bombay Light Cavalry and stationed in Baroda, opened a lock hospital in the mid-1820s, to control rising cases of VD and bore its cost for six months. He claimed that 'upwards of twenty unfortunate creatures were restored to perfect health who a few months before were labouring under a most loathsome disease'.[9] In the Bombay Presidency, however, measures were more of a temporary nature. Lock hospitals were established only when the situation was reported to be grave, and were closed when VD was supposed to have abated. Sometimes steps were taken to improve the system. Following the advice of the Superintending Surgeon of Poona Division—which was supported by the Commander in Chief—the appointment was sanctioned of 'two steady old police peons' or constables who knew the regulations and the 'haunts of prostitutes' was sanctioned.

In Bengal the reforming Governor-General William Bentinck abolished lock hospitals in 1830 and pointed out to the Court of Directors that this would save Rs. 30,000 a year. The Bombay and Madras governments refused to follow suit because according to them VD had decreased after the establishment of lock hospitals.

The system seemed to be disagreeable to the Indians. Indian women found these hospitals 'horribly revolting to their habits and customs'. Moreover, it was also suggested that 'the native curative means' which they had known 'for years or ages' were 'held in estimation by them which would not be the case if they found them less efficacious than those employed by the Europeans'. Alternative means for controlling VD were sought. It was suggested by Dr. Murray, Deputy Inspector-General of Hospitals for the Queen's troops, that VD patients should be treated in general civilian hospitals with a VD ward under an

Indian matron who must be acquainted with the 'haunts and habits' of prostitutes. Another device proposed was that wherever European troops were stationed a public dispensary should be opened with a ward attached to it for 'female patients' and each regiment should have an Indian matron to supervise the prostitutes. Yet other measures suggested by the Medical Board were the introduction of regular medical inspections in European regiments and the expulsion of diseased or suspected females from cantonments. But it was not possible to forcibly subject women to inspection or expulsion. Experiments began with the appointment of matrons at public dispensaries, but that arrangement too was severely disliked by Indian women. In the years following the abolition of the lock hospital, Madras was gradually moving towards restoration of the system, but not Bombay and Bengal.

A widespread discussion about the health of the British army was prompted by the Crimean war and the revolt of 1857. According to many witnesses before the Royal Commission on the Sanitary State of the Army in India, venereal disease was a very serious problem and it was suggested that to tackle the problem more soldiers should be allowed to marry. Many expressed distrust of the lock hospital system because it could never be applied to all Indian prostitutes. In India, only 12 per cent of a regiment could marry. In Madras the Principal Inspector General calculated, on the basis of statistics of the previous eighteen years, that the annual average proportion of VD cases amounted to 25.5 per cent among European troops but only 3.5 per cent among Indian troops. Almost all Indian soldiers were married. Medical opinion within the army, however, was divided. Some put forward the suggestion that lock hospital system should be re-established as a pragmatic measure.[10] There was no unanimous medical opinion regarding the effectiveness of mercury in treating diseases like syphilis and thus it seemed that confinement of patients to hospitals would prove to be effective in preventing the spread of infection.

The Royal Commission on the Sanitary State of the Army in India, appointed in 1859, submitted its report in 1863. It strongly recommended that the lock hospitals be reintroduced with greater vigour. The Military Cantonments Act of 1864 provided for the medical inspection and regulation of brothels. This was followed by the Contagious Diseases Act of 1868 (partly modelled on a similar Act passed in Britain in 1866), which strengthened the system by providing compulsory registration of brothels and prostitutes and the detention and medical examination of prostitutes. The Act could be introduced in any locality specified by a local government with the prior approval of the Governor-General.

After 1868, the overall responsibility for lock hospitals was entrusted to the provincial sanitary commissioner. The application of the Act differed from one area to another. According to the superintendent of lock hospitals in Calcutta in 1870, the enforcement of the Indian Act was more strict than that of the British Act since it applied to the 'whole civil population of Calcutta where the military element is very small'.[11] In 1876 the Lieutenant-Governor pointed out that the Commissioner of Police did not always enforce the penal provisions of the Act. Out of a total of 2,359 women arrested for not registering in 1870, only 301 were presented in court.[12] In the city of Bombay, on the other hand, women were employed to act as spies to detect attempts at extortion by the police.

Contradictory opinions were expressed regarding the effectiveness of the Contagious Disease Act. According to the Lieutenant-Governor of Bengal, it was successful to a great extent: 'the working of the lock hospitals has, on the whole, been productive of good, both in diminishing the prevalence and mitigating the severity of venereal diseases among registered prostitutes and European soldiery'.[13] On the other hand, the Government of India Sanitary Commissioner wrote in his annual report of 1881 that 'year by year it becomes more and more evident that, however, successful such a system promised to be when it was originally adopted, its practical working has proved a complete failure'.[14]

A controversy gradually developed regarding the continuance of the Act. In 1875, the *Hindoo Patriot* expressed pleasure[15] that the Act appeared to have led to some improvement in the health of European troops. In contrast the *Bengalee* expressed the opinion that it was useless and its implementation was accompanied by 'a degree of oppression . . . which far outweighed any benefits which might be supposed to have accrued from it'. It was pointed out just how the Act became an engine of oppression in the hands of the police and unscrupulous neighbours.

Agitation by evangelical organizations in Britain and India against the Act, and the repeal of the Act in 1886 in Britain increased pressure for its abolition in India. Ultimately in June 1888 a resolution was passed by the House of Commons condemning the compulsory examination of women and calling for an end to lock hospitals and the licensing of army prostitutes. This, however, did not mean the end of the medical inspection of Indian prostitutes. The Military Cantonments Act of 1889 provided for 'the prevention of the spread of contagious diseases within a cantonment, and the appointment and regulation of hospitals . . . for the reception and treatment of persons suffering from any disease'.

Directed by the liberal government in Britain, the Indian government passed legislation in 1895 stating that medical examination or compulsory treatment of prostitutes suspected of suffering from venereal disease could not be carried out under the Cantonment Act (1889).[16] Case of venereal disease continued to show an upward trend, reaching a peak in 1895 when 522 out of 1,000 men were found to be affected. In 1897 the Conservative Secretary of State, Lord Hamilton, instructed the Viceroy, Lord Elgin, to repeal the 1895 Act and to reintroduce medical inspection in cantonments. Despite protest from Indians,[17] the cantonment legislation was passed and, as pointed out by scholars, 'the military authorities won the last battle in a long campaign'.[18]

In the 1860s, piecemeal and sporadic efforts were made by individuals to provide medical care to Indian women, mainly maternal care and training centres for midwives and also trained female doctors from abroad, and the opening of the doors of medical education for women. For the white elite, apparently the basic initial stimulus for this interventionist role came from writings of people like James Mill who articulated a version of gender relations prevalent in India that justified the British presence.

The condition of women is one of the most remarkable circumstances in the manners of nations. . . . The history of uncultivated nations uniformly represents the women as in a state of abject slavery from which they slowly emerge as civilization advances. . . . As society refines upon its enjoyments . . . the condition of the weaker sex is gradually improved till they associate on equal terms with the men and occupy the place of voluntary and useful coadjutors. A state of dependence more strict and humiliating than that which is ordained for the weaker sex among the Hindus cannot easily be conceived.[19]

As the nineteenth century progressed, traditional Indian practices of childbirth and maternal care were discredited. An important target of attack was the unhygienic condition under which birthing was taking place, in a special room called the *aturghar* or *sutikagriha*. Dr. Stewart, a professor of the Calcutta Medical College was horror-stricken to see the 'filthy, smoky, and crowded hovels, to the straw of which the unfortunate Bengalee females are condemned by native usage in the hour of suffering'.[20] Meridith Borthwick has shown that early medical literature on the horrors of childbirth in India emphasized the place of birth as the main cause of the high rates of mortality among both mothers and infants. Childbirth was attended by aged women and *dais* or midwives, described as 'experienced and courageous women of advanced age and with clean cloths before whom she [the birthing mother] may not feel shy, who have cut their nails and who cheer her with friendly words'.[21] The *dai* generally was a lower caste Hindu or poor Muslim woman. In Bengal, she could be a *hadi* or *dom*,[22] in north India she was generally a *chamar* or of the sweeper caste and,[23] in the south, she was of the barber caste. There was little incentive for the *dai* to hone her skills or for the midwife to develop professional skills as in Europe, even before the coming of medical men. This was a traditional occupational grouping, devoid of professionalism. The recruits were a caste-based restricted hereditary group and often the calling was handed within a family from mother to daughter.

The duties of the *dai* could extend from cutting the umbilical cord, washing the puerperal garments, and burying the placenta, to living in the home of the parturient mother and giving her and the infant regular massages and warm fomentations. Thus they had to perform menial tasks and received meagre pay. Many studies of the medical profession have a tendency to glorify the role of midwives. It has been emphasized that they had more empathy for their patients than the doctors. The *dai* appeared to symbolize rites exclusive to women which involved non-Sanskritic rituals. Among the Calcutta elite there was a mixed attitude towards the midwives. People like Lal Behari Day portrayed a kindly and well-intentioned village midwife.[25] On the other hand, many accounts highlighted the ignorance of the *dai,* jeopardizing the health of both mother and child by interfering with the process of natural labour. Many accounts recorded cases of how midwives began pulling the heads of infants as soon as they appeared or of how placenta was removed by hand immediately after the infant had fully emerged. According to Geraldine Forbes many treatises written by the Bengalis were actually translations of English works and condemned the traditional method of birthing.[26]

Many travellers' and medical men's accounts in the nineteenth and early twentieth centuries suggested that the birthing process was no less hazardous in India than it was in England. Little was actually known about prenatal and postnatal care.[27] In fact, 26 per cent of children died in England in 1868 before the age of five.[28] Undoubtedly in most parts of the world in the nineteenth century, child delivery was a hazardous process and was also painful until the discovery of chloroform in 1847. According to Meredith Borthwick, some Bengalis thought that European women had overcome the difficulties of childbirth by observing the rules of health; but as shown by P. Branca, puerperal fever was actually a frequent cause of maternal mortality in England. At the same time, new medical techniques were developing in England and raising the standard of midwifery.[29]

According to Geraldine Forbes, 'the Indian way of giving birth would probably not have received attention it did, had it not been for the increase in the numbers of missionary women in India in the latter half of the century'.[30] It had been pointed out by Balfour and Young that the stimulus for the first *dai* training schemes was provided by those missionaries who had gained access beyond the *purdah* as teachers, and were grieved to see their young pupils dying in childbirth.[31] In the 1860s a number of missionary societies came forward to sponsor women's associations to recruit and train female missionaries. These associations tried to focus on education in the *zenana*.[32] Conditions in the *zenana* in fact became a focus of critical attention: it was considered to be a place of dirt, darkness, and disease.[33] A vast body of Anglo–Indian discursive writing was produced in the second half of the nineteenth century focusing on Indian women, and one most striking thing about this was its preoccupation with the *purdah*. In the opinion of Indrani Sen,

> while 'native' men evidently sought to control their women's sexuality and bodies by shutting them up inside the *zenana* away from the colonial gaze, the efforts of the Anglo-Indians were geared towards bringing these very women out in the open through the gradual eradication of the *purdah* system. In this conflict between the two groups, the 'native' woman's body became, in effect, the terrain upon which the cultural contest was staked out.[34]

The hegemonic ambitions of Western medicine were represented by the activities of women missionaries from England and the United States, who came to India from the 1860s and established hospitals, dispensaries, and training centres for midwives and nurses. A member of the American Methodist Episcopal Church established a medical training school for women in Nainital in 1869, but it did not last for more than three years. The same mission sent Dr. Clara Swain, a graduate of the Women's Medical College of Pennsylvania, to Bareilly, in 1870. She started a small class with three married women and fourteen girls from the mission's orphanage. Before long, missions on both sides of the Atlantic began to champion the ability of women to apply the 'double cure': healing the body as well as the spirit.[35] Fanny Butler was the first British Medical woman to come to India in 1880. Swain, Butler, and other missionaries opened dispensaries, hospitals, and training centres for midwives and nurses.[36] A civil surgeon in Punjab launched a scheme in 1866 in Amritsar to train *dais*.[37] In 1886 the project was revived by a medical missionary of the Church of England Zenana Missionary Society.

The role of medical missionaries was important in more ways than one. It was they who made a small but influential group of

Indians desire medical attendance in childbirth, thus creating the climate later for more widespread work. The missionary women commented on what seemed to them the horrors of childbirth and the appalling customs and beliefs surrounding it. It was pointed out by a missionary author, who was not herself a doctor but who described the practices of Bengali village midwives in the mid-nineteenth century, that 'unscientific' and 'unchristian' customs seemed to have been indistinguishable.

In the course of time, public opinion in England as well as India came to realize that Indian women were in urgent need of female physicians. Dr. Frances Hoggan (one of the first British women to qualify as a doctor) argued that the Indian Medicine Service (IMS) had failed to fulfil the medical requirements of Indian women and that only women doctors could best serve the interests of Indian female patients.[38] She pointed out that the 'shrinking' of women patients from male doctors was not a 'prejudice' to be disregarded, but a 'natural' attitude and it was the 'right of Indian women' to expect this sentiment to be 'respected and not outraged'. She also suggested that the government should produce two classes of women practitioners; ordinary practitioners for the country and the more accomplished section capable of teaching and practising in the bigger towns.

George A. Kittredge, an American businessman in Bombay, was touched by this plea. In his view, 'to be successful in this country, women must be recognized as the equals of men in the medical care of their own sex. This recognition could never be gained by those who have taken a partial medical course and an inferior degree in an Indian University.'[39] The Medical Women's Fund was set up to raise public subscription for the promotion of female medical education. Kittredge began to raise funds among his Indian friends to bring in female physicians from Europe and North America. He was helped in his endeavour by a rich Parsi, Sorabji Shapurji Bengali, and was able to collect Rs. 40,000 within two months. The effort gained further momentum when Pestonji Hormusji Cama came forward to offer a donation of more than Rs. 1.5 lakh to meet the cost of establishing a hospital for women and children. The Cama Hospital was opened in August 1886.

The first systematic attempt to provide Western medical care to Indian women began in 1885 with the establishment of the National Association for Supplying Female Medical Aid to the Women of India, or Dufferin Fund, under the direct patronage of Queen Victoria. According to a frequently repeated story, the Queen-Empress came to know about the health condition of Indian women from Elizabeth Bielby and Mary Scharlieb. In 1881, Bielby, a medical missionary, engaged in medical work in Lahore was asked by the Maharaja of Punnah or Panna to treat his wife, who was severely ill. Bielby helped her to get well. When Bielby was ready to depart for England for further medical studies, the Maharani requested her to meet Queen Victoria and convey a special message:

> I want you to tell the Queen and the Prince and Princess of Wales, the men and women of England, what the women of India suffer when they are sick. Write small, Doctor Miss Sahib; for I want you to put it into a locket, and you are to wear this locket round your neck till you see our Great Queen and give it to her yourself. You are not to send it through another.[40]

On completion of her medical training at the London School of Medicine for Women, Bielby went to see the Queen, who reportedly replied: 'We had no idea it was as bad as this. Something must be done for the poor creatures. We wish it generally known that we sympathize with every effort made to relieve the suffering of the women of India.'[41] She gave her photograph to Scharlieb to take to Indian houses to show how much the queen felt for her subjects.

When Lady Dufferin went to India in 1883, the Queen reportedly asked her to find out the possibilities of providing medical care to Indian women. Lady Dufferin surveyed the work done by the Government of India and other institutions and came to the conclusion that 'taking India as a whole, its women were undoubtedly without that medical aid which their European sisters are accustomed to consider as absolutely necessary'. In 1885, she wrote to the wives of the various Governors and Lieutenant-Governors and being assured of their support, drew up a prospectus. This was published in various languages from Simla, the summer headquarters of the Government of India. It was met with warm response and the press almost unanimously gave approval. During the Viceroy's autumn tours, favourable allusions were made to it in municipal addresses. Finding the responses satisfactory, Lady Dufferin finally inaugurated the Fund.

The explicitly stated aim of the Dufferin Fund was 'To bring medical knowledge and medical relief to the women of India' by providing women medical workers.[42] The prospectus mentioned three areas of intensive work. First, it would provide medical teaching and training of women as physicians, hospital assistants, and midwives. Second, medical relief would be provided to Indian women by way of dispensaries and hospitals under female superintendence. Third, the aim was also to provide for trained female midwives and nurses to take care of women patients and children.

It was stated in the prospectus published on 18 August 1883 that the Queen Empress had become the Patron of the Fund. She was the 'Supreme Patron', Viceroy Dufferin the 'Patron in India'. The Fund was administered by a Central Committee that met regularly. The first Central Committee consisted of high-ranking officials of the government and two Indians closely connected with the government. The Countess of Dufferin served as Lady President. It was categorically stated that despite government personnel and support, the association was not intended to be a government institution and would be free of official interference in all respects.[43]

By 18 January 1886, approximately Rs. 148,344, and £235 respectively had been collected in India and England.

The Fund organizers made it clear that the Fund was non-sectarian. No employee of the Association was allowed to 'proselytize or interfere in any way with the religious beliefs of any section of the people'.[44] It was suggested that the masses could be more readily reached through the influence of the 'moulvis and Pundits'. Lady Dufferin unequivocally stated her wish that the Fund should not employ medical missionaries.

The Western-educated urban Indian elite was favourably disposed towards the Dufferin Fund that seemed to promise modern, scientific medical assistance to women. It was welcomed by the richer sections of the educated Indian elite who had played a collaborative and enthusiastic role in spreading Western science and medicine since the inception of colonial rule and contributed to the funds for running dispensaries, hospitals, and colleges. Historians have pointed out that medical philanthropy and patronage in India sprang from a variety of motives.[45] It was seen by the Indians as a means of extending leadership over one's own community and society. It must be remembered that India had a strong cultural tradition of philanthropic and charitable works, which found fresh outlets in the colonial era. Medical philanthropy was also partly a political strategy. The Indian elite wanted to utilize it as a means of gaining prestige and influence, and recognition from the colonial regime.

It would appear that in Bengal elite (which included the English educated Indian intellectuals) patronage and philanthropy for Western science as well as medicine sprang from the belief that Western science which was integral to Western culture, was an important modernizing force.[46] People like Ramkamal Sen,[47] Dwarakanath Tagore[48] and Motilal Seal[49] were great patrons of Western science, philanthropists who generously contributed to the promotion of Western medicine.

In general the Bengali intelligentsia welcomed the institutionalization of Western science. The Hindu College established through the efforts of the *bhadralok* in Calcutta in 1817, had to teach, apart from the humanities, subjects like 'geography, astronomy, chemistry and other sciences'.[50] The college received a donation of scientific equipment from the British India Society of London for teaching astronomy, optics, mechanics, and chemistry in 1823.[51] In 1823, when the General Committee of Public Instruction decided to spend government money to set up a Sanskrit College in Calcutta, Rammohun Roy strongly protested. In a letter to Lord Amherst, he wrote:[52]

> If it had been intended to keep the British nation in ignorance of real knowledge, the Baconian philosophy would not have been allowed to displace the system of schoolmen, which was the best calculated to perpetuate ignorance. In the same manner the Sanskrit system of education would be the best calculated to keep this country in darkness, if such had been the policy of the British Legislature.

He urged the government to

employ European gentlemen of talent and education to interest the natives of India in Mathematics, Natural Philosophy, Chemistry, Anatomy, and other useful sciences, which the natives of Europe have carried to a degree of perfection that has raised them above the inhabitants of other parts of the world.

A small hospital was established in Calcutta in 1831 for the treatment of patients and for imparting clinical instruction to students at the Native Medical Institution.[53] The moving spirit behind the establishment of the hospital was Ramkamal Sen, who made a liberal donation and had taken a very keen interest in the development of medical education from the very beginning. The hospital was planned on a small scale and accommodated thirty patients. It had an outpatients department as well. Naba Krishna Gupta, the senior and most advanced student of the medical class, was appointed Apothecary. He was assisted by other students by rotation each month.

The Calcutta Medical College,[54] from its very inception, was enthusiastically supported by the Calcutta elite. Many luminaries of Calcutta including Dwarakanath Tagore, and Ram Kamal Sen wholeheartedly welcomed it and tried to promote its medical education programme by instituting scholarships and prizes for students. Dwarakanath instituted scholarships of Rs. 2,000 per year for three years.[55] Already in 1833 he was elected a member of the committee of the Leprosy Society.[56] In 1835 he joined Motilal Seal and Rajah Protab Chandra Singh to plan the establishment of a fever hospital for the poor.[57] Four students of the Medical College were sent to England with financial help from Dwarakanath Tagore, Professor Goodeve, and (partly) the government.[58] Three of them, including Dwaraka Nath Bose, Bhola Nath Bose, and Gopal Chunder Seal qualified as Members of the Royal College of Surgeons (MRCS) in 1846 and returned to India to join the uncovenanted Medical Service. Surjee Coomar Chukerbutty remained abroad and obtained the MD degree of University College London, becoming the first Indian to qualify for the Indian Medical Service. He also became a distinguished professor of the Medical College holding the Chair of Materia Medica from 1864 till his death in 1874.

In 1837 Madhusudan Gupta[59] suggested that Calcutta Medical College start a programme to train midwives and build a lying-in hospital for poor women. In 1840, a large lying-in hospital was constructed on the Calcutta Medical College grounds by a liberal public subscription. It contained 100 beds and imparted instructions in midwifery.[60] In 1841 Dr. H. Goodeve, who had been appointed in 1837 to the Chair of Anatomy and Medicine, took charge of Anatomy and Midwifery.[61] It seems that from the middle of the nineteenth century the *bhadralok* in Bengal showed an eagerness to adopt new medical skills in case of childbirth though it was accessible to a few who could afford high cost. The report of the Midwifery Hospital stated in 1848 that all of their college graduates—six or more in number—practising in Calcutta were 'habitually called to take charge of the women of the families they attend during their confinements'.[62]

In the 1870s an attempt was made in Bengal to open full-fledged medical training to women students. In 1875, Neel Kamal Mitra submitted a petition seeking admission for her granddaughter, Biraj Mohini, to the Bengali course at the Calcutta Medical College. Mitra, however, insisted that Biraj Mohini be given a separate curtained-off seat and taught dissection in the presence of her husband or himself. This proposal was turned down.

In 1874, Mary Scharlieb, the wife of an English barrister in Madras, who later became a gynaecologist, persuaded the authorities in Madras to allow women students to attend classes at the Madras Medical College.[63] A new chapter in the history of medical education in India opened in 1875 when four women (including Mrs. Scharlieb) were admitted to the three-year certificate course.

In 1881–2, Abala Das and Ellen D'Abrieu, who had passed the entrance and the F.A. examination respectively of the Calcutta University, went to Madras with government scholarships to study medicine since they were barred from entering the Calcutta Medical College. The Council of the Medical College was opposed to the idea of admitting female students to the Medical College although the government was in favour of it. Brahmo reformers of nineteenth-century Bengal were enthusiastic supporters of medical education for women. The admission of female students who had passed the First Arts Examination to the

Calcutta Medical College was permitted by a resolution of 23 June 1883. The first beneficiary of this new rule was Kadambini Basu, one of the first female graduates in India, who later married the Brahmo reformer Dwarakanath Ganguly. In 1883, Kadambini entered the Calcutta Medical College and was awarded the GBMC degree in 1886.[64]

Rules were relaxed in 1886 by providing for admission of girls who had passed either the entrance or a special preliminary examination to a three-year course in the Medical College. The Maharani of Cossimbazar made a donation of Rs. 1.5 lakh for the construction of a hostel for the female students.[65] The government offered scholarships of Rs. 2 a month to all female students. Bidhu Mukhi Bose and Virginia Mary Mitter received these scholarships and became the first Indian women to graduate in 1888–9.

Indian middle class support for female education, including medical training, was no doubt partly prompted by an urge to providing an answer to the imperialist critique of Hindu male behaviour towards women. (This was implicit in the Dufferin Fund advocates' reference to Indian men as a 'population which . . . makes small account of the health or of the lives of women'). There was also an urge to create a compatriot wife who would bring about discipline, order, and hygienic practices in middle-class home. The Bengali male agenda of social reform in the nineteenth century aimed at improving the lot of women; but it was also addressed to the need of reshaping norms and functions of middle-class family life as a site of moral and cultural restructuring of nation. Education was the core of the reform of women's condition and education and had little to do with economic functions or the pursuit of employment and the growth of professional expertise. The main social utility and commitment of women's education lay in the constructive role expected to be played by the new women in family welfare.[66]

These new women were expected to develop as companions to men, scientific nurturers, and members of civil society; they were to remain a socially distanced class from the common or lower class women, inhabiting a world of unrefined, coarse, popular culture.[67] The large number of pedagogic texts written in this period produced a normative discourse on the family and framed new rules and guidelines for the housewife for ideal home and family management: nurturing of children, regulation of dietary habits, creation of hygienic environment, etc. Parents, particularly mothers, were advised to educate themselves to be able to understand and execute the experts' instructions. 'For the proper rearing of children, for proper knowledge about the child's nutrition and health, you should read appropriate books, you should take advice from expert physicians . . .'.[68] Women were accorded a crucial role in this standardized, remodelled domesticity. A lack of knowledge and education would harm not only the family but also the nation. (Women who were ignorant of the rules of the body would not only harm themselves, but by producing weak and deficient children would also destroy the nation.)

Thus with the emergence of the family as the locus of nationalist restructuring, women were awarded an augmented status. The Indian middle class internalized some of the imperialist critique of Indian/Hindu domestic life but at the same time modified some of the received ideas about bourgeois domesticity, privacy, and individuality to suit their own needs. This became part of the discourse on modernity that evolved in the nineteenth century. Comparisons between 'Indian' and 'European' homes were made by Bengali authors and the latter was described as an epitome of perfection while the former seemed to suffer from a grievous lack of order, discipline, and cleanliness. The Bengali *bhadralok* was anxious to counter the notions of bourgeois privacy by defining freedom as freedom from selfishness. Even when women were educated, they should be good wives. The anxiety was partly reflected in the critical outlook expressed in the print media about educated women who seemingly neglected other duties. A *bhadramahila* was expected to master the technique of becoming a *sugrihini* by acquiring elementary knowledge of all sorts of medical remedies for common ailments and save the family expenses in doctors' fees.[69]

Brahmo reformers like Durgamohan Das and Dwarakanath Ganguly were enthusiastic supporters of medical education for women.[70] Brahmo *Public Opinion* wrote in 1883.

> If there be any one country where, more than at another, the want of lady doctors is most keenly felt, it is no doubt India. The system of zenana seclusion makes it nearly impossible for male doctors to be very useful in treating female patients. Consequently, a very large number of our women come to premature death from want of proper medical attendance. . . . Besides, there are diseases peculiar to them, which it is simply impossible for male doctors to diagnose or treat.

The *Bamabodhini Patrika* wrote, 'Everyone with prudence will admit that as for men, medical education is equally necessary for women. There are certain types of female diseases which can only be appreciated by women and their treatment by males cannot be as effective as by females.'

There seemed to be also an awareness among Indian men that a vigorous female domain of healing actually existed, which pre-dated British rule.[71] The vast majority of Indian women relied on these traditional systems of healing which existed both within the domestic sphere and among certain specialists. These traditions had come under attack in the nineteenth century by people influenced by the British civilizing discourse on science. Western medicine in fact contributed to, and appropriated to itself, the cultural arrogance of post-1857 colonial rule. Disdainful attitudes towards indigenous medicine, and faith in the superiority of Western medicine buttressed by the bacteriological discoveries of Pasteur and Koch became the hallmark of imperial rule. As feared by nationalists like Bal Gangadhar Tilak, racial discrimination meant that medical education in colonial India developed on a tiered basis. Supervision remained in the hands of English physicians and Indian women were deemed fit only for subordinate posts. While it is true that the available choices of medical treatment for women widened with the promotion of Western medicine, it had very limited appeal. Most Indian women continued to depend on non-Western practices of healing, deprived of state patronage and discredited in the civilizing agenda of imperial rule. Efforts at promotion of Western health care for Indian women, as represented by Dufferin Fund, on the other hand provided ideological support for imperialism by putting emphasis on the benevolent and philanthropic aspects of British rule. This could be politically exploited by the British government. It has been pointed out that the Dufferin Fund was a 'distinctive example of what became a characteristic effort of the British government of India, to graft onto India the English model of philanthropy, which colonial rulers believed represented a progressive civic ideal'. Following the example of the Dufferin Fund, other organizations were established in subsequent years. These provided a new role for white women in India by promoting a new ideology of love and sympathy between ruler and ruled and at the same time contributed towards the reproduction and maintenance of the empire.

NOTES

1. The first regular hospital was founded in Calcutta in 1707 for the treatment of East India Company soldiers and sailors. The Presidency General Hospital, established in 1768, was also for Europeans only. In 1787 the government built-up another hospital at Dum Dum, chiefly for vaccine inoculation of soldiers and their families. The first hospital for treatment of Indians was opened in 1793 by Sir John Shore in Bengal.
2. It seems that the establishment of charitable dispensaries from the 1830s was one of the earliest attempts at providing Western medical care to Indians. Dispensaries became local centres for vaccination against smallpox and for conveying Western ideas about sanitation and hygiene.
3. It is generally held that until the mid-1820s there was friendly co-existence of Indian and Western systems of medicine in colonial India. In the Native Medical Institution (NMI) established in Calcutta in 1822, in Calcutta Sanskrit College and the Calcutta Madrasa, parallel instructions were carried out in indigenous and translated Western medical texts. This ended in the triumph of the Anglicists in introducing English language and European Sciences in India, which led to the abolition of the NMI and medical classes at the Sanskrit College and Madrasa and foundation of the Calcutta Medical College in 1835 where only Western medicine was taught. See Poonam Bala, *Imperialism and Medicine in Bengal: A Socio-historical Perspective*, New Delhi, 1991; Anil Kumar, *Medicine and the Raj: British Medical Policy in India, 1835-1911*, Delhi, 1998; David Arnold, *Colonizing the Body: State Medicine and Epidemic Disease in Nineteenth Century India*, Delhi, 1993.
4. Ibid., p. 254. According to Radhika Ramasubban, Western medicine, at least till 1900, remained confined to a small enclave of white residents

and soldiers. See R. Ramasubban, 'Public Health and Medical Research in India: Their Origins Under the Impact of British Colonial Policy', *SAREC Report*, Stockholm, 1982; 'Imperial Health in British India, 1857–1900', in Roy Macleod and Milton Lewis, eds., *Disease, Medicine and Empire: Perspectives on Western Medicine and the Experience of European Expansion*, London, 1988.

5. Relevant studies on gender and medicine in Colonial India are: Maneesha Lal, 'The Politics of Gender and Medicine in Colonial India: The Countess of Dufferin's Fund, 1885-1888', *Bulletin of History of Medicine*, vol. 68, 1994, pp. 29–66; Malavika Karlekar, 'Kadambini and the Bhadralok', *Economic and Political Weekly* (hereafter *EPW*), vol. 21, no. 19, 1986, pp. WS-25–31; Judy Whitehead, 'Modernising the Motherhood Archetype: Public Health Models and the Child Marriage Restraint Act of 1929', in P. Uberoi, ed., *Social Reform, Sexuality and the State*, New Delhi, 1996, pp. 187–209; Barbara N. Ramusack, 'Embattled Advocates: The Debate over Birth Control in India, 1920-40', *Journal of Women's History*, vol. 1, 1989, pp. 35–64. The work of Geraldine Forbes includes: (i) 'Medical Careers and Health Care for Indian Women: Patterns of Control', *Women's History Review*, vol. 3, no. 4, 1994, pp. 515–30, (ii) 'Managing Midwifery in India', in D. Engels and Shula Marks, eds., *Contesting Colonial Hegemony: State and Society in Africa and India*, London, 1992; (iii) 'Colonial Imperatives and Women's Emancipation: Western Medical Education for Indian Women in Nineteenth Century Bengal', in *Modern Historical Studies,* vol. 2, 2001, pp. 83–102. See also Mridula Ramanna, 'Control and Resistance: The Working of the Contagious Diseases Act in Bombay City', *EPW*, vol. XXXV, no. 17, 2000, pp. 1470–6; Indira Chowdhury, 'Instructions for the Unconverted: Marie Stopes, Indian Women and the Making of a Pamphlet, 1920–1955', in *From the Margins*, vol. II, no. 1, 2002, pp. 212–42. Sujata Mukherjee: (i) 'Women, Medicine and Empire: Female Practitioners and Patterns of Health Care in Colonial Bengal', *Modern Historical Studies*, vol. II, March 2001, (ii) 'Disciplining the Body? Health Care for Women and Children in Early Twentieth Century Bengal', in Deepak Kumar, ed., *Disease and Medicine in India: A Historical Overview,* Delhi, 2001, pp. 198-214. See sections in Geraldine Forbes, *The New Cambridge History of India, IV.2, Women in Modern India*, Cambridge, 1998. Also Meredith Borthwick, *The Changing Face of Women in Bengal, 1849-1905,* Princeton, 1984; Dagmer Engels, *Beyond Purdah? Women in Bengal 1890-1939,* Delhi, 1999; Kenneth Ballhatchet, *Race, Sex, and Class Under the Raj: Imperial Attitudes and Policies and their Critics, 1793-1905*, New York, 1980; Sabyasachi R. Mishra, 'An Empire "De-Masculinized"! : The British Colonial State and the Problem of Syphilis in Nineteenth Century India', in Deepak Kumar, ed., *Disease and Medicine*, pp. 166–79.

6. Quoted in Ballhatchet, ibid., p. 10, Memorandum, October 1886, pp. 1888, LXXVII (158), 235 ff.
7. Cited in Mishra, op. cit., p. 171, Letter from Brigadier C. Troup, Agra, 27 May 1861.
8. Govt. resolution 22 April 1805, MMP 22 April 1804, 26, 86ff.
9. Cited in Ballhatchet, op. cit., p. 14, Weeks to Govt., 27–2–1805, Bom MP 29 June 1825, 29.
10. Ibid., p. 35
11. Mark Harrison, *Public Health in British India: Anglo-Indian Preventive Medicine 1859–1914*, Delhi, 1994, p. 73.
12. IOR P/1002. GOI (Sanitary), extract from the Procedgs. Govt. of Bengal (Indicial Dept.), 6 August 1873.
13. IOR P/1203. GOI (Sanitary), resoln. no. 357, 21 December 1877.
14. Cited in *Indian Medical Gazettee*, 1883, 18: 102.
15. *Hindoo Patriot*, 17 May 1875, p. 234.
16. IOR P/4753, GOI (Sanitary), no. 19, May 1895.
17. *Bengalee*, 20 February, 1897, p. 89.
18. K. Ballhatchet, op. cit., p. 95.
19. James Mill; quoted by Dipesh Chakrabarty, 'Postcoloniality and the Artifice of History: Who Speaks for the Indian Pasts?'; *Representations*, 37.
20. Dr. Stewart, *Report on the Statistical History of the Female Hospital*, GRPI for 1846-47. Appendix E, no. 10, p. clxxv.
21. Dr. J. Jolly, *Indian Medicine,* tr. C.G. Kushikar, Delhi, 1977, p. 69.
22. See Borthwick, op. cit., p. 155.
23. G.W. Briggs, *The Chamars,* Calcutta, 1920, pp. 24–6, 53–4 quoted in Charu Gupta, *Sexuality, Obscenity, Community: Women, Muslims, and the Hindu Public in Colonial India,* Delhi, 2001 p. 177.
24. Janet Chawla, *Child-Bearing and Culture—Women-Centred Revisioning of the Traditional Midwife: The Dai as a Ritual Practitioner*, Delhi, 1994, pp. 2–3, 80–2.
25. G. Forbes, 'Managing Midwifery in India', in D. Engels and Shula Marks, eds., *Contesting Colonial Hegemony*, p. 156.
26. Ibid., p. 157.
27. B. Ehrenreich and D. English, *Complaints and Disorders: The Sexual Politics of Sickness*, London, 1976, p. 23.
28. P.H. Chavasse, *Counsel to a Mother*, London, 1869, p. 3.
29. P. Branca, *Women in Europe Since 1750*, London, 1978.
30. G. Forbes, 'Managing Midwifery in India', pp. 157–8.
31. M. Balfour and Ruth Young, *The Work of Medical Women in India*, London, 1929, pp. 13–14.
32. G.H. Forbes, 'In Search of the "Pure Heathen": Missionary Women in Nineteenth Century India', *EPW,* XXI/17 (26 April 1986, WS2–WS8).

33. Janaki Nair, 'Uncovering the Zenana: Visions of Indian Womanhood in Englishwomen's Writings, 1813-1940', *Journal of Women's History* vol. 2, no. 11, 1990, pp. 88–94.
34. Indrani Sen, *Woman and Empire: Representations in the Writings of British India (1858-1900)*, Delhi, 2002, p. 53.
35. Geraldine Forbes, 'Colonial Imperatives and Women's Emancipation: Western Medical Education for Indian Women in Nineteenth Century Bengal', in *Modern Historical Studies*, vol. 2, 2001, pp. 83–102.
36. Balfour and Young, op. cit., pp. 13, 16–18, 29.
37. Forbes, 'Colonial Imperatives', op. cit., p. 86.
38. Arnold, op. cit., p. 261.
39. G.A. Kittredge, *A Short History of the "Medical Women for India" Fund of Bombay*, Bombay, 1889, p. 3.
40. Balfour and Young, op. cit., p. 20.
41. Ibid., pp. 20–1.
42. Harriot Dufferin, 'The National Association for Supplying Female Medical Aid to the Women of India', *Asiatic Quarterly Review*, vol. 1, 1886, p. 259.
43. Counters of Dufferin's Fund, First Annual Report, pp. 8, 13–14.
44. Ibid., p. 15
45. Arnold, op. cit., pp. 269–74.
46. Members of this new elite were in many cases descendants of the early indigenous collaborators of European merchant communities who ventured new opportunities and became rich by serving the mushrooming European commercial houses in Calcutta as *banians* and *dewans* or business intermediaries. With the introduction of the permanent zamindari settlement they became powerful proprietors of landed properties in Bengal, gaining further wealth and prominence. It appears that in early eighteenth century the Tagore family lived on the swamps of Calcutta and acted as suppliers of goods and provisions to the captains of ships and free merchants. Darpanarayan Tagore (d. 1793), son of Joyram Tagore who served as *amin* to the East India Company, was a prominent merchant and bania of Calcutta. See Pradip Sinha, 'Social Changes', in N.K. Sinha, ed., *The History of Bengal (1757-1905)*, Calcutta, rpt. 1996, p. 391.
47. Ram Kamal Sen (1783–1844), founder of the famous Sen family of kalutola, initially worked as a compositor in a Hindustani Press on Rs. 8 a month. He attracted the attention of Dr. Wilson, who later became Professor of Sanskrit in the University of Oxford. He was first promoted to some 'subordinate situation' in the establishment of the Arabic Society, which brought him to the notice of distinguished members of the European society. He then became a member of the Committee of the Calcutta School Book Society and helped in the compilation and translation of several useful works. He was placed

by Wilson at the head of the Native Establishment of the Mint. Later, he accepted the office of Treasurer to the Bank of Bengal (ibid., p. 397).

48. Dwarakanath represented a unique example of far-flung and sophisticated entrepreneurial activities. The range of Dwarakanath's business interests included Calcutta house properties, zamindaris, indigo factories, silk filatures, insurance and banking, commercial houses, coal mines, tea and even shipping. For a discussion of his career and activities see Blair B. Kling, *Partner in Empire: Dwarakanath Tagore and the Age of Enterprise in Eastern India,* Calcutta, 1981.
49. Motilal Seal attained great prominence in the 1830s and 1840s. He has been described as a successor of Ramdulal De, the Bengali millionaire who died in 1825 and showed great talent for combining commercial activities with strong speculative skill.
50. Kapil Raj, 'Knowledge, Power and Modern Science: The Brahmins Strike Back', in Deepak Kumar, ed., *Science and Empire: Essays in Indian Context (1700-1947)*, Delhi, 1991, p. 120.
51. Aparna Basu, 'The Indian Response to Scientific and Technical Education in the Colonial Era', in ibid., p. 126.
52. S.N. Sen, *Scientific and Technical Education in India 1781-1900,* Delhi, 1991, p. 162.
53. The Native Medical Institution or the School for Native Doctors was established in Calcutta in June 1822 to train Indian doctors. It imparted three years' training in medicine in the local languages. Major English works were translated and the trained doctors were to serve in the army or civil department. The institution became operational in 1824 and existed till 1835.
54. The Calcutta Medical College, established by an order of 20 February 1835, ushered in a new era in the history of medical education in India. It was a pioneer institution for systematic education in Western medicine. Its stated purpose was to train native youths aged between 14 and 20 irrespective of caste and creed in the principles and practices of medical sciences in accordance with the mode adopted in Europe.
55. Kling, op. cit., p. 48.
56. Brajendra Nath Bandopadhyaya, ed., *Sambadpatrey Sekaler Katha 1818-1840*, Calcutta, 1949–50, II, p. 315.
57. Kissory Chand Mitra, *Memoir of Dwarakanath Tagore*, Calcutta, 1870, pp. 26–7, 64.
58. Sen, op. cit., pp. 272–3.
59. Madhusudan Gupta was a student of the Native Medical Institution and later became a superintendent of the Hindustani class at Calcutta Medical College. In 1836 he, along with four others, performed dissections on human bodies.

60. *Centenary Volume of Calcutta Medical College*, Calcutta, 1935, p. 23.
61. Ibid., p. 24.
62. M. Borthwick, *The Changing Role of Women in Bengal, 1849-1905*, pp. 156–7.
63. Mary Scharlieb, *Reminiscences*, London, 1924, pp. 29–30.
64. Karlekar, op. cit.
65. Sen, op. cit., p. 514.
66. Himani Bannerji, 'Fashioning a Self: Educational Proposals for and by Women in Popular Magazines in Colonial Bengal', *EPW*, 1991, 26–43.
67. Geraldine Forbes, *Women in Modern India*, p. 41. Also Sumanta Banerjee, 'Marginalization of Women's Popular Culture in Nineteenth Century Bengal', in Kumkum Sangari and Sudesh Vaid, eds., *Recasting Women: Essays in Colonial History*, Delhi, 1989.
68. Pradip Kumar Bose, 'Sons of the Nation: Child Rearing in the New Family', in Partha Chatterjee, ed., *Texts of Power: Emerging Disciplines in Colonial Bengal*, Calcutta, 1996, p. 123.
69. In 1878, a native doctor's fee varied from Rs. 3 to 10. A European doctor charged Rs. 16. Borthwick, op. cit., p. 217.
70. See Malavika Karlekar, 'Kadambini and the Bhadralok'.
71. For further discussion see Mukherjee, 'Women, Medicine and Empire', op. cit.

CHAPTER 6

Plague and Prophylactics: Interrogating Colonial Medical Intervention in Eastern India

ARABINDA SAMANTA

My paper seeks to engage with two things. First, it will try to delineate the different perceptions of people of the plague epidemic that ravaged eastern India in the closing years of the nineteenth and early decades of the twentieth centuries. It will also explore the interface between plague, people, and the medical world, all three of which were enmeshed in a tragic saga that merits fresh historical interrogation. Second, the paper will explore the multiple sites where the rhetoric of Western medical intervention was contested by a popular construction of the disease. It will view the disease and its remedies as functions of power and knowledge informed by a relationship between ruler and ruled in a colonial society. The colonial administration lacked a homogeneous perception about how to deal with the epidemic, and reacted differently in different situations. Nor did all categories of people react to the epidemic in a similar way and their opposition to the colonial state was not uniform either. The study seeks to interrogate how these multiple layers of perception were imbricated.

Extant work on epidemic plague in India generally concentrates on Bombay, and scholars emphasize the panic and the obtrusive nature of state intervention. To David Arnold, for instance, the upsurge of public resistance to state-sponsored medical intervention in Bombay was due to cultural differences and a repugnance of the oriental body to Western medicine.[1] For Raj

Chandavarkar, on the other hand, the widespread hostility was due to the coercive manifestation of a brutally intrusive state.[2] To Ira Klein, again, it was a conflict between Western anti-plague measures and popular culture.[3] But what about eastern India, which also shared with Bombay the phenomena of opposition to Western medicine and preventive state measures, widespread rumours, panic and scare, riots and strikes? Did it attract the same kind of government intervention and the same form of popular resistance? Were the rumours generated during the epidemic purely 'elite discourse'?[4] Was the popular construction of plague based purely on rumour and religion rather than fact and reason? What exactly was the material base of the panic it generated? In fact, at the end of the century, Western medical practitioners knew very little about the disease; nor did they have a specific remedy for it. In such moments of blissful ignorance of the aetiology of the disease, what exactly did the native medical practitioners have at their disposal to prescribe as remedy and what did the people do to circumvent the epidemic?

THE RAT, THE FLEA, AND MAN

Any contagious and fatal epidemic disease was originally called a plague (from *plaga*, a stroke), but the term is now applied to the appearance of high fever with inflammation of lymphatic glands or bubo, and is therefore called the Bubonic Plague. In Sanskrit medical books it has been described as *Vidradhi* and *Visharpa*, and in Yunani medicine as *Taoon*.[5]

By the end of the nineteenth century medical tracts emerging from Bombay sought to locate the aetiology of the disease in climate and social position. The climate and season, they argued, had a 'special influence' on the onset of plague.[6] A moderate amount of heat was found to be favourable to its occurrence, whereas a very high or low temperature was unfavourable to it. But, they argued, there had been exceptions. It had prevailed during the severest cold of winter on the Volga and in Moscow, and in extreme heat, as in Smyrna.[7] As regards social position, it was argued that poverty had been found to be the chief predisposing cause of plague. The Great Plague of London of

1665, called the *poor's plague* and of Baghdad and Mesopotamia in 1878, where it was called by Cabiadis *miseriae morbus*, are cases in point. But in Bombay, it was argued, poverty had had nothing whatever to do with the outbreak of the disease. There had been neither famine nor poverty. It broke out there among the well-to-do classes.[8]

A disease so fatal in its nature and against which human power was so futile was in ancient times naturally attributed to wrath of the gods. Supernatural, astrological, and in some instances rationalistic, causes were assigned to it. In the fourteenth century, the College of Physicians of Paris ascribed it to the influence of constellations in India. But more natural explanations followed. The putrefaction of dead animals was assigned as a cause in Egypt. Poisoning of water supply was also believed to be the origin. Undue heat, rain, damp grain, and absence of the Etesian winds were thought to generate plague.

Scientists of the late nineteenth century were also a divided house: some who believed in the germ theory attributed plague to a specific germ, holding that germs could never arise *de novo*; others believed that atmospheric changes and certain telluric conditions or insanitary surroundings engendered the seeds of pestilence which were carried through air, water, or other media.[9] The arguments in favour of the first theory were, however, so strong that it was then almost universally accepted that plague was due to a specific poison which grew and multiplied under favourable conditions, and that wherever it occurred it was caused by the implantation of those germs in a suitable soil. The soil best suited to the plague seed was one where insanitary conditions prevailed. Dirt and filth, bad ventilation, and overcrowding were supposed to be its manure.

By the closing years of the nineteenth century, it was maintained that plague was a miasmatic or soil-bred disease, and that the germs found in earth, water, or in fermenting or decomposing material, a suitable nidus for growth. In this sense it was like malaria, which was endemic in a particular area, appearing and disappearing according as climate or other conditions were favourable or unfavourable. The Chinese had a peculiar idea of the infection of plague. They considered that it rose from the soil, and believed that it first attacked small animals with breathing organs near the soil, such as rats, then animals with breathing organs a little higher, such as poultry, pigs, dogs, goats, cows and so on until it reached man.[10]

Rats are undoubtedly attacked with plague before and during its prevalence among human beings, and they play an important part in the spread of the infection. In Kumaon, the rat plague was observed by the people, and was recognized as a forerunner of the plague. *The Times of India* observed on 30 September 1896:

> It was known more than a month ago to all the people of Mandavi and to all municipal sweepers in the district that the rats were dying in thousands all over the districts. They were found dead and dying almost everywhere, and in places where dead rats were never found before.

The causative agent, a bacterium, was also discovered and the intimate relation between rat plague, the flea, and human plague was also elucidated. During the Hong Kong epidemic, the great Japanese bacteriologist S. Kitasato, who formerly worked with Koch in Germany, described a bacillus in plague-stricken patients, and showed by experiments that those bacilli if injected into lower animals produced in them symptoms of plague. A. Yersin simultaneously discovered the same germs. They argued that these germs must be considered to be the specific poison that produced the symptoms of plague. The bacilli were found in the blood, in the buboes, and in all internal organs of the victims of the plague.[11] Thus in 1894, at Hong Kong, A. Yersin and S. Kitasato described the plague bacillus. In 1897, M. Ogata isolated the organism from rat fleas. In 1898, P.L. Simond demonstrated the flea transmission.[12]

It was, however, in 1908 that the whole problem was analysed in a complete and rigidly scientific fashion by a Plague Commission appointed by the Government of India. The Report of this Commission presents, first, a survey of various areas of plague in relation to the prevalence of the disease in rats as determined by the trapping and examination of many thousands of rats in the same localities. This study showed a close and roughly quantitative relation between the epizootic among rats and the epidemic among human beings, the course of the former preceding that of the latter by one or two weeks.

Later sections of the Report give epidemiological evidence to show that direct transmission of plague from man to man is not so important a factor, that insanitary conditions are significant only in so far as they favour the breeding and harbouring of rats, and that the spread of plague from one locality to another is generally due to imported rat fleas on the bodies or in the luggage of persons who themselves are often in good health. It is probable that the infected rat commonly plays a major role in place-to-place dissemination. In transport of the disease by sea, the rat is undoubtedly the chief agent. This 1908 study of the Indian Commission laid a sound basis regarding the rat–flea–man links.[13]

PLAGUE MORTALITY IN EASTERN INDIA

The first known case of plague in Calcutta was discovered in April 1898, and at the end of that month the city was declared infected.[14] This continued until October, when cases ceased to be reported, and the city was accordingly declared free of plague. During these six months, attacks and deaths reported were 230 and 192 respectively. The outbreak therefore cannot be said to have been epidemic in the usual understanding of the word.

During the next four months, the disease remained quiescent. Isolated cases were reported which bacteriological examination showed to be plague, and there were a considerable number of cases in which the cause of death not being sufficiently apparent, it was considered advisable to disinfect the concerned houses. Mortality during the whole period was above average, and in the opinion of the Health Officer the greater part of the excess was in fact due to plague.

It was not however till the end of February 1899 that the existence of plague was sufficiently demonstrated to warrant a declaration that Calcutta was again infected. This outbreak lasted until the end of June 1899, and during these four months 1,336 cases and 1,223 deaths from plague were reported.[15] There was extensive concealment of cases, and the estimates of the authorities often varied, the largest being slightly over 3,000. The highest mortality was reached in March and April. From April onwards it fell rapidly, and in June mortality was again normal. During the second half of 1899, 1,131 cases were returned as plague, and 898 as 'suspicious'.

In 1900 the disease assumed very different proportions, and became for the first time a virulent epidemic. From January to the end of June, 7,897 cases and 7,373 deaths were reported.[16] Of these, 3,667 cases or nearly half the total occurred in the month of March, and 1,996 in April, after which the virulence of the epidemic abated rapidly.

The total mortality due to plague in Calcutta from its first outbreak down to 30 June 1900, a period of two years and three months, had been according to the official returns, 10,997. If in accordance with the opinion of the Chairman and Medical Authorities, suspicious cases and excess mortality not otherwise accounted for were put down to plague, the total might be rounded off to 13,000.

PROPHYLACTIC INTERVENTION REVISITED

Plague in India was primarily a disease of the long-tailed house rat, *rattus*. Other rodents were of comparatively little importance. They were less numerous, they tended to live in sewers or fields rather than in houses, some varieties did not harbour many plague fleas. In the vast majority of cases man was infected more or less accidentally by hungry, infected rat-fleas, which strayed from the carcasses of house-rats that died of plague.[17] When there was a plague epidemic among rats, the severity of the human epidemic depended on the closeness of the association of rats and man. Human plague would cease to be a serious problem, it was believed, when we no longer provided food and shelter for rats in our houses. The real anti-plague measure, it was argued, was therefore 'rat-exclusion'.[18] This entailed a radical change in the habits and housing conditions of the people, which it was argued would come about eventually through education and social uplift.

It had been the practice to deal with plague outbreaks by certain time-honoured temporary measures, namely evacuation, inoculation with anti-plague vaccine, and rat trapping and poison-baiting. When plague attacked Calcutta, Mungher, and Patna in 1898, certain measures had been formulated by experts and prescribed for general adoption throughout India by the British government. Their leading features were carefully enforced death registration, house-to-house visitation, segregation of the sick and minimizing contacts, the evacuation of infected localities, disinfection, and the enforcement of general sanitary precautions.[19]

Right of Entry and Examination

The owner and occupier of any house should permit the Health Officer to enter his premises and examine any person to believed to be infected with bubonic plague. If the person be a female who according to the custom of the country did not appear in public, the examination would be made by a female doctor or female hospital assistant.[20]

Segregation

If on examination a Health Officer suspected that someone was infected, he might cause such person to be removed to a hospital, and might arrange for his detention, diet, and medical treatment. He might also cause the other occupants of the house to be removed to a segregation camp to be detained under observation for ten days. The relatives, friends, hakims, vaids, and priests of sick persons would be allowed free access to them during the day, subject only to such precautions as the Health Officer might consider necessary.[21]

Disposal of Dead

The friends of the deceased were permitted to dispose off the corpse by cremation or burial in accordance with their religious practices, but were to obey the directions of the authorities as to the time, route, and method of removing the corpse. The corpses of Europeans or Mohammedans were to be buried, if possible, at least 6 ft deep, at a place far from habitations and so situated that there was no risk of contaminating sources of water supply. Where cremation was the custom, the body should be completely burned at the usual burning ghat or other isolated locality, the clothes in contact with the body being either burnt or disinfected.[22]

INOCULATION

In 1900 the Government of Bengal had ordered the District Officers to resort to Mr. Haffkine's system of anti-plague inoculation. But by 1903 it modified the order and argued that inoculation should be *encouraged* wherever possible, and that in places threatened with a visitation of plague, no efforts should be spared to *explain* to the people the advantage of becoming inoculated before the disease was among them.[23] It was at the same time strictly enjoined that compulsion must on no account be resorted to, and that the choice whether or not they would be inoculated must entirely be left to the people.

PROPHYLACTICS IN PRACTICE

In eastern India, provisions regarding segregation and evacuation was modified by offering inoculation as an alternative, and by permitting isolation in private houses of the better classes. People in Bukhtiarpur, Danapore, and Chakberia in Patna district, Chaitantola in Monghyr and elsewhere were up in arms in June 1900 in connection with regulations regarding evacuation and disinfection.[24] The Lt. Governor, therefore, directed that local officers abstain from employing force regarding evacuation and disinfection. They should encourage people to adopt them in every way, by advice, by persuasion, and by supplying them with materials or labour that might be desired. If any person or class of persons nevertheless finally refused to comply, force should not be used.[25] In spite of this, the first appearance of plague was signalled by an unprecedented panic and exodus from the town, and by several disturbances. Fortunately, the outbreak was a slight one, and meanwhile it served the purpose of indicating how far the general policy till then prescribed was practicable in Calcutta and how far not. When plague reappeared in the succeeding year, the policy of the government was stated in a resolution of which the following is an extract:

> In order that effective measures may be taken to prevent the spread of infection, it is of the first importance that every case, which occurs, shall be promptly brought to notice of the authorities. Foreign Governments would have just cause for complaint if the regulations in force were repugnant to the sense of the people affected as to drive them to withhold information and hide away their sick instead of bringing them forward for treatment.[26]

Compulsory removal to hospital and segregation were accordingly abolished. The policy followed from that time was to make every effort to obtain information about cases, and to induce the people to accept the measures which experience had indicated to be of most avail in preventing the spread of the disease. Of these, the principal, in fact practically the only one, generally used in Calcutta, had been disinfection.

DISINFECTION

The sheet anchor of the plague policy observed in Calcutta was the disinfection of every room or apartment in which a case had occurred. Such measures as house-to-house visitation and the offer of rewards for information had never been suggested in Calcutta; nor had penalties for not volunteering information been enforced, except in occasional instances. The procedure throughout had been to disinfect every room in which either a case of ascertained plague or a case of suspected plague had occurred. Thus between January and June 1899, only 1,336 cases of ascertained plague were reported, but 4,750 premises were disinfected. From February 1900 this policy was extended to the disinfection of whole *bustees* or streets in any part of which a case of plague occurred.[27] In carrying out these measures, the Chairman observed that the value of the influence of men of local standing was enormous. He cited three instances in which local opposition was readily overcome by invoking the aid of the landowners Raja Binoy Krishna Bahadur, Maharaja Sir Narendra Krishna, and Babu Sarat Chandra Mukherji respectively.

In many places human plague disappeared in the hot weather only to recrudesce in the following cold season. It was then realized that temperature and humidity in the depths of rat burrows were such that plague might continue among rats and fleas over the hot weather. Not only so, but it had been shown in some experiments in Madras Presidency in the 1930s that infected rat fleas were capable of conveying infection after as long as 29 days' starvation in rat burrows.[28]

By the 1930s certain methods of destroying fleas were found to be of very little value. The disinfection of grain by exposure to the heat of the sun or in hot air chambers was given up. The use of kerosene oil emulsion to spray plague-infected houses was of some value, but could not affect fleas hidden in burrows or crevices or under rubbish.

A note on 'trapping and baiting as anti-plague measures' prepared by Dr. G.D. Chitre was circulated in 1919, and these methods were widely employed. Various fumigants such as sulphur dioxide and carbon bisulphide had been used from time to time. Hydro-cyanic acid gas was the fumigant of choice as far as the lethal effect on rats and fleas was concerned. By the 1930s, experiments had been carried out in the Cumbum Valley in the Madras Presidency, and also in Bombay and Lahore under the auspices of the Indian Research Fund Association, the Travancore Durbar, and the Bombay and Punjab governments, with the hope that HCN fumigation of rat burrows might prove a more effective means of arresting plague and might even be successful in obliterating endemic plague.[29]

SERUM THERAPY

Serum therapy was in 1897 at an interesting stage of development from the scientific point of view, but as a practical method of treatment it was of uncertain value. Laboratory prophylactic experiments with anti-plague serum had been so successful that its use as a preventive or protective measure might perhaps ultimately allow some modification of the segregation thought so necessary in dealing with suspects.[30] Even a weak anti-toxic serum gave good results. But as to its use as a curative agent in practical medicine, it had been a comparative failure both in

Bombay and in China. This failure was scarcely to be wondered at because the serum employed was admittedly of weak anti-toxic power. Moreover, the practical difficulty of bringing the patient under treatment adequately early to give him a fair chance of recovery had been against the success.[31]

INOCULATION

Inoculation was carried on by Haffkine in Bombay Presidency, but with what success, it is difficult to say. Figures given of what occurred in the House of Correction show that its use was of benefit, but such figures are very incomplete and much more information is required to warrant a conclusion. James A. Lowson, Plague Commissioner, Bombay, argued in 1897 that this form of protection would never come up to that provided by an anti-toxic serum. The reason was that the pain and discomfort of the inoculation was great, and unless inoculation could be 'sprung' on a whole community at once, the results of inoculating a few were generally sufficient to cause the remainder to fight shy of any such vaccination. Laboratory work on the vaccine, according to Lowson, had so far not given very successful results, and the results on man had not proved so striking as one could wish.[32]

At first sight the figures of 1907 with their improvement on the previous year indicate that inoculation was at last commencing to emerge from the shadow of the popular disfavour under which it had languished so long. It was argued that an increase from 101 to 3,122 inoculations might very well be taken as an indication that some of the educated at last were beginning to realize the benefits and safety of plague prophylaxis.[33] It is, however, to be noted that a few Europeans and one or two educated Bengalis had had themselves and their children inoculated; but with these extremely rare exceptions the whole of the inoculations were done on people attracted by the reward of eight annas: beggars, drunkards, victims of the opium, hasheesh or cocaine habit. Moreover, the same persons used to present themselves for inoculation two or three times within the same month. Many probably did a round of the different districts and were inoculated several times. It was impossible to verify this from the registers, as nearly all the names in them were false.[34]

THE EXTERMINATION OF RATS

During the year 1907, as many as 1,22,090 rats were killed. The number seems large, but one might doubt whether it had at all materially affected the rat population in the city, particularly when we know that in Tokyo, after the destruction of 40,00,000 in three years, the result, according to Kitisato, was disappointing. In northern Calcutta the number of rats killed was greater because the northern district was the one in which the chief grain centres were situated. Many of the grain men or *dalwallahs* had found rat collecting sufficiently paying to take it as a regular calling.[35] In other parts of the city the rats were too scattered to make the task worthwhile. In the European parts of the town, where active measures were taken by private individuals and carefully kept up, there was some diminution in the rat population.

The result of trapping was educative rather than practical. Rewards for live rats brought in were lowered from two annas to six pice from January 1907. The attention of people was directed to the importance of rat destruction, and perhaps in the future, it was anticipated, a time would come when each householder would take care of his own house. In fact, it was suggested in 1906 that the first step would be to insist that all grain godowns be rat-proof. As a matter of fact, some of the most recently erected *dal* godowns had as the result of Dr Crake's (District Medical Officer) persuasion been built *pukka* and rat-proof.

Rat destruction had often been unpopular with the people for religious reasons, and had more often than not received little or no cooperation from the public. Any reduction of the population by trapping could be only temporary, because rats bred so rapidly that they soon replaced the casualties. A serious objection to rat destruction by trapping and baiting was that the infected fleas were not destroyed. It might be argued that to kill rats when plague was raging merely released plague-infected fleas and increased the chances of man being infected.

INTERROGATING THE MEDICAL INTERVENTION

The treatment of this pestilence might be said to have been hitherto altogether ad hoc. The great number of medical men who had since the beginning of the nineteenth century visited the East and practised for many years in countries where plague was most prevalent had thrown no light upon the cure.[36] Leibermeister argued that the treatment of plague was principally the use of prophylaxis. The treatment of individual cases could only be symptomatic.[37] Aitken regretted in 1880 that recent experience had not in any degree advanced the successful treatment of the plague.[38] Dr. J.F. Payne who had personal experience of the disease wrote in 1885, 'No special line of treatment has proved efficacious in checking the disease once established.'[39] Dr. B. Scheube, Government Physician and Sanitary Councillor at Greiz, formerly Professor at the Medical College in Kyoto, admitted that the treatment of the plague focused on the symptoms; unfortunately, he argued, it held out little prospect of success.[40]

One could multiply such testimonies, but it is needless to do so. There could not be the slightest doubt that the disease by its very nature was of the most serious kind, and attended with heavy mortality. But whether the treatment adopted by the old school, especially in the olden days, did aggravate the disease and add to the mortality was a question which merits historical interrogation.

This much was certain: some members of that school had admitted the injudiciousness of that treatment. Thus Dr. Gavin Milroy, one of the greatest authorities on epidemic diseases, argued that there was little in medical writings at all satisfactory or encouraging in respect of the recovery of the sick. The perusal of recorded histories of cases of plague as observed in Malta in 1813 and in Egypt in 1835 left the impression that patients would have fared better had they been treated with light nourishing food and cordials frequently administered, together with simple saline or acid medicines, and without active purgation or blood-letting, or other energetic measures.

Dr. Dyson, Sanitary Commissioner of Bengal and member of the Plague Commission, had condemned the use of antipyretics such as antifebrine, and phenacetin drugs as they had been found to produce unfavourable results. Plague was essentially a disease with adynamia of the greatest description as its pronounced condition, and therefore nothing should be done to aggravate that condition.

Dr. Mahendra Lal Sarkar, the celebrated homeopath, admitted in 1899: 'We have as yet had no testimony from members of our own school.' Indeed homeopathy had not yet held trials for this disease. And this for the simple reason that there had been no occasion for it, neither in Europe where plague had ceased to exist before homeopathy developed into a workable system, nor in America where the disease was happily unknown. Hence in homeopathy books with but few exceptions the disease was not mentioned at all. Dr Marcey and Hint, in their *Homeopathic Theory and Practice of Medicine*, argued: 'If we may be allowed to judge of its nature from those phenomena which seem to be characteristic we suppose the following remedies will correspond to its manifestation, and prove to it homeopathic, namely: Arsenic, Acid-nitr, Rhus-tox, Veratrum, Merc. Bell, Ipecac, Carbo-veg.'

Dr Richard Hughes in his admirable little *Manual of Therapeutics According to the Method of Hahnemann*, stated about the plague, 'Homeopathy has no practical knowledge of its therapeutic, and happily none of us are likely to have any occasion to treat it. If we had, Arsenic and Lachesis are the two medicines on which I should feel disposed to rely.'[41]

Dr. Constantine Hering stated in 1879 that Lorbacher had proposed Lachesis, Arsenic, Carbo-veg, Phosphorus and Secale as the main remedies for plague.[42] To this he added Badiaga. To him, what Lachesis would do was uncertain. Still more uncertain was Arsenic. Hering supported Raue in administering Kali phos, which he considered very promising. Stramonium, he argued, had more similarity to the plague symptoms than Belladonna, and Silica more than Hepar. Loimine, a preparation of the pus of the plague brought to India by Dr Theuille, had allegedly cured cases of the greatest intensity.

Raue, mentioned by Dr. Hering as having proposed Kali phos as remedy, contented himself with giving the quotation from Hering for the article about the plague in his *Special Pathology and Therapeutics* (1895 edition) without making any additional suggestions of his own. Winterburn, in his article on plague in Arndt's *System of Medicine*, says

> In addition to the remedies mentioned by Hering, I would like to suggest Crotalus, as possessing analogically a more intimate relation to the plague than Lachesis, Arsenicum or Phosphorus; though here again we cannot not know what is the right remedy until the patient is before us, or what a remedy will do until we have clinically tested its efficacy.

Thus one can see that the killer disease drew interested attention in different schools of medical intervention, but with little or no results. This only exhibited mounting tension in the medical world. Plenty of fresh air and absolute rest in the horizontal position were supposed to be the first essentials of treatment in the plague hospitals.[43] No plague patient was allowed to sit up in bed until the temperature had remained normal or sub-normal for four or five days. In the majority of cases, a preliminary calomel purge was given, while from the onset almost free stimulation by alcohol and by food was supposed to be necessary. Where many patients were brought in a condition of collapse, an ounce of stimulant mixture was administered as a routine procedure. Pyrexia was treated by tepid sponging, since the usual chemical antipyretics were found to produce dangerous collapse. Ice bags applied to the head were found the best remedy for the delirium, although an occasional dose of bromide was given. For persistent vomiting, nothing succeeded better than liquor morphiae internally, with ice to suck, and a mustard leaf to the epigastrium.

A diet of milk, rice, congree, sago, arrowroot, egg-flip, chicken or goat soup and soda water were used during the acute stage.[44] Delirious patients frequently had to be fed by enema or the nasal tube. Subsequently, as convalescence advanced, the diet was gradually increased.

Serum treatment as advised by Drs. W.M. Haffkine and Yersin was also practised. But since the publication of Dyson and Calvert's report, Dr. G.S. Thompson, Civil Surgeon of Satara, who had subjected cases under Yersin's serum to a critical analysis, came to the conclusion that the results gave 'no better hopes of recovery than ordinary empirical treatment'. Serum treatment, he argued, did not fulfil the expectations and laudations that had preceded its trials in the present epidemic. Moreover, it was not sufficiently perfected in details of strength and dosage to justify its application to patients suffering from plague. He was therefore inclined to think that the serum treatment, despite its temporary reputation, would not stand the test of time.

As regards Haffkine's prophylactic serum inoculation, Indian doctors too expressed almost the same opinion. Mahendra Lal Sarkar, for instance, argued in 1898 that it was uncertain how long the immunity would last.[45] It was uncertain whether, when the immunity had been lost, the inoculated might not become more susceptible to the disease. And it was still more uncertain whether the inoculated, while immune themselves, might not be the carriers of infection to others. Serum inoculation introduced into the human system a virulent septic poison, and Haffkine had thought that the degree of immunity was faithfully proportionate to the strength and dose of the prophylactic. Hence he had proposed to increase the virulence of the poison. In other words, he had sought to introduce into the healthy human organism as virulent a culture as he could possibly make in the laboratory. Sarkar argued that this was far from pleasant to contemplate for those who would like to be inoculated. Nevertheless, George Lamb, Director of the Pasteur Institute of India, carried out an extensive series of researches on plague, both epidemiological and experimental, which gained the respect and admiration of bacteriologists all over the world. He was responsible for carrying a detailed enquiry into the mechanism of the epidemic spread of plague in India, the results of which had been published in five reports. He initiated and throughout bore a prominent part in the long series of experiments and observation, which resulted in the conclusive proof of the transference of plague from rat to rat, and from rats to man by the agency of fleas.[46]

PATHOLOGY OF POPULAR MIND

The prophylactic measures adopted by the government were well intentioned, but the motives were often misconstrued. Rumours were afloat that the government desired to poison the people, to reduce the superabundant population, or to deter foreigners from invading India, even to seize the people's money and property. Some said it was to propitiate the plague demon, for the permanent consolidation of British rule in India.[47] Among other absurd rumours that were circulated was one that the intention of the government was to interfere with the religion and caste of the people, and to destroy caste and religious observances with the ultimate design of forcing Christianity on the natives.

Dr. Mahendra Lal Sarkar argued in 1898 that if all the *golmal* about the plague had produced any other effect it had produced one thing which must be regretted, and that was the scare.[48] There had been a tremendous exodus from the town in consequence of the panic caused by the report that everybody suffering from the disease would be taken to the segregation camp for treatment. Crowds of men and women were seen thronging the streets of Calcutta; the thoroughfares presented a shocking scene; and the railways stations were so overcrowded that many of those who had gone there were obliged to return home, as the carriages were not sufficient to carry them away.[49] The cart men, coolies, *mehters*, and *dhangars* had struck work and gone back to their village homes. Home servants and cooks were also leaving the town for fear of being inoculated or segregated; it was becoming more difficult to get washer men to wash clothes. The mercantile community was meeting the greatest difficulty in moving goods and working godowns for want of sufficient coolies.[50] The landholders were suffering from the loss of their tenants who were leaving the city, and the general public from want of menial servants of all classes.

Although panic would gradually abate, the segregation fears was increasing. The government under the rules of the Venice Convention was bound to observe them to prevent the spread of the plague. (The Venice Convention held that segregation was one of the best means that could be adopted to prevent the plague taking an epidemic form.) The question was whether the Convention meant that by segregation the patient was to be separated from the other members of the family, and placed in a separate apartment of the house or whether he was to be taken for treatment to a plague hospital. That the latter was not the intention of the Convention seems to be evident from the concession already made by the government that the members of a family attacked by plague might be lodged in a separate apartment of a house, subject to the approval of the authorities. People wondered, then, 'Why remove patients of the poorer classes to the Manicktollah Hospital, if they dread to be taken there? Then why not allow them to be treated in their own houses?'[51]

People also argued that in order to disabuse the minds of the masses about segregation, they should be treated in separate apartments of their own houses by their own doctors, where this was possible, under the supervision of the Health Officer and his Assistant. But in cases where the accommodation was not considered sufficient, the patients of the poorer classes might be treated in the Ward Hospitals that the Commissioners had opened in their respective wards, under instructions from the authorities. This, the people argued, would be in consonance with local feelings, habits, and customs, and would in no way to raise unnecessary alarm.[52] People were leaving the city in large numbers as they considered their hearths and homes sacred, and did not like to be invaded by police officers.

The *Times* had also taken up the subject of segregation. 'The Indian Press', it argued, 'continues week after week protest against the stringency of oppressive measures, which could with difficulty be forced upon a European population, and which are so opposed to the most sacred traditions of Eastern life. . . . Europe understands this perfectly well in regard to Turkey. It realizes that to compel the Sultan to apply to his Oriental provinces the same that may with safety be applied in London or Berlin would shake the Ottoman Empire to its foundation.'[53] As regards the measures adopted in Bengal, the *Times* asked: 'How far is a Government justified in introducing measures at the instance of foreign advisers which can only be enforced at the cost of panic?'

An alarm spread that a person suspected to be suffering from an attack of plague had been carried away to the isolation hospital at Manicktollah, and that it had no provision of separate wards for the treatment of Mohammedan and Hindu plague patients. To allay the fear, it was suggested that the *Baithakkhanas* or ground floor rooms of the outer block of Bengali gentlemen's dwelling houses might be allowed to be used as segregation hospitals for the use of plague patients.[54]

Ambulance vans were also matters of controversy. Rumour was afloat that people dreaded the ambulance van more than death itself. They believed that when a patient was put into it, his sufferings were great, and the disinfectants made the van more uncomfortable.[55] People argued that the *palki* instead of ambulance van should be introduced to carry the patient to plague hospital. If *Oriya* bearers could not be found to carry plague patients in *palkis*, *Romoni* or up-country bearers might be employed on the duty. *Palkis*, it was argued, which had no cloth covering inside, might be used as they could be easily disinfected.

CONCLUSION

Evidence from eastern India indicates that the plague epidemic spread consternation throughout the province not because people were scared by the alarming tolls of death, but especially because in the hour of crisis, the government behaved in a way that was marked by both proactive haste and brute indecision. Segregation camps, inoculation, ambulance vans, ward hospitals, plague precautions and many more measures the government resorted to were the right therapeutic statements in the wrong places at their worst conceivable forms. Admittedly, therefore, whatever the government did, generated adverse public reaction, bred suspicion, and fanned rumours (presumably because people were not adequately briefed about the benefits they might derive from such measures). Nor had people experienced any appreciable benefits from the measures in the past when vaccination was enforced to mitigate epidemic smallpox. Government measures might have produced the desired results had they amounted to an approximation of the traditional therapeutics and the rhetoric of Western medical intervention.

What the people did attempt instead as prophylactics largely relates to an appropriation of the accumulated wisdom of a tradition-bound society, which enjoins that what cannot be ended, should be mended through divine intervention. They turned to providence. When the epidemic was at its worst, the urban literati suggested that a day of general prayer be observed in all churches, chapels, masjids, Brahmo Samajes, temples, and all places of public worship for offering humble prayers.[56]

> Let us then make pujah to propitiate the Deity according to the customs of our various nationalities. Let the Hindu worship his idol, utter the mantras, which his Shastras have taught him, and make *homa* and observe other propitiatory rites. Let the Christian pray in his Church. Let the Brahmo pray in his Samaj. Let the Mahomedan utter his prayers in his Masjid. I have no doubt that the united prayers of His children would be heard and we would be freed from the plague by the Almighty Giver of all good and blessings.[57]

In the previous years, it was argued, when an epidemic of cholera or small pox had appeared in the city, *sankirtan* parties were formed in different parts of the town. They chanted hymns, and went from door to door in groups, singing the name of *Hari* in loud chorus with a view to driving away the epidemic.[58] There was no reason why similar parties should not be formed at the then crisis when the fear of plague was driving them mad.[59]

From the contemporary literary tracts and vernacular periodicals it appears in retrospect that the incidence of plague was popularly viewed as a visitation of fate. One had to submit to it with patience, but without an effort to do what was humanly possible to mitigate the calamity. People were also distrustful of Western methods for the protection of public health and extirpation of epidemic disease. Added to this, both Hindus and Mohammedans viewed with the greatest dislike any intrusion into their homes, and especially any possible interference with the privacy of their women. Among Hindus, again, the caste system and its elaborate rules pr evented the close association, and interdining between superior and inferior castes. In the case of the city of Bombay, the greatest opposition was experienced from the Sunni Mohammedans, and especially the Konkani Sunnis. In the case of eastern India, riots, strikes, and disturbances were actions of the marginalized municipal coolies, scavengers, sweepers, carters, butchers, durwans, menial servants, railway coolies, and mill workers—people who were hit the hardest by the plague regulations.

Where the incidence of plague had been really severe, the people had readily given information to the plague authorities and had sought their aid. They had also to some extent taken refuge in flight. But once the virulence of an epidemic had abated and the fear of death was no longer before their eyes, their attitude had throughout been one, if not of actual hostility, at least of indifference to precautionary measures and unwillingness to undergo the inconvenience incurred in order to guard against future calamity. It was this attitude that was the perpetual obstacle in dealing with plague in eastern India.

NOTES

1. David Arnold, 'Plague: Assault on the Body', in *Colonizing the Body: State, Medicine and Epidemic Disease in Nineteenth-Century India*, Delhi, 1993, pp. 200–39.
2. Rajnarayan Chandavarkar, 'Plague Panic and Epidemic Politics in India, 1896–1914', in Terence Ranger and Paul Slack, eds., *Epidemics and Ideas: Essays on the Historical Perception of Pestilence*, Cambridge, 1992, pp. 203–40.
3. Ira Klein, 'Plague, Policy and Popular Unrest in British India', *Modern Asian Studies*, vol. 22, no. 4, 1988, p. 739.
4. Chandavarkar, 'Plague Panic', p. 223.
5. A. Mitra, *The Bubonic Plague*, Calcutta, 1897, p. 1.
6. *Bubonic Plague in Bombay*, a paper read by Dr. A.G. Viegas at the Meeting of the Grant College Medical Society on 24 November 1896.
7. Ibid., p. 23.
8. Ibid., p. 24.
9. Mitra, op. cit., p. 8.
10. Ibid., pp. 9–10.
11. C.E.A. Winslow, *Man and Epidemic*, Princeton, 1952, p. 200.
12. Mitra, op. cit., p. 12.
13. Winslow, op. cit., pp. 200–1.
14. Report on the Epidemic of Plague in Calcutta during 1898 and 1899 and up to June 1900, by E.N. Baker, Secretary (hereafter Secy) to the Government of Bengal (hereafter GOB). Home Department, Sanitary A, February 1901. National Archives of India (hereafter NAI), Delhi.
15. Baker, op. cit., p. 7.
16. Ibid.
17. 'A note on hydro cyanic acid gas fumigation of rat borrows as an anti-plague measure', in *Health Bulletin*, no. 21, by Major W.J. Webster, Assist-Director, Central Research Institute, Kasauli, New Delhi, 1935,

p. 1. Oriental and India Office Collection (hereafter OIOC), British Library (hereafter BL). V/25/850/94.

18. Ibid., p. 1.
19. *A Compilation of Regulations issued by the Government of India and Local Governments in Connection with Plague*, Calcutta, 1898. V/27/856/6A, OIOC, BL, London.
20. Ibid., p. 89.
21. Ibid., p. 90.
22. Ibid., p. 91.
23. *The Bengal Plague Manual, Being a Collection of the Extant Regulations and Executive Orders in Connection with Plague Issued by the Government of India and the Government of Bengal*, Calcutta, 1903, p. 68.
24. Ibid., p. 30.
25. Ibid.
26. Baker, op. cit., p. 8.
27. Ibid., p. 10.
28. Webster, op. cit., p. 2.
29. Ibid., p. 3.
30. *Report on the Epidemic of Plague from 22nd February to 16th July 1897*, by James A. Lowson, Plague Commissioner, Bombay, Bombay, 1897, p. 40. OIOC, BL, London, V/27/856/13.
31. Lowson, op. cit., p. 40.
32. Ibid., p. 41.
33. *Report on Plague in Calcutta for the year ending 30th June 1907*, by T. Frederick Pearse, ed. W.C. Hossack with Appendices by W.C. Hossack and H.M. Crake, Calcutta, 1907. NAI, Delhi.
34. Pearse, op. cit., p. 3.
35. Ibid., p. 4.
36. Copland, *Dictionary of Practical Medicine*. Cited in *Therapeutics of Plague, Being Suggestions for the Prophylactic and Curative Treatment of the Disease*, by Mahendra Lal Sarkar, Calcutta, 1899, p. 2. OIOC, BL, London.
37. Zimssen's *Cyclopaedia*, cited in M.L. Sarkar, op. cit., p. 3.
38. Aitken, *Science and Practice of Medicine*, 1880.
39. *Encyclopaedia Britanica*, 1885, 9th edn.
40. B. Scheube, *The Diseases of Tropics: A Manual for Medical Men.*
41. Sarkar, op. cit., p. 6.
42. *North American Journal of Homoepathy*, August 1879.
43. Drs Dyson and Calvert's Report, para 24. Cited in Sarkar, op. cit., p. 45.
44. Drs Dyson and Calvert's Report, para 25. Cited in Sarkar, op. cit., pp. 46–7.
45. *Calcutta Journal of Medicine*, June 1898.

46. Lamb Papers. Mss Eur D 893. OIOC, BL, London.
47. *The Bengal Plague Manual: Being a Collection of the Extant and Executive Orders in Connection with Plague Issued by the Government of India and the Government of Bengal,* Calcutta, 1903, p. 29.
48. *Calcutta Medical Journal*, May 1898.
49. *The Indian Mirror*, 5 May 1898.
50. Ibid., 11 May 1898.
51. Ibid., 10 June 1898.
52. Ibid., 11 June 1898.
53. Cited in *The Plague*, by Gobin Chunder Dhur, Calcutta, 1898, p. 18.
54. *The Indian Mirror,* 23 June 1898.
55. Ibid., 17 June 1898.
56. Gobin Chunder Dhur, *The Plague: Being a Reprint of Letters Published in the Indian Mirror for Allaying Popular Alarm and Conciliating the People to the Action of the Authorities*, Calcutta, 1898, p. 5.
57. *The Indian Mirror*, 30 May 1898.
58. Ibid., 29 May 1898.
59. Ibid., 1 June 1898.

CHAPTER 7

Coping with Epidemic: Responses in Bombay Presidency, 1900–1919

MRIDULA RAMANNA

In the nineteenth century British officials tended to ascribe periodic outbreaks of the epidemics of cholera, smallpox, and 'fevers' (which latter term embraced a whole gamut of inexplicable complaints) to the lack of a sense of sanitation and hygiene among Indians. The promotion of a colonial sanitation policy in Bombay Presidency began in the middle of the century, primarily because, as surgeon-general William James Moore put it, 'the improvement of the public health of natives was generally held as one of the most important measures by which the health of Europeans sojourning in their midst was to be maintained'.[1] Bombay's health officials, Thomas Gilham Hewlett (serving between 1865 and 1873) and Thomas Stephenson Weir (1873–97) showed unmistakable zeal in cleansing and sanitizing that city. Their annual health reports speak of the need for cholera hospitals in every locality, for a cleaning of *gullies* (trenches) and alleys with carbolic acid, for the removal of tiles from the roofs of houses which had cholera cases, for the white washing of walls, and for sanitary raids. The other culprit in the colonial perception was the pilgrimage, whether the *simhasta* at Nasik, or a *jatra* to Alandi, Jejuri, or Pandharpur within the Presidency, or the annual *haj*. These only meant a fresh supply of 'contagion' to Bombay, Hewlett had asserted. Sanitary Commissioner John Lumsdaine gave a vivid description of what he thought happened at the jatras: there were the excesses and irregularity of endless festivals, the feasting and fasting, the strong emotions during religious observances, the fatigue and exposure, the overcrowding and filth and subsequent dispersion, with an 'utter disregard of all ordinary prudence and the voluntary courting of danger by the fanatical adoption of noxious practices'.[2] It was thus felt that these events should be controlled; yet, even in cantonments, it was found that enforcement of strict quarantine had not been possible because supplies had to be brought in from outside. The difficulties were regarded insurmountable, especially with regard to fairs like those held at Hardwar, where regulation was an occasion for extortion and oppression.

There then came a shift in the attitudes of the colonial state, in the late nineteenth century, the turning point being the plague epidemic of 1896–7, when drastic controls were introduced to combat the epidemic and Indian protests were brushed aside. The outrage over this unparalleled interventionism turned to vehement opposition, which together with the spreading national consciousness, led to the abandonment of the policy. After 1900 greater caution was adopted and intrusive measures were given up. But epidemics continued to regularly appear, and in 1905, after three years of high mortality, the Government of India sought the people's support to fight the disease.

In this context, was there a change in Indian attitudes? This article will explore Indian responses to controls in relation to cholera, smallpox, and plague outbreaks between 1900 and 1919, by drawing on evidence from different parts of the Bombay Presidency. An attempt is made to answer the following questions: what was the record of sanitary progress and smallpox vaccination? What were the reactions to the regulation of pilgrims, who were seen as carriers of disease? How were the less interventionist plague combating measures perceived? At the end of the two decades, in 1918–19, came the influenza epidemic, when the state threw up its hands in despair, neither knowing how to treat it nor having enough personnel to deploy, in view of doctors being away on war duty. It was the relief provided by voluntary organizations in the cities of the Presidency that enabled the authorities to cope with the challenge, thus indicating growing public involvement in health care.

CHOLERA AND SMALLPOX

The peaks of cholera deaths, reached in the last quarter of the nineteenth century, have been linked to famines. The containment of the latter and other stimuli like sanitary improvements led to lower cholera mortality in the early 1900s. In Bombay city, the advocacy of hygiene was linked to sanitary reform and led to the inauguration of the Vehar, Tulsi, and Tansa water works, (these lakes supplying water to the city) and discussions about various drainage schemes in the 1870s and 1880s. These measures were funded partly by the civic body, where Indians had some representation, and so the plans and their implementation became the subject of much debate. Indian responses to sanitary reform were ambivalent: while some supported and funded the efforts, others objected to tap water and underground drainage as ritually polluting.[3] The Bombay government proceeded with these plans in an ad hoc manner, primarily because of its reluctance to make financial commitments and the limits placed by the Government of India.[4] What was the state of sanitation in towns and villages? There were 157 city and town municipalities, 25 district local boards, and 212 *taluka* local boards, and, by 1914, a total population of over 18 million under the administration of local boards. The Bombay government sanctioned certain allotments from provincial funds and from a grant of Rs. 7 lakh from the Government of India. Priority was given to pilgrim centres, seen as breeding grounds of epidemics.

While officials in the early 1900s still subscribed to the view that the condition of the soil was inimical to the growth of cholera, adequate water supply was recognized as a priority. It was found that the disease severely attacked towns and villages located on river banks and that a plentiful mango crop exacerbated the problem. But a constant refrain in the sanitary reports was the lack of funds and complaints against municipalities for not spending money to their best advantage. In Dharwar and Belgaum, water supply schemes were drawn up only to be shelved; water closets were nowhere in use outside Ahmedabad and Karachi. Those cities which had no piped water supply depended on wells, tanks, and *nullahs*. In times of famine or scanty rains people had to depend on subsoil water: and the lower and scantier it became, the chances of cholera increased. Pamphlets in English and local languages were issued, emphasizing the importance of water supplies and advising villagers not to wash their clothes near the source of the supply. As for drainage, it was noted in 1906 that in almost all towns, night soil was removed by manual labour, and cities like Poona, Sholapur, Surat, Hubli, and Hyderabad were badly in need of 'scientific drainage'.[5] The Deputy Sanitary Commissioner of Sind averred that non-officials on these boards had little interest in sanitary matters. In view of the expenses on plague other taxes were regarded as a burden; also, the appalling calamity of plague had distracted the attention of the general public from the dangers of accumulating waste in the neighbourhood of dwellings. Elaborate surveys and reports were made, but their recommendations were seen as impractical and utopian.

The Village Sanitation Act became a reality and, by 1910, was in force in 300 villages. A quarter of the contributions to the village sanitation committee came from government, a third from the local board, and the rest from the people. However, the sanitary committees were unpopular and contributions were voluntary only in name. The *Sanitary Commissioner's Report* of 1908 concluded that, 'with an income so uncertain and so variable from year to year it is small wonder that few works of permanent utility are carried out'.[6] The Collector of Bijapur felt that they only wasted money; the Deputy Sanitary Commissioner of Gujarat saw no improvement in villages where the Act was in force; and the Collector of Karachi observed that sanitary committees were 'anaemic' institutions, collective sanitation not appealing to the people.[7] The inducement that appealed most to villagers to establish a sanitary committee seems to have been the fact that street lamps would be provided. Expenditure under the head of sanitation comprised of the salaries of sweepers, who only cleaned the houses of 'leading' residents.

The year 1915 was exceptional in so far as cholera deaths were concerned. 'Only once in the past fifty years have the inhabitants been so fortunate in the incidence of the disease.'[8] But again the rates rose.

Although high death rates from smallpox were recorded in 1900, 1905, and 1913, there is also noticeable an overall decline.

Vaccination had varied responses. Dr. Shantaram Kantak, Superintendent of Vaccination, had protested strongly against 'omissional infanticide' caused by apathy and indifference.[9] The sanitary department was advised to obtain early information about outbreaks, encourage vaccination and re-vaccination, and supplement existing staff. Birth data could aid smallpox vaccination campaigns, but those who wished to avoid vaccination gave wrong information or concealed births. The collection of data was viewed as a precursor to taxation. In the early twentieth century there was no marked improvement and very low birth rates were recorded in Sind, where the work was left to illiterate subordinates, while in the southern regions of the Presidency, registration was considered 'untrustworthy' because the hamlets were scattered over forest areas. Even in Bombay city and Ahmedabad, high rates of 'stillborn' were recorded, which were ascribed to 'second hand' information. The refusal of adults to submit to re-vaccination was seen as being responsible for its introduction into certain areas. In 1916, registration was termed 'defective' in many towns of the Presidency. It was felt that the municipalities were too timid to enforce the bye-laws. The vaccinators, who were expected to check birth and death registers and report omissions to the Deputy Sanitary Commissioner, performed this duty in a perfunctory manner, for fear of offending and bringing to punishment police patels, who were the registrars.

The difficulty in promoting smallpox vaccination in the talukas of Sirsi and Sidapur was compounded by the reluctance of parents to allow vaccination in the first year of an infant's life. Besides, the vaccinators themselves suffered malaria in these highly malarial districts, and there was the problem of maintaining a potent lymph supply.[10] In Sind it was found that where there was opposition to vaccination, mortality was 70 per 100,000, while in the protected areas it was only 10 per 100,000.[11] Sanjay Bhattacharya has shown that caste and religious considerations still mattered in the early 1900s, in the choice of the vaccinifier. Though the Municipalities of Sukkur and Rohri had passed resolutions adopting compulsory vaccination, the authorities deferred to popular wishes that arm-to-arm vaccination of Hindu children be done exclusively from

Hindus, and of Muslim children from Muslim vaccinifiers.[12] Hyderabad city was recorded as a place notoriously opposed to vaccination, while in Malegaon, the opposition came mainly from *momins* (weavers). In Sholapur, extreme laxity in the registration of births was reported in 1909, therefore the vaccinator had no reliable information to guide him as to the number of children available for vaccination, and a large proportion of the infant population escaped his notice. On the other hand, the Collector of Dharwar reported that three-fourths of the pilgrims were persuaded to undergo vaccination. As for the success rate of primary vaccinations, it was in the range of 90 per cent and above. However, it was found that re-vaccination figures were small and were confined to large towns and places severely affected by smallpox. The record of sanitary progress and responses to vaccination were thus mixed.

PILGRIMAGE AND EPIDEMIC

The authorities reckoned that the 'primitive habits' among inhabitants of a small village may not influence the health of people outside, but similar 'habits' on the part of 150,000 pilgrims were very likely to adversely influence the public health of a whole district. Yet there was hesitation in implementing regulations on pilgrims. David Arnold has referred to the Government of India cautioning the Bombay government in 1905 about the political 'inadvisability' of invoking the Epidemic Diseases Act of 1897 when cholera broke out at Pandharpur.[13] The disease was constantly tracked. Thus, in Sind, in 1906, the first cholera patient came to Karachi from Calcutta in a steamer and died at the harbour. Then three months later, there were cases among emigrants from Sukkur, and among a number of persons who had congregated at Sehwan for the *Lal Shabaz* fair. From there, it spread to Hyderabad, Jacobabad, and again to Karachi.[14] Death rates in pilgrim towns like Pandharpur, Dakor, and Gokarn remained high. While the use of permanganate of potash was advocated, this was not always accepted; brahmin priests in Nasik objected and stopped the 'permanganation' of the Ram Kund and Lakshman Kund, the 'holy of holies', in 1906.[15] During the *simhasta* fair of 1908, elaborate precautions included a thorough inspection of all incoming and outgoing pilgrims at the railway station and the provision of a cholera hospital near the station, both at Nasik and Trimbak. Fairs were sometimes prohibited during epidemics, as at Pandharpur in 1904, 1905, 1908, 1915 and 1916. Manjiri Kamat has shown that these prohibitory orders were not planned actions, but reflections of the indecision and anxiety of the government.[16] Though the figures rose again, there is perhaps better public awareness, for there are references to the fear of cholera reducing the number of pilgrims.

The press constantly campaigned for better facilities at pilgrim centres. *Kesari* held that impure water, unwholesome food, a lack of rest houses, and exposure to rain had led to the outbreak of cholera among the pilgrims of Pandharpur in 1914 during the *ashadhi* fair; it was suggested that these could be removed with cooperation between the people and the government. The paper felt that the latter's priorities were all wrong. 'If the sanitary department turns a little from the rat campaign, destruction of mosquito larvae and counting of plague germs on the plague flea to such work as providing pure water and places of shelter for the pilgrims it will prevent cholera.'[17] Sub-assistant surgeons who were attached to the different *palkhis* on their inward and outward marches to Pandharpur inspected the *palkhis* and weeded out all infectious cases even before they reached the destination. For the protection of the existing water supply, recommendations included the conversion of step wells into draw wells. All medical personnel were deployed for the regulation of pilgrimages, and their other duties suspended. Trenches were dug, dustbins provided, extra sweepers employed, foodstuffs inspected, and sheets of muslin provided to sweetmeat dealers to cover their wares. In addition to the stocks with the municipality, the collector of Sholapur reported in 1919 that he had arranged for 175 tins of pesterine, 3 drums of bleaching powder, 17 night soil carts, incinerators for burning refuse, and provisions for lodging 23,382 persons in private houses.[18] It was undoubtedly the involvement of Indian doctors like Anthony D'Mello, Deputy Sanitary Commissioner, which made the regulations more acceptable.

Even in Bombay city, where vaccination had been compulsory

since 1877, John Andrew Turner, the activist health officer (1901–19), deplored the practice of taking children suffering from smallpox in a public conveyance to a temple (obviously a *shitala* temple), and to smallpox and plague cases travelling along with healthy passengers on the railways. In 1915, a pilgrim committee, comprising of the sanitary commissioners of Bombay and India and two Indian representatives, K.R. Godbole and S.G. Haji, was appointed to enquire into the possibility of improving sanitary arrangements at pilgrim centres, camping grounds, lodging houses, and better arrangements on railways. They visited nine pilgrim centres and five railway stations, and recommended better co-ordination with the railways.[19]

Varying reactions are found in newspapers to measures regulating the movement of pilgrims. In 1916, the sanitary commissioner's justification for stopping the Siddheswar fair at Sholapur was that there had been plague deaths. The *Kalpataru* printed a letter from a person allegedly a victim showing that he was alive. More stridently, the *Kesari* warned that the public was 'rapidly ceasing to be dumb driven cattle bending their heads at any freakful order'.[20] On the other hand, the *Bombay Samachar* supported the order prohibiting people from visiting Pandharpur because of the prevalence of plague, and observed that hundreds of pilgrims had deliberately violated orders prohibiting visits during the *karthiki* fair in the previous year, in defiance of the police. Government had refrained from action because of consideration for their 'ignorance and religious zeal'.[21]

As for pilgrims going abroad, in accordance with the Government Resolution no. 1746, dated 11 June 1875, weekly reports regarding cholera outbreaks and, after 1896, of smallpox or any epidemic disease, 'especially one that is likely to affect foreign trade in Bombay or Karachi', had to be submitted by the Municipal Commissioner of Bombay, and by the collectors of districts, to the government.[22] To 1891, weekly reports of the numbers of deaths from cholera in Bombay were sent to the British Consul in Alexandria. The inspection of clothing, beds, linen, and boxes being carried on to ships was done under the supervision of the health officer of the port. Cases of plague, cholera, and smallpox were sent to the Arthur Road infectious diseases (AR) hospital (established in 1888).[23] Vessels, crew, and bedding were inspected and disinfected, with a 5 per cent solution of carbolic acid or a one in thousand solution of a corrosive sublimate. Strict examination and disinfection had become so well known by the 1900s that the port health officer claimed that few persons actually suffering from plague attempted to travel by sea, knowing they would not be allowed to sail.[24] Plague inspection at ports was also objected to; it was argued that the coastal traffic need not be subjected to medical inspection since there was no connection with Europe or America and 'no imperial question' was involved. Thus there was no justification for the maintenance of a costly establishment, 'without any other result than the harassment of the passengers'.[25] Governor Lord George Sydenham Clarke (1908–13) conceded to this demand, and abolished inspection in 1908.

To the hardworking and zealous Turner, epidemics were brought into the *urbs prima in Indis* by pilgrims. He ascribed the introduction of cholera to the water brought from the wells of Mecca. As part of his campaign against smallpox, he suggested that all hajis on board an infected vessel prior to disembarkation, not previously vaccinated against smallpox, should be compulsorily vaccinated, and should be housed in temporary camps before their departure. While this move was welcomed, the collection of the name and address of every haji disembarking was termed 'uncalled for'.[26] The authorities were assisted by the pilgrim committee, who supervised the vaccination of haj pilgrims to immunize them against smallpox, which, during some years, was already prevalent in Arabia, when they embarked on their journeys. Separate arrangements were made for ladies in purdah, vaccination being carried out under the supervision of the woman Superintendent of Pilgrims by a Woman Municipal Vaccinator. Yet there were obstacles. Some pilgrims would not allow the vaccination to take effect, turning down their sleeves or washing their arms. Some even pressed out their vesicles or applied ash or irritants to quickly dry them. Since it was voluntary vaccination, and most were illiterate, the relative success was regarded a credit to the department. The statistics of pilgrims who underwent vaccination in the second decade of the twentieth century, shows a rise in the percentage of those vaccinated, from 45.8 per cent in 1910 to 93.89 per cent in 1920 (Table 7.1), with a fall in the numbers of persons going on haj during the war years (1915–18). Taking care of Shia pilgrims was reported to be easier. They were regarded 'not only superior in respect of cleanliness' to average pilgrims going on haj, but were also numerically smaller, and went to Kerbala at any time, not just during Ramzan. The pilgrim department's expenses included paying for interpreters and compensation for grain destroyed during anti-plague operations. The Protector of Pilgrims also reported to the government that the services of a hakim had been employed to treat sick pilgrims, since those coming from the north preferred unani medicine.[27] The action by the Commissioner of Police in taking heed of the objections of hajis to the use of Western medicines on the grounds that they contained alcohol, was appreciated.[28] Another problem was the unwillingness of some pilgrims to return from Bombay to their homes on completion of their pilgrimage. The Haj Committee had to raise money from philanthropists to create a fund for this purpose. It is significant that in 1912 the Begum of Bhopal promised an annual contribution of Rs. 2,400 besides an initial

TABLE 7.1: SMALLPOX VACCINATION AMONG PILGRIMS GOING ON HAJ

Years	Number of pilgrims	Number vaccinated	Percentage
1911	22,856	15,140	66.24
1912	15,464	10,587	68.46
1913	15,319	11,198	73.09
1914	11,700	9,102	77.79
1915	2,432	2,308	94.90
1916	2,632	2,449	93.00
1917	1,339	1,198	89.46
1918	1,361	1,169	85.89
1919	9,218	9,001	97.64
1920	18,008	16,908	93.89

Source: GD, 950, 1916–18, Protector of Pilgrims/Commissioner of Police, no. 45, P.P/6 of 1919, 15 February, 1919; GD, 950, 1920, no. 312/6 P.P of 1921, 24 February, 1921.

donation of Rs. 7,200.[29] On some occasions the pilgrims would be sent upcountry directly to their homes, without being allowed to stop in Bombay.

PLAGUE POLICY

Plague appeared regularly during these decades, the maximum mortality being registered in the months of February and March. The unpopular measures of the previous century having been given up, attempts were made in the city of Bombay to promote sanitary consciousness and to disseminate information about diseases. Thus was set up the Bombay Sanitary Association, in 1903, at the combined initiatives of Turner and Nasserwanji Hormusji Choksy, the Medical Superintendent of the AR Hospital. The first public lecture held under its auspices delivered by Choksy, was entitled, 'Some Common Sense Views on Plague'. The Bombay Sanitary Association arranged several other lectures, with visual aids including lantern slides, films and exhibits, in subsequent years. Lectures were held on the occasion of the *karthiki* fair at Pandharpur in 1912, and pamphlets, written in the vernaculars, on plague and cholera were distributed.

The 1905 resolution calling for public cooperation to fight plague was hailed by the *Jam-e-Jamshed*, for having the courage to admit 'the futility of making war on such an enemy without the cooperation of the people'.[30] *Maratha* saw it as a 'climbing down' from 1897. The policy regarding plague 'as bound to be amenable to their laws and administrative measures . . . offers a pitiable contrast with the almost fatalistic tone of the present resolution'.[31] From the outset, the funding of the anti-plague campaign had been a sore point with Indian opinion. It was observed that the cost of the measures for the suppression of plague in 1896 was Rs. 25 lakh for the Presidency, of which the Government of India had offered only Rs. 5 lakh.[32] The *Voice of India* pointed out that 'of the crores and crores of money recovered from our rural masses every year not a fraction is given back to them'.[33] The departments swallowed up the bulk of the 'miserable' allotments and the people had no voice in managing their affairs, nothing in reserve to fall back on when famine or plague struck. The paper concluded that this was the 'weakest armour' of the British administration and warned, 'God knows how the officials will answer on the day of reckoning'.[34]

The policy now was to promote the prophylactic and the extermination of rats. The former was received with mixed responses; some contemporary newspapers endorsed its efficacy, and others reported that it produced weakness in the limbs and sudden pains in the joints, and associated it with leprosy and continuous fever.[35] The authorities hoped that since Indians were familiar with the smallpox vaccination, they would accept inoculation. They had to be told that inoculation was not a certain prophylactic but reduced the chances of infection with plague; an inoculated person, when attacked, had increased chance of recovery and the process, if properly done, was harmless. It had been found that where people were not warned, they declined to have anything to do with it. While one view contended that it offered protection for only six months, another held that to get oneself inoculated year after year was troublesome.

Acceptance or rejection depended on who was providing it. The successful propagation of the prophylactic in Poona in 1908 was particularly significant since that city had been in the forefront of agitation against it in the previous decade. Under the leadership of Gopal Krishna Gokhale and his team of volunteers led by Gopal Krishna Devadhar, the Poona committee coordinated with the municipal authorities not only in propagating inoculation (persuading 15,250 persons to get themselves inoculated) but also in rat destruction, evacuation, and the disposal of the dead.[36] They compiled statistics showing 1,250 deaths among the un-inoculated, and only 4 deaths among the inoculated, and concluded that there had not been a single instance in which 'evil effects' had resulted from inoculation.[37] In fact, their work was so effective that a writer in the *Oriental Review* asked why Indian leaders in Bombay could not take a leaf out of the book of their Poona counterparts.[38] Bal Gangadhar Tilak's *Kesari* had changed its stand on the grounds that the serum had become more pure, and claimed that had the press not protested initially, the quality of the serum would not have improved.[39] At the same time, the paper deprecated efforts to pressurize the people to accept inoculation. *Maratha* maintained that the plague administration had fully justified the popular demand made, after the first epidemic, that local bodies should be entrusted with the task of carrying out all preventive measures, under the guidance of the government.[40] Endorsement of inoculation by the Aga Khan had enabled the Khoja community in Karachi to procure the vaccine from Bombay in April 1898, even before the activities of the plague committee commenced. S.M. Kaka, health officer of Karachi, recorded that in the five inoculation stations established in 1898, actually 3,756 males and 2,063 females were inoculated once and 155 males and 132 females twice, women being inoculated in their own houses.[41] While the prophylactic was advocated in Broach and Kaira, newspapers from Satara and Belgaum objected to compulsion. The repeated failure of inoculation in Bijapur was ascribed to the lukewarm attitude of lower 'native' revenue officials.[42] On the other hand, the endorsement by Tilak that the vaccine was prepared with utmost care made it acceptable in Dharwar. Doctors played an important role in the promotion; the Bombay Medical Union (established in 1884) endorsed it, as did a prominent vaid, Marishankar Govindji Shastri. Dr. Bhalchandra Krishna Bhatavadekar explained that it was unsuitable only in cases of tuberculosis, rheumatism, kidney disease and general debility. The success of its promotion in Wai was attributed to the efforts by Dr. D.A. Turkhud.[43] In 1915, the Sanitary Commissioner reported that 29,925 persons had been inoculated at civil hospitals and dispensaries.[44]

As for rat destruction, resistance came from the Jain community, who objected to it on religious grounds, and consequently the death rate among them was 53.9 per 1,000, against 27.9 among Lingayats and 20.7 per 1,000 among brahmins during the years 1901–6.[45] Inspired by Turner's success in Bombay, the Bombay government issued a directive to collectors in other cities to provide an incentive of Re. 1 for every twenty rats killed. The endeavour was to prevent the spread of plague from village to village by destroying rats in all adjoining villages.

INFLUENZA, 1918–1919

When the influenza epidemic broke out in Bombay city in June 1918, there were differences among the colonial officials about the source of the infection. Turner contended that the crew of a ship that had docked in the port at the end of May had brought the fever. The Government of India gave it the usual explanation, unsanitary conditions in the city. The epidemic soon spread to other parts of the Presidency. The inadequacy of governmental resources to provide immediate relief was soon apparent, and the Sanitary Commissioner admitted as much.

An idea of the number affected could be gauged by the fact that a total of 220,000 influenza cases were treated in the Presidency's hospitals and dispensaries. In Bombay city, accommodation for seriously ill patients was made available in the government run Jamsetjee Jejeebhoy (established in 1845) Goculdas Tejpal (1874) and the AR hospitals, and at a military camp in the suburb of Dadar. The Bombay Municipal Corporation formed a medical sub-committee under the chairmanship of Rahimtulla Currimbhoy, with Dr. Kavasji Dadachanji, President of the Bombay Medical Union, Sir Cowasji Jehangir, and Dr. M.C. Javle as members. They visited dispensaries and brought to the attention of the Controller of Prices the need to reduce the high prices of medicines. Letters were sent to the 'large employers of labour' suggesting the steps to be taken, while mill managers were summoned to discuss preventive measures to be adopted in mills. Posters were put up in English, Marathi, and Gujarati, advising the people to call in medical help or go to hospitals. Arrangements were made for the distribution of milk and pneumonia jackets. By way of affording additional facilities for the poor in the disposal of their dead, cremation and burial fees were suspended at the municipal cemetery. The military authorities provided orderlies and ambulances for the removal of patients to hospital.

These efforts were not enough, and Turner appealed to the public, observing, at the same time, that the ignorance and superstition of the people, their different social surroundings, and modes of life had made it difficult for the authorities to check the spread of disease.[46] The response was immediate. The Hindu Medical Association took the lead. Under the guidance of a committee of doctors, volunteers made house visits and directed those who had no one to nurse them to hospitals. The patients were mainly poor mill workers, 'who were ill fed and lived in badly ventilated dark rooms with smoky atmosphere full of coal and dust'.[47] Initially, the people seemed to have been uncommunicative and had to be persuaded to avail of treatment. 4,195 out-patients were treated at various centres maintained by the Association. The figures collated showed that the very young and those in their prime were the most affected.

The other body which responded to Turner's appeal was the Social Service League (SSL). Founded in 1911 by social reformers N.M. Joshi, N.G. Chandavarkar, Bhatavadekar, B.N. Motivala, and G.K. Parekh, the SSL had the provision of medical relief and the promotion of sanitation and hygiene among its aims. It had conducted anti-tuberculosis campaigns, and provided training in first aid. Now, the SSL set up the Influenza Relief Committee comprising of industrialists, judges, and doctors, and collected a fund of Rs. 52,148, which included contributions from charities (N.M. Wadia Charities, Tata Sons), professional associations (such as those of cotton brokers, cotton merchants, and mukadams) and individuals giving amounts ranging from Rs. 500 donated by Parbhuram Popatram a well-known vaid, to small sums given by anonymous donors.[48] Twenty centres, in different parts of the city, were opened to distribute the stock mixture, milk, and clothing, reaching out to 17,684 persons. Temporary hospitals for castes and communities, Jain, Lohana, Marwari, Bohra, and Pathare Prabhu were started, as had been done during the plague epidemic of 1896–7. Beds in Parsi hospitals, and a ward of twenty beds at the A.R. hospital, were set apart for influenza patients of that community as were two wards in the Maratha hospital for Hindus. To overcome objections regarding possible pollution, nurse volunteers from the St. John Ambulance Brigade Overseas (1904) represented Hindus, Parsis, and Jews.[49] Two hundred volunteers were provided by the community and caste associations of the Dawoodi Bohras and Ismaili Khojas, who had done exemplary work during the plague epidemic of 1896. The other bodies were the Pathare Prabhu Social Samaj, Kitte Bhandari Aikyawardhak Mandali, Kshatriya Bhandari Dnyati Samaj, Gaud Saraswat Brahmin Mitra Mandal; professional associations like the Kamgar Hitwardhak Sabha, St. George's Nursing Association; youth organizations like Radiant Club, Students' Brotherhood, Students' Social Union, Young Men's Hindu Association, Young

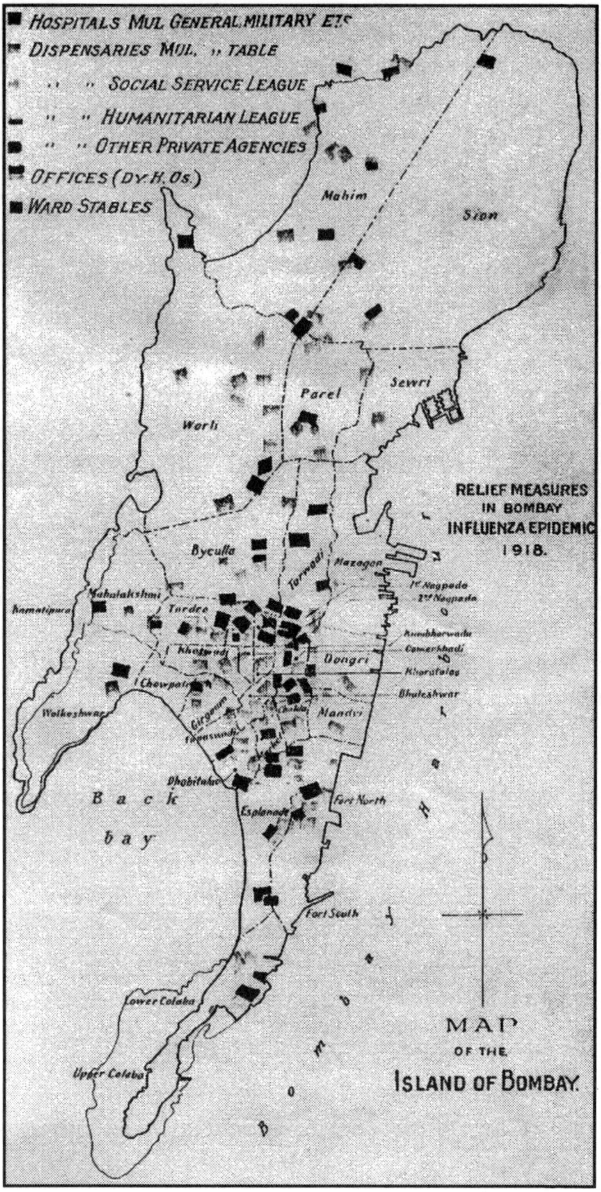

Source: General Department Volumes, Maharashtra State Archives, 232, 1919.

Men's Mohammedan Association, Young Men's Khoja Association, Presidency Students' Federation, and professors and students of Wilson College; political groups like the Home Rule League; and miscellaneous bodies like the Temperance Association, the Bombay Humanitarian League, Matunga Residents' Association, and Swajan Hitachintaka Samaj, Telugu Free Library, and the Church Mission Society.[50] Providing medical relief were twelve doctors, two of whom were women, and senior students of the Grant Medical College. The Western India Turf Club lent its premises as an office, while the Japanese and Shanghai Piece Goods Association operated two relief stations. Blankets were provided by mills, cologne water and tubes of menthol by Messrs Gobhai & Company.[51] While arrangements were made for the temporary boarding and lodging of patients discharged from hospital, a special corps arranged for the cremation of the dead.

The *Hindustan* rightly observed that the principal burden had fallen on the philanthropic sections of the public.[52] Turner, for his part, paid credit to all these organizations and also acknowledged the support of officials such as the Vaccination Superintendent, the Deputy Health Officers, and the sanitary and conservancy staff who had been relieved of their regular duties, and the personnel of the registration branch, birth *karkuns*, disinfecting staff, and death *ramoshis*.[53]

Relief was organized in the other cities of the Presidency on similar lines. In Poona, H.N. Apte, the President of the Municipality, supervised the opening of a temporary hospital and under the leadership of Vithal Ramji Shinde, the Depressed Classes Mission assisted by the Arogya Mandal and the Poona Volunteer Corps provided relief.[54] In Sholapur, a few doctors and citizens supplied free medicines. The Municipal Commissioner of Surat established two travelling dispensaries and an infectious diseases ward at the local civil hospital, while volunteers visited houses, distributed medicines, including a 'medicated mixture' prepared by Dr. Gajjar and supplied free to the Municipality.[55] The Parsi orphanage, Jain hospital, and Mitra Mandal did relief work in this city.[56] In Karachi, aid came through a fund collected at the initiative of Jamsetji N.R. Mehta and twelve mobile dispensaries and bands of volunteers. Led by

Dr. Haji Gulam Husein Kassim, Municipal Councillor, doctors provided free medical treatment. Information regarding rest, nourishment, and preventive measures was disseminated through the press and by the distribution of leaflets, in English and local languages.[57] A citizens' influenza committee assisted by the Salvation Army took up the task in Hyderabad.

In Ahmedabad, it was the Gujarat Sabha, an association concerned with social and political reform since 1884, that set-up the Influenza Relief Committee. The Municipality here seems to have been obstructionist to relief efforts, having refused the use of one of its school buildings as a hospital, which was subsequently opened in a building donated by Seth Jamnabhai Baghubhai, a businessman. In fact, this indifference to the suffering of the people and the 'chronic insanitary condition of the city' were the subject of six editorials in *Praja Bandhu.* The paper pointed out that while the burden of taxation had been increasing, municipal expenditure had been on the high salaries of the Municipal Commissioner, the Engineer and the Health Officer. 'We do not know what would have been the condition of the suffering poor, if the Gujarat Sabha had not come to their rescue.'[58] Subsequently, the *Praja Bandhu* started a fund to aid the efforts of the Sabha. The St. John's Ambulance Division transported patients to the hospital, while the Salvation Army provided the nurses, and medical students of Byramjee Jejeebhoy Medical College and schoolteachers volunteered their services, for house-to-house visits. As in Bombay, funds came from Ahmedabad Advance Mills, mill owners including Ambalal Sarabhai, and from associations such as the *Panch Kuva Mahajan,* a body of cloth traders, *Maskaty Mahajan*, comprising of cloth merchants, and the Yarn Merchants Mahajan. Twenty-four dispensaries were established and around 3,000 patients were given relief. The Gujarat Sabha also provided funds to the Salvation Army to reach villages on the outskirts of Ahmedabad, and in the districts of Kaira and Panch Mahals. The district local board met the expenses of travel, board and lodge of the medical students who worked in these villages, and supplied them with thermometers and stethoscopes.[59] David Hardiman has mentioned the Patidar Yuvak Mandal distributing medicines during the epidemic to the 'Kaliparaj', as the Chodhris, Dhodiyas, and Gamtis, Adivasis of South Gujarat, are referred to.[60] This public involvement in dealing with the influenza, which was a worldwide pandemic, reduced casualties in the cities of the Presidency, even as nationalist voices were asking whether there was no remedy for influenza in Indian medicine.

CONCLUSION

While there is no denying that the official figures of mortality may be unreliable, and many more fatalities may have occurred during epidemics, there was a perceptible change in the official view about the health policy to be promoted in the twentieth century. In his presidential remarks to the Bombay Sanitary Conference in 1911, S. Harcourt Butler specified that the basis of all sanitary knowledge must be knowledge of the people, the conditions under which they live, their ways of life, their social customs, their habits and surroundings, and even their financial means. The *Memorandum on Indian Sanitary Policy* (1914) reiterated this view. Pardy Lukis, Director General of the Indian Medical Service, acknowledged many shortcomings in the policy that had hitherto been pursued, by the grafting on India of Western sanitary measures and systems, without due regard to the fundamental differences of Western and Eastern civilizations. He contended that conditions of life in temperate and tropical climates were different, and that 'uneducated and suspicious people' were unprepared for reform or any kind of change. Their apathy, fatalism, and resentment of interference constituted a formidable obstacle to sanitary progress. For sanitary reform to be successful, was required 'a knowledge of the people' and a recognition of the diversity of local conditions in a country of numerous castes, communities, and creeds, a variety of climates, temperatures and even levels of sub-soil water. After the ravages of plague and the discoveries of the results of medical research, which the resolution contended, had made some real impression on the educated communities in India, the demand for sanitary surroundings and preventive measures had been made.[61] Micro focal analysis such as the above show that in the efforts of individual officials like John Turner in Bombay there was an attempt to implement the policy regarding cooperation with the people. Doubtless, he subscribed to the colonial assumption about epidemics being caused by the 'habits' of the people, but he did try to search for solutions at the ground level, by successfully liasing with Indian doctors, philanthropists, civic leaders, and established local bodies, like the Social Service League.[62]

As for Indian responses, ambivalence is perceptible in resistance to epidemic combating efforts and in the support extended to relief measures. The contemporary press expressed these mixed attitudes. The opinions of the nationalists, during these years when national consciousness was spreading, are important. Initially, during the 1896 plague epidemic, Tilak had accepted segregation as necessary, and alluded to the superstitious folly of persons who regarded hospitals as chambers of death; at the same time, the *Kesari* and *Maratha* regularly reported the havoc caused in Poona by the intrusive measures. Ian Catanach has seen Tilak's opposition to all anti-plague measures from late 1899 as a reaction to the rise in the fortunes of Gokhale.[63] It was not just that; to Tilak, British medical intervention could be effective only if it worked with the support of Indian leaders. The endorsement of the plague regulations and prophylactics by community leaders like the Aga Khan led to their acceptance among the Khojas. Gokhale's stewardship caused the anti-plague campaign to be more effective in Poona. Similarly, hospitals organized on caste and community lines made the concept of hospitalization acceptable. However, the fact that these institutions catered to their own groups would have meant that pollution prescriptions were maintained. There is also a notable change in pilgrim behaviour, showing a willingness to comply with regulations, attributable to greater awareness.

While other public health challenges like tuberculosis and malaria were ever present in the Presidency, the purpose of studying responses to four specific diseases in this chapter, was to show the gradual change in Indian response, from outright opposition to cooperation in coping with epidemics. The public's readiness to contribute generously to relief efforts, especially in the cities, was in keeping with the practice started in the previous century, when several medical institutions were founded and funded by them. The caste and community organizations and professional associations involved in this effort in Bombay represented the cosmopolitan social composition of the *urbs prima in Indis*. It is significant that a number of these bodies had health care and dissemination of knowledge about sanitation and immunization among their aims. This is indicative of growing civic consciousness and of the understanding that health was as much a public responsibility as the state's. Last, it was the promotion of anti-epidemic measures by Indian doctors, whether the prophylactic, or better hygiene or information about the rat as a plague carrier, which made their fellow Indians adopt them. Trained in Western medicine, their conviction in the system was total and both private practitioners and doctors in municipal or government employment were mainly responsible for whatever success was achieved in reducing casualties.

NOTES

* The author wishes to acknowledge the assistance of the Wellcome Trust, London, for a grant that enabled her to consult the India Office and the Cambridge University libraries in 2001.

1. *Transactions of the VIIth International Congress of Hygiene and Demography*, London, 1892, vol. IX, p. 123, in General Department Volumes (hereafter GD), 29,1892, Maharashtra State Archives.
2. *Report of the Sanitary Commissioner, Bombay Presidency* (hereafter *RSC*), Bombay, 1875, p. 223.
3. Mridula Ramanna, *Western Medicine and Public Health in Colonial Bombay, 1845-1895*, Delhi, 2002.
4. Ira Klein, 'Imperialism, Ecology and Diseases: Cholera in India, 1850–1950', *Indian Economic and Social History Review*, vol. 31, no. 4, 1994, pp. 515–16.
5. *RSC*, 1906, p. 6.
6. *RSC*,1908, pp. 23–4.
7. Ibid., pp. 21–2.
8. *RSC*, 1915, p. 7.
9. Shantaram Kantak, 'Vaccination Systems in India', *Indian Medical Congress*, Calcutta, 1895, p. 256.
10. *RSC*, 1907, p. 10.
11. *RSC*, 1908, p. 14.
12. Sanjoy Bhattacharya, 'Re-devising Jennerian Vaccines? European Technologies, Indian Innovation and the Control of Smallpox in

South Asia, 1850–1950', in Mark Harrison and Biswamoy Pati, eds., *Health, Medicine and Empire*, Delhi, 2001, p. 221.
13. David Arnold, *Colonizing the Body: State Medicine and Epidemic Disease in Nineteenth Century India*, Berkeley, 1993, p. 197.
14. *RSC*, 1906, p. 9.
15. *RSC*, 1908, p. 11.
16. Manjiri Kamat, 'The Palkhi as Plague Carrier: The Pandharpur Fair and the Sanitary Fixation of the Colonial State in British India, 1908–1916', in Harrison and Pati, eds., *Health, Medicine and Empire*, pp. 299–316.
17. Extract from *Kesari*, 21 July 1914, in GD, 967, 1915.
18. GD, 826, 1920, Report of the Deputy Sanitary Commissioner on the *ashadhi* fair, 1919.
19. Govt. of Bombay Order no. 2712, 23 April 1917 in GD, 967, 1918.
20. *RN*, *Kalpataru*, 23 January 1916, *Kesari*, 30 January 1916.
21. *RN*, *Bombay Samachar*, 25 January 1916.
22. GD, 33, 1900, Govt. resolution no. 4562, 15 August 1898.
23. GD, 33, 1900, Statement no. 1068, 1900, 6 April 1900, signed by Acting Health Officer, Port of Bombay.
24. *RSC*, 1902, p. 6.
25. *Report of Native Papers, Bombay Presidency* (*RN*), *Sookree*, 13 July 1907.
26. *RN*, *Akhbar-e-Islam*, 13–15 June 1904.
27. GD, 131,1912, Protector of Pilgrims/Police Commissioner, no. 771-PP-11, 10 June 1912.
28. *RN*, *Akhbar-e-Islam*, 27 July 1915.
29. Ibid.
30. *RN*, *Jam-e-Jamshed*, 30 January 1906.
31. *RN*, *Maratha*, 28 January 1906.
32. *RN*, *Hindustan*, 18 July 1897.
33. *RN*, *Voice of India*, 23 April 1904.
34. Ibid.
35. *RN*, *Jam-e-Jamshed*, 20 April 1905, *Rajahavsa*, 1 March 1899, *Kaiser-i-Hind*, 27 August 1916.
36. *Gopal Krishna Devadhar*, ed. H.N. Kunzru, Poona: Servants of India Society, 1939, pp. 134–9.
37. *RN*, *Bombay Samachar*, 26 August 1908.
38. *RN*, *Oriental Review*, 26 August 1908.
39. *RN*, *Kesari*, 30 May 1911; ibid., 7 November 1911.
40. *RN*, *Maratha*, 8 September 1907.
41. S.M. Kaka, *Report on the Effect of Preventive Inoculation with Prof. Haffkine's Plague Prophylactic*, Karachi: Municipal Commissioner's Office, 1899.
42. *RN*, *Broach Samachar*, 28 September 1905; *Kaira Times*, 22 October

1905; *Shri Shahu,* 20 September 1905; *Parikshak*, 27 July 1905, *Karnatak Vaibhav,* 14 March 1908.

43. *RN, Indian Spectator*, 4 November 1905,*Vrittasar*, 3 April 1911.
44. *RSC*, 1915, p. 15.
45. J.A. Turner, 'Plague in Bombay', *Indian Medical Gazette*, July 1906, p. 302.
46. GD 232, 1919, Health Officer's Letter, 25 November 1918, p. 18
47. GD 232, 1919, *Report of the Relief Work of the Hindu Medical Association.*
48. 'Report of the Influenza Epidemic', *The Social Service Quarterly*, IV, 1918, pp. 148–54.
49. GD 232, 1919, pp. 34–8.
50. GD 232, 1919, Health Officer's Letter, 25 November 1918, p. 20.
51. Ibid., p. 21.
52. *RN, Hindustan,* 9 October 1918.
53. GD 232, 1919, Health Officer's Letter, 25 November 1918, p. 23.
54. GD 232, 1919, President, Poona Municipality/Collector, 29 September 1918; *Times of India*, 15 October 1918.
55. GD 232, 1919, Municipal Commissioner, Surat/Collector, 11 December 1918.
56. GD 232, 1919, A.E.L. Emmanuel Collector /Commissioner, Northern Division, 14 January 1919.
57. Ibid. Report by Dr. E.D. Shroff, HO, Karachi, 2 September 1918.
58. *Praja Bandhu,* 20 October 1918.
59. 'Gujarat Sabha Influenza Relief Committee', *Social Service Quarterly*, 1918, p. 156.
60. David Hardiman, *Gandhi in his Time and Ours*, New Delhi, 2003, p. 138.
61. GD 1074, 1914, Government of India Resolution on Sanitation, 23 May 1914.
62. See Mridula Ramanna, 'Local Initiatives in Health Care Bombay Presidency 1900-1920', *Economic and Political Weekly*, vol. XXXIX, no. 41, October 2004, pp. 4560–7.
63. Ian Catanach, 'Poona Politicians and the Plague', *South Asia*, vol. 7, 1984, p. 8.

CHAPTER 8

Seeing Like a River: The Bengal Presidency's Hydraulic Transition

ROHAN D'SOUZA

In the twilight of Empire Bengal's rivers began to be increasingly recognized as an indisputable 'water problem'. For the official colonial imagination, the delta's fluvial arms were temperamental and snaked their way across the capacious flood plains only to wastefully empty 'millions of tons' of their valuable hydraulic burden into the Bay of Bengal. Usually a swollen rage during the monsoon, and an irrelevant trickle by the winter, such hydrographical quirks, it was authoritatively held, regularly depressed the Bengal peasant. Around December of 1945 in a radio broadcast, R.G. Casey, then Governor of Bengal, forcefully argued for a definitive response to the region's perplexing hydrology:

> . . . the water problem of Bengal necessitate(s) our so handling (of) the great rivers that their flow is equalised and controlled as between summer and winter in order that they may provide an adequate and balanced output. . . . This would avoid disastrous flooding in the monsoon and would cure the dry or stagnant state of many of our rivers in the winter.[1]

The Governor's confident declarations, undoubtedly held both technological seduction and promise. By 'equalizing' the rivers and moderating their discharge, the fluvial tapestry of the delta could be rationalized, and its troubling variability finally rendered consistent, becalmed, and tractable. The Governor's broadcast, however, was not restricted to simply declaring a quest for engineering elegance. His proposal about 'balancing' the Bengal rivers, in the remaining portion of the radio address, was dramatized in terms of a productivist cornucopia: controlled hydrology, he claimed, through a slew of large dams and barrages across the Teesta, Ganga, Damodar, and mighty Brahmaputra, could be harnessed for the simultaneous realization of irrigation, electricity, flood control, and navigation. In effect, 'rationalized' rivers could lubricate the needed economic thrust to propel Bengal's millions into an altogether higher orbit of development. Casey's hydraulic utopia was not in any way exceptional for its time. The 1940s saw the emergence of a new era in the promotion of a multi-purpose river valley development, the national quest to replicate the triumph of the Tennessee Valley Authority (TVA), and a pronounced science and technology hubris in the economic realm.[2] These robust tropes, in fact, were rapidly combining to underpin and energize the soon to be formulated Truman doctrine of 1949—the grand master text, according to Arturo Escobar, that laid out and announced the rearrangement of the post-war world along a new hierarchy based on 'development'.[3]

Perhaps the properties of the delta—a colloidal compression of land and river—made it a particularly receptive site for mammoth fantasies about organizing grand societal schemes and control around water. But 'training' rivers through large dams and barrages was not the only type of fluvial vision that had been anticipated for Bengal. In the early summer of 1930, Sir William Willcocks (1852–1932), one of the Empire's most celebrated hydraulic engineers, delivered a series of lectures on the rivers of Bengal at Calcutta University. In the lectures, he argued for abandoning the familiar colonial route for hydraulic manipulation in the region, and advocated many sharp and unexpected divagations; not unlike, perhaps, the unpredictable deltaic rivers themselves. He had voiced some of these claims earlier,[4] but in Calcutta University for the first time he cogently brought together an entirely fresh template for interpreting the region's hydraulic potential. This by implication rejected almost a century of colonial engineering interventions to harness the rivers. The profound change was especially unusual given the fact that Willcocks had launched his career through the well furrowed hydraulic orientations of the Indian Irrigation Service. During the course of a long and eventful career, he had earlier ably espoused the ideals and virtues of modern canal irrigation in different theatres such as Egypt and Mesopotamia (modern-day Iraq).[5] But in the winter of his life, as if in an act of deep contrition, Willcocks at Calcutta University chose not only to argue against the need for another grand canal scheme, but even went so far as to suggest that the colonial government had effected, since its rule in Bengal, the physical erasure of the delta's once vibrant practice of inundation irrigation. According to him, a large network of 'overflow canals' traversed the deltas of the Ganges and Damodar basins and irrigated almost 70,00,000 acres of land, all this much prior to the great civilizing and modernizing impetus of British rule. The broad and shallow inundation canals were designed when the deltaic rivers were in flood, to tap the silt-laden crest waters that were rich in fine clay. These canals were, furthermore, long and continuous and ran almost parallel to each other. The most striking feature of flood irrigation, however, was its importance as a fertilizing agent and not merely a source for water. Willcocks argued that the 'rich red water of the river and the poor white water of the rainfall' had in fact been made to combine in order for agrarian production to be sustainable and successful.

> . . . if your rice fields have been irrigated by rain water alone, they are weak and cry for irrigation in October with excessive and costly supplies of poor river water. . . . If however you have irrigated your rice fields with rain and river water mixed together in the early months of the monsoon when the river water is rich and full of mud, you so strengthen the plants of rice that they resist the hard conditions of an early failure of the monsoon in a way rice irrigated by rain water along has no knowledge of the River water in the early months of the floods is gold.[6]

The muddy waters, moreover, combated malaria as they carried fish eggs which entered the canals, floated into the subsidiary channels, tanks, and rice fields and then soon became young fish who instantly fell upon the larvae of the mosquitoes and 'lived on them'.[7] The other significance of overflow irrigation, he claimed, was that it did not obstruct the build-up of the delta; by diffusing silt deposition over the flood plain, it allowed the annual inundations to actively function as geomorphic agents which through time, by raising the land, blunted the ferocity of flood currents.[8]

As far as the technique of inundation irrigation was concerned, Willcocks surmised that since 'overflow irrigation' hinged on the ability to tap the muddy crest of water of the river when the latter was in spate, the effort was concentrated on leaking or diverting its flow at certain points along its channel. Long shallow inundation canals were therefore led from the river's main channel and the actual irrigation was performed by the cuts in the banks of these canals, which were then closed when the floods receded. To keep the canals functioning, however, required a considerable amount of routine effort on the part of the cultivators, as the structures needed to be de-silted (known as *pulbandi*) and their banks regularly repaired. According to Willcocks, by 1815 in central Bengal, a combination of Maratha–Afghan wars (1794) and years of neglect by the cultivators finally doomed overflow irrigation. For Willcocks, this decay meant that the once vibrant inundation canals could not leach the rivers of their silt laden waters and these bottled-up flows in the main stem caused the entire deltaic system to become even more hydraulically volatile. Nonetheless, Willcocks held, cultivators in the delta continued to effect breaches along the banks of the river. Such breaches or the practice of *kanwa*[9] (as it was popularly referred to) were supposedly carried out 'when the fear of inundation had passed or a breach had occurred elsewhere' and had thereby reduced the current along that stretch of the river.[10] Though carried out in an ad hoc manner, these breaches were an improvization evolved from a still older era of overflow irrigation and enabled the cultivators to 'more or less perfectly' irrigate their lands. Willcocks called this practice of *kanwa* (i.e. the breaching of natural levees) the zamindari-bank irrigation.[11]

Clearly, Willcocks and Governor Casey based their readings of the delta on contrasting, if not entirely opposed, hydraulic principles. Willcocks believed that overflow irrigation enabled cultivators to harness the rivers by adapting to the latter's natural run and tumble. For him, inundation irrigation, by tapping the varied tempers of the flood line, periodically invigorated the cultivator's fields with nourishing silt-laden turbid flows. Thus the silt followed the river onto rice fields and this fertilized land, in turn, by regularly soaking these leaden currents aggraded the delta and blunted the ferocity of the floods; regular deposition, in this way, accumulatively built-up the region over a period of time and graduated it towards a calibrated equation between the flood plain and its fluvial endowments.[12] Thus Willcocks wanted the restoration of inundation irrigation so that as a technique and a set of practices it would weave the rivers and surrounding alluvial plains into a mutually beneficial embrace. In effect, the Willcocks thesis was premised on a notion of hydraulic manipulation that was based on flood dependence and a cultivation strategy that was oriented towards consolidating the delta through geomorphologic process.

In contrast, Governor Casey viewed the delta's rivers as essentially poised against the surrounding alluvial plains. The rivers, according to him, appeared in two opposed moods; they either burst their banks and injured cultivation or so weakened their flows that crops withered from lack of water. In short, understood only from some distant higher ground, the river appeared unpredictable, volatile, and unstable, a fluid phenomenon that was temperamental and constantly reworking its boundaries with powerful and yet yielding qualities. A conception of river hydrology, in effect, premised as the unstable fluid opposite to settled agriculture. For Casey, therefore, the delta's rivers had to be conceived only through the language of control. Rivers needed to be 'equalized', 'balanced' or 'tamed' and their flow harnessed through the science of engineering. The Great Bengal Delta was conceptualized, in this official imagination, not as a composite of land and water but elementally as separate, discrete, and often times antagonistic, spheres of land and water.

These contending notions about the properties and dynamics that made up the delta in their turn drew on several legacies of hydraulic interventions in the colonial period. That is, colonial rule in the attempt to make the rivers tractable, legible and calculable to its various imperatives for rule, administration and drives for surplus appropriation, sought to re-plumb the delta as a hydraulic entity. This paper will argue that the various colonial attempts to recast the delta through river control, by the early decades of the twentieth century, effected a hydraulic transition. I broadly describe this as a *flood vulnerable landscape imposed over a kaleidoscope of hydraulic events.* Put in another way, a clutter of structural indentations such as drain sluices, embankments, and canals transformed the delta from an organic geomorphologic process to instead becoming a collage of disconnected fluvial events that were regulated principally as practices of political economy. The Bengal delta, in effect, was to be galvanized into a new type of colonial productive order through an assemblage of technological artefacts such as drainage, flood control, and irrigation. But first, above all else, let me situate this account in its material or rather, in this case, its gelatinous ecological, context—the unique entanglement of land and river.

BETWEEN RIVER AND LAND

The Bengal delta physically manifests as a vast alluvial tract crisscrossed by the Ganga, Meghna, and Brahmaputra River systems. At some point in the southern portion of what was then the colonial district of Dacca, these three enormous rivers collide, intermingle and weave their way towards the Bay of Bengal in a sort of irregular funnel. While these masses of flow hurtle tortuously in the direction of the sea, innumerable distributaries and channels erupt from the main stems of the giant rivers and cut fresh paths onto the surrounding soft alluvial plains. These several fluvial arms, especially when charged with monsoonal torrents, range wildly over a mesh of marshes, swamps, wetlands, ox-bow lakes, and innumerable tiny rivulets and streams, energizing some and enervating others. To a geologist, the delta is a geomorphologic process: a vortex of dynamic equilibriums; an inestimable number of rapidly evolving relationships between land, river, and fluvial energy.

The rivers' peculiar volubility in the delta has been explained to have essentially brought about by the fact that after they debouche onto the 'la' alluvial plains, they witness a steep subtraction in their channel gradients. Thus a loss in channel depth evens out the rivers' currents and thereby weakens their carrying capacity. The detritus comprising soil, silt, rocks and other organic matter is therefore rapidly dropped onto the bed of the river, which when rapidly clogged in this fashion is then unable to transport its water in the same channel. Consequently, the river repeatedly bursts its banks, overflows, or shoots off effluents to unburden itself of its waters. The rivers' instability is actually a quest for the necessary gradients that will permit them to discharge their flow.[13] In the recurring process of erosion and deposition, the rivers, however, end up building the delta; that is, it is slowly raised above that of the sea. Put differently, the delta is a collection of dynamic drainage, flood and wetland ecologies.

To unravel the colonial impact on this complex ecological context, I will, in order to prevent the paper from becoming unwieldy, discuss the wide ranging hydraulic transformations with indicative examples rather than attempt an exhaustive treatment of the subject. This paper will, therefore, be directed at uncovering some aspects of the dynamics of the colonial interventions in drainage, irrigation, and flood control during the nineteenth century. Exploring the political debates and bureaucratic differences that erupted following the adoption of many of these schemes, I argue, will throw light on the multi-sided hydraulic transition of the Bengal delta.

PLUGGING LEAKS AND SHORING UP THE LAND: THE DRAINAGE CONUNDRUM

As pointed out earlier, the delta is riven by innumerable drainage lines that leak waters into surrounding basins and rivers, and ultimately into the Bay of Bengal. A kind of encompassing fluvial tapestry of crisscrossing rivulets, lean streams and circuitous flows are vital to calibrating the overall hydraulic momentum in the delta. Traditionally, it appears, 'natural drainage' in the deltaic tracts was a valued element in the calculus of both habitation and agrarian production. Some of this orientation is indicated in an account of the 'general system' of village drainage in Bengal in *Bengal Drainge Committee Report of 1907* (quoted in turn from the *Epidemic Commission Report of 1864*):

> The drainage of all villages . . . in Lower Bengal is effected by the water first turning in the nearest paddy-fields lying in the direction of their slope, then it collects in the bheels from which it rushes through

khals into larger streams, which again communicate with navigable rivers.[14]

This description suggests that, historically perhaps, villages and their contexts in the delta were strategically situated with regard to drainage. Water was kept circulating between different types of depressions and basins and was finally led unhindered to a flowing stream or river. Such an equilibrium, however, was fragile and tenuous: a single disturbance or obstruction could upset an entire network of carefully maintained circulation. It is perhaps akin to the 'butterfly effect' in chaos theory. Despite the delicate nature of the drainage pattern, colonial rule had during the course of the nineteenth century inaugurated a number of projects for road, railway, and embankment construction in the region. These modes of transport, with their emphasis on permanent all-weather structures and mostly built in unrelentingly straight lines, marked a sharp break from movement in the earlier era, which was predominantly based on circuitous rough paths and 'crooked' routes.[15] The colonial transport networks in Bengal, in fact, radiated along an east-west axis, while the region's natural drainage lines dropped from north to south.

By the second half of the nineteenth century, official concern built-up over what was rapidly discussed as the problems caused by the obstruction to drainage. Several enquiry committees were commissioned to deliberate on issues of waterlogging, the impacts of stagnant water bodies on malaria, and the question of the role of railways and roads in interrupting natural drainage. One of the first of these Commissions which debated the impacts of drainage congestion on the spread of malaria was the Epidemic Commission of 1864. In the course of its enquiry, the Commission had to contend with a somewhat comprehensive response on the issue of drainage from Raja Digambar Mitter (more about him later). In the years 1867 and 1868, another set of elaborate enquiries were held on the subject and the conclusions were summed up by Colonel Nicolls, then Chief Engineer of Bengal. According to Nicolls, in a note of 4 March 1869,

> . . . roads and railways in Bengal have not obstructed the drainage of the country so far as to cause or to aggravate sickness. But some obstruction is inevitable and should be remedied as soon as possible.[16]

Colonel Nicoll's mixed assessment, which firmly dismissed claims about the impact of roads and railroads in causing malaria through waterlogging (while recognizing that the latter did occur) marked the general tenor of the findings and conclusions by a succession of similar investigations. It appears that, for the authorities, there was an inability to conclude that a straight correspondence existed between malaria and the construction of roads and railways. On the other hand, there could be no doubt that all-weather transport structures would logically interrupt and cause several scales of drainage impact. In fact, it was precisely over this contradictory colonial assessment that two sharply opposed ideas on the issue of drainage began to congeal. At one end, formulated chiefly by Raja Digambar Mitter, it was argued that Bengal's drainage problems were entirely man-made, in that the 'natural drainage system' had been systematically interrupted by obstructions. The solution, according to this view, was the restoration of the 'natural' pattern by reconnecting the circuits for water circulation, i.e. drainage should flow from 'villages to the arable lands, from paddy fields to beels and from beels through khals and water courses to the navigable rivers . . .'.[17] In contrast, the other proponents argued that drainage congestion was characteristic and intrinsic to the Gangetic delta, and that only man-made interventions through artificial drainage schemes could cure the delta of stagnant malarial swamps. In effect, in this latter assessment, the deltaic tract was not perceived as a single organic bloc made up of intricate patterns of drainage but instead appeared as a disconnected aggregation of marshes and waterlogged sites.

This divide between those arguing for the restoration of natural circuits against those insisting that the delta had to be cured of its naturalness was most dramatically played out during intense discussions over the Bengal Sanitary Drainage Act that had been introduced at the meeting of the Governor's Council on 9 February 1894. In a subsequent debate by the Council in the same month, the members broke out into various kinds of dissenting voices over the Bill. One of the members, L. Ghosh, averred that the solution to the drainage congestion in the delta lay in carrying out schemes that facilitated 'surface drainage' in the villages. Ghosh, however, contrasted his advocacy of village level surface drainage schemes as being in sharp opposition to those who were arguing for 'subsoil drainage', which would mean the draining of maidans, paddy-fields, and beels.[18] In fact, the differences between the proponents of surface and sub-soil drainage respectively mirrored larger disagreements over the question whether the delta's unsanitary conditions could be relieved through the restoration of the natural drainage pattern (surface schemes) or by removing its naturalness (subsoil schemes). Along the same curve of differences was the other fear that the Bill had evolved a string of procedures for addressing the issue of drainage only through the narrow optic of propertied interests. According to the format specified in the Bill, the government would act on a drainage 'complaint' only if it was moved, at the local level, through a Drainage Commissioner who had been qualified by his 'interests in the land'. The Drainage Commissioner was then expected to conduct a survey of the afflicted locale and submit a report based not only on the physical nature of the undertaking but, more specifically, with an eye towards the 'particulars as to the estate and tenure holders and cultivating raiyats of the local area with the annual value of the estates, tenures and holdings . . .'.[19]

This procedure for addressing how drainage was to be assessed and solved, was clearly weighed in with an obvious bias towards the landed element of the delta. That is, drainage was to be perceived only as an adjunct to the land question, with the reasoning for relieving congestion based entirely on its 'value' in terms of meeting the revenue demand. Thus draining was treated as a localized affair, which too, though not obviously stated, preferably in only remunerative zones. In effect, the Drainage Bill, in terms of its operative part, was slanted clearly in favour of those who had suggested that the delta had to be delivered from its naturalness and the government should, in the main, push for subsoil drainage, while only carrying out financially remunerative surface schemes. The Bengal Sanitary Drainage Act finally passed in the late 1895, however, deleted section 3 which had urged for the general drainage of marshy lands and even rice lands. The Bill instead concentrated on stating that drainage measures were to be taken only when obstructions caused deterioration of sanitary conditions in the locale. The

Bill was a little too late and not enough in terms of reversing the congestion that almost a century of colonial rule had brought about in the delta. More specifically, to highlight just one example, in the extreme fringes of the delta, in the districts of the 24-Parganas and areas abutting the Sunderbans, a rapid period of land reclamation had been ongoing from the earliest decades of the nineteenth century.

In the normal course of tidal action, lands were continuously raised with deposition of silt by slack ebb currents from the Bay of Bengal. After a point, these raised lands became high enough to encourage the growth of vegetation, which after some years could be cleared for settled agriculture after an embankment was thrown around it. The reclamation of land from the sea, however, if prematurely carried out—without the lands being sufficiently raised—could result in drainage complications such as waterlogging or the silting of river beds tapped behind the embankments. It is more so from the fact that the deltaic rivers when approaching the sea are extremely sluggish, prone to rapidly dissipate into marsh, creek, or just stagnant lake. To explain the consequences of such premature reclamation enterprise, I quote C. Addams Williams, then Executive Engineer in the Public Works Department, who details a telling example:

> On the south-east of Calcutta are situated what are called the salt lakes. On the flow side, the water finds its way from the Bidyadhari River through side branch khals into the lakes and deposits its silt; on the ebb tide the water flows back into the river and the flow and ebb keep the river open. At the northern portion of the lake called the Panchannagram basin was some years ago cut off from tidal spill into the lakes by an embankment and brought under cultivation. In the meantime, the tides have continued to spill into the lakes by an embankment and brought under cultivation. In the meantime the tides have continued to spill into the lakes to the south of the embankment: the land has continued to rise in consequence. . . . We have therefore a low depression . . . from which it is almost impossible to drain the rain water; it is quite possible that these low lands will become entirely waterlogged before long and cultivation will entirely cease. . . .[20]

The pursuit of premature reclamation was, in fact, the dominant response amongst the colonial authorities to drainage in the region. Keenness to extend the cultivable frontier was backed in equal earnestness towards realizing revenue from settled cultivation. The conundrum over drainage in the Bengal delta and its often times baffling consequences, nevertheless haunted British engineering expertise. By the latter decades of the nineteenth century, as I have all too briefly alluded to above, colonial officialdom found itself sharply divided over the issue of drainage. The split essentially occurred over the question of whether to restore the 'natural regime' of the delta or, in contrast, command the flow of the waters through a spectrum of drainage schemes. For various reasons of political economy, the colonial dispensation overwhelmingly chose the latter path by taking up often costly works such as the Mograhat Drainage Scheme, Dankuni Drainage Works, and Rajpur Drainage Scheme and many others.[21] But shadowing these elaborate undertakings was a fairly consistent critique by both local opposition and officials within the administration and engineering branches as well, who argued that the delta's hydraulic regime would only be further complicated and aggravated by such initiatives.

The congealing of these two opposed views on the question of drainage, in fact, mirrored similarly contrasting opinions on the issue of irrigation and flood control in the Bengal delta. In short, the divide over interpreting and understanding the colonial response to the delta's hydrology, throughout the nineteenth century, was indicative of general discomfiture with the gradual and sometimes abrupt hydraulic transition that was being then effected in the Bengal Presidency. Natural deltaic river regimes increasingly were made to give way to an altogether new hydraulic grid that was being assembled and shaped in the image of the colonial imperative. Thus the hydraulic differences between Willcocks and Governor Casey, sprang from a debate over a century of flood control, drainage, and irrigation legacies in the delta. In my opinion, they stated the manifestos of the two antagonistic and incommensurable sides of the great Bengal hydraulic transition.

NOTES

1. R.G. Casey, 'Poverty or Plenty', Broadcast Speech by the Right Honourable R.G. Casey, Governor of Bengal on All India Radio, Calcutta, on 8 December 1945, in *Personal Diary R.G. Casey,* Mss

Eur F 48/4, May 1945–February 1946. Oriental and India Office Collection (hereafter OIOC), The British Library, London.

2. Rohan D'Souza, 'Damming the Mahanadi River: The Emergence of Multi-Purpose River Valley Development in India (1943-46)', *The Indian Economic and Social History Review,* vol. 40, no. 1, 2003.
3. Arturo Escobar, *Encountering Development: The Making and Unmaking of the Third World*, Princeton, 1995.
4. Sir William Willcocks, 'The Restoration of the Ancient Irrigation of Bengal', Lecture delivered in the British India Association hall, Calcutta on 6 March 1928, The Calcutta General Printing Company Limited, Bowbazar Street, Calcutta.
5. For a brief account of Willcocks' experience in Egypt, especially his bitter fight with Sir Murdoch Macdonald on the question of the flow data records of the Nile, see Herbert Addison, *Sun and Shadow at Aswan,* London, 1959, pp. 69-78. For an autobiographical sketch, see William Willcocks, *Sixty Years in the East*, London, 1935.
6. Sir William Willcocks, *Ancient System of Irrigation in Bengal*, rpt., Delhi, 1984, p. 32.
7. Ibid., p. 60.
8. 'The ancients by increasing the supplies of muddy water, steadily improved the lands as time went on, and also decreased the danger of inundation.' Ibid., pp. 35–6.
9. According to Willcocks, *Kanwa* is derived from the Persian or Arabic word meaning to dig. Ibid., pp. 7–8.
10. Some of the claims in Willcocks' thesis, however, were challenged by C. Addams Williams, then Chief Engineer in the Bengal Irrigation Department. The most substantial of these was one directed against the assertion that several of the rivers and distributaries in the Ganges delta (the Bhagirathi, the Gorai and some of the off-takes of the Bairab, the Jelingi and the Mathabhanga) were originally canals lined out and dug parallel to each other. Instead, to Addams Williams, these rivers were natural stream courses and showed typical deltaic meanders. Their features, he contended, were designed by natural factors rather than human ingenuity. See *Note by C. Addams Williams, on the Lectures of Sir William Willcocks, on Irrigation in Bengal Together with a Reply by Sir William Willcocks*, Calcutta, 1931, pp. 4–6. See also Willcocks' reply, especially in pp. 37–9.
11. Sir William Willcocks, *Ancient System of Irrigation*, pp. 21–2.
12. In a recent study, Arabinda Biswas, upholding Addams Williams, pointed out that the 'remarkable straightness' of the Gangur River in the area between the Damodar and Banka rivers, indicated an element of truth in Willcocks' conjectures. Arabinda Biswas, 'The Decay of Irrigation and Cropping in West Bengal, 1850-1925', in Boudhayan Chattopadhyay and Pierre Spitz, eds., *Food Systems in Eastern India,* Geneva, UNRISD Report No. 87.3, 1987, quoted in Anil Agarwal

and Sunita Narain, eds., *Dying Wisdom: Rise, Fall and Potential of India's Traditional Water Harvesting Systems*, Delhi, 1997, p. 101.

13. On deltas see James M. Coleman, *Deltas: Processes of Deposition and Models for Exploration,* Boston, 2nd edn., 1982; Robert Kay, ed., *Deltas of the World*, New York, 1993 and Michael N. Oti and George Postma, eds., *Geology of Deltas*, Rotterdam, 1995.
14. *Report of the Drainage Committee, Bengal (Presidency Division)*, Calcutta, 1907, pp. 20–1.
15. See Jean Deloche, *Transport and Communication in India: Prior to Steam Locomotion*, Delhi, 1993, 1994.
16. Report of the Drainage Committee of 1907, p. 33.
17. W.A. Inglis, *A Review of the Legislation in Bengal*, Calcutta, 1911, p. 69.
18. Ibid., p. 68.
19. Ibid., p. 67.
20. C. Addams Williams, *Drainage Problems in the Ganges Delta: A Series of Six Lectures Delivered at the Shibpur Engineering College*, Calcutta, 1913, p. 6.
21. Excellent debates on the issue of drainage are available in the following sources. W.A. Inglis, *The Canals and Flood Banks of Bengal*, Calcutta, 1909, pp. 465-561; *Papers from 16th March to 14th April 1914 Relating to the Mogra Hat Drainage System in the 24 Paragana District*, Selections from the Bengal Government, Calcutta, 1915; G.C. Machnonchy, Superintending Engineer, Public Works Department, *Problems Regarding Flood Drainage*, Calcutta, 1905; C. Addams Williams, *History of the Rivers of the Gangetic Delta, 1750-1918*, Calcutta, 1919.

CHAPTER 9

Technology and Defence Production Personnel under Colonial Dependence: Cossipore Factory in the 1890s

ARUN BANDOPADHYAY

The single most important feature of modern warfare is perhaps its domination by modern weapons. The industrial underpinnings of the nineteenth-century revolution created a massive change in the organization and technology of war, which were quickly mastered by the European countries on a highly competitive basis. Tools or weapons were made in close connection with modern science and technology, and if only the right ones could be discovered, they could ensure victory to a great extent. 'Extended warfare between the great powers intensified efforts to apply scientific knowledge and inventiveness to the business of destruction.'[1] The high superiority of weapons rapidly became the desired objective of all, and other aspects of military preparation turned into things of secondary significance.

Revolutionary changes inevitably set in the technology of defence production, particularly in small arms and artillery, from about the middle of the nineteenth century.[2] Between 1815 and 1870, the weapons of both infantry and artillery in Europe were transformed. The muskets of the Napoleonic era were smoothbores, accurate at 50 yards, with a maximum range of 200 yards. The guns were also smoothbore muzzleloaders, as all guns had been since the fifteenth century. 'But by 1870, the principal armies of Europe were equipped with rifles accurate up to 500 yards and with rifled guns with ranges between 2000 and 3000 yards.'[3] Between 1875 and 1914, the consumption of firearms and artillery increased enormously. The British field artillery was inferior to that of the European Powers in range and rapidity of fire.[4] In 1889, the field artillery of the French and German armies could have barrels that alone recoiled against a progressive buffer action instead of the whole carriages. 'This made possible a faster rate of fire and used shrapnel shells which burst in the air above the enemy.'[5]

Under the conditions of general competitiveness among the European countries, the artillery and the engineering branches of the armed forces were the first to acquire a professional outlook, linking military technology to the rise of professionalism. This outlook spread from one country to another and by the end of the nineteenth century, a loose term, Military–Industrial Complexes, came into being. These comprised the 'totality of the institutions involved with developing, manufacturing and manning the hardware of war'.[6] The very requirement of war in such a changing scenario presupposed a state of the art technology in the production of arms and ammunitions, failing which there was always a risk of the product being obsolete. 'In war, especially in modern wars, wars in which weapons change(d) rapidly, one thing (wa)s certain; no army of 50 years before any date selected would stand "a day's chance" against the army existing at that date.'[7]

Herein lay a basic dichotomy between the technology of defence production and the general dependence of that production as promoted by colonialism. Technology did not permit backwardness and laxity in that world of rapid change, but colonialism often failed to rise to the task at the right time and in sufficient measure. There was an inherent limitation within colonialism for the development of truly independent, modern, and dynamic technology of ordnance production. The very nature of the contradiction usually led it to sacrifice the interests of the colonized country for the sake of the colonizer. However, the paradox of the situation was that the technology of defence production and the demands of new personnel as understood in modern times had also an inherent tendency to get out of the limitations generated by colonialism for its own sake. The problem of technology transfer in defence production under these strange conditions of colonial dependence is thus doubly complicated. This dichotomy and deviation, pursued so dramatically in British India, in the use of technology and the recruitment of qualified scientific personnel in Cossipore Factory (known at that time as the Foundry and Steel Factory) in the 1890s, form the main theme of my essay.

COSSIPORE UNDER THREE DISTINCT PHASES OF TECHNOLOGY

The early years of Cossipore's technological experience were promising, particularly after the establishment there of the Gun Foundry in 1830. The most notable characteristic of Captain Hutchinson's experiments of the 1830s was the display of engineering feats that were high by contemporary world standards. There is also reason to believe that production technology did not lag a behind. We have seen a certain degree of optimism during this period, so much so that with introduction of the cupola furnace by the middle of the 1830s, Hutchinson could write that Cossipore would require no more cast iron articles in future for this purpose. This phase of promise and hope continued to the onset of the Mutiny, when Captain Broome made some notable additions and experiments and Cossipore, 'beside producing brass ordnance, turned out a very large stock of iron work of all descriptions including machinery and carriages of a Light Field Battery'.[8] It is probable that Cossipore was given a relatively free hand in its production, which was reflected in the use of its technology.

For about thirty years after 1858, however, Cossipore's technological experience was discouraging. There was obstruction rooted in the change of imperial policy of ordnance in India. Cossipore was not given the sophisticated machinery available in this period of technological change. The mode of construction of even the brass ordnance became the copy work of the mode practised in Woolwich. It was even stated that the Ordnance Select Committee at Woolwich would decide what Cossipore should produce. The respective Governors-General of India, from Canning to Northbrook, did not try to check the process, despite appeals from influential quarters. Northbrook even paved the process by his decision of virtual stoppage of gun making in Cossipore in July 1873. However, the ongoing manufacture of 9-pounder muzzle-loading rifled projectiles could not be immediately halted, and Northbrook had to write to the Secretary of State for India in August 1873 for the import of 'such quantities of (pig) iron, in addition to that to be locally purchased and under supply, as may be deemed sufficient after consideration of Colonel Maxwell's statement' for the work.[9] Known as the Foundry and Steel Factory (FSF) at that time, Cossipore of course tried its best to get out of the compulsions imposed from above. For example, as early as 1875 Major T. Nichol, the Superintendent (1874–82) of the Factory, complained to the Inspector-General of Ordnance and Magazine that the steam drop in the shell factory was only capable of forming bevels of the shrapnel shells up to 80 pounders but was entirely occupied to the production of 9-pounder projectiles. For the extra production of 9 × 7 pounder shells, a second steam drop similar to the one in use in the Royal Laboratory at Woolwich was necessary. Nichol believed that Cossipore on its receipt would be able to turn out shrapnel shells for the 7-inch, 8-inch, 9-inch, and 10-inch guns and also for Palliser shells.[10] Such complaints were not seriously considered and there was no immediate reflection in Cossipore's production. The FSF was only able to project considerable re-direction of its operation and technology in the early 1880s.

The requirement of the new items of defence along with a changing technology in the last quarter of the nineteenth century made considerable inroads on the policy of colonial dependence of ordnance production in India. The cost factor of production was another cause for the modification of the imperial policy towards Cossipore. Cossipore was henceforth given more active assistance in terms of equipment and personnel. Consequently, a process of relative 'freehand' in technological matters was initiated, first in the 1890s, then in the 1900s and finally after the formation of the Munitions Board during the First World War. The first of these changes is discussed in this paper. However, the change of policy was hinted as early as November 1887 when India Office wrote on behalf of the Secretary of State to the Governor-General of India, requesting the latter to supply information about the plants required by the Cossipore Factory for the manufacture of French time fuses by the latest technology. Of course, more important changes were in the offing, and in the next ten years Cossipore's technological experiments in the production of steel and the deployment of highly qualified technical personnel would open up the possibility of a major reshuffle of preference and priority in ordnance production in the 1890s. We shall discuss the two aspects of change in some detail in the following two sections. A short, interrogative note instead of a conclusion will follow on the 'limited' prospects and problems of technology transfer under colonial dependence.

NEW TECHNOLOGY AND THE MANUFACTURE OF STEEL IN COSSIPORE

In the history of modern iron and steel production in India, the success story of Tata Iron and Steel Company (TISCO) in the early years of the twentieth century is well known and widely appreciated.[11] This is quite legitimate if we consider the range and scale of operations and the developments that were followed in this unit of production at what is now known as Jamshedpur. At TISCO, actual production started in 1911–12.[12] However, steel had been successfully produced in the FSF, Cossipore by modern technological methods more than fifteen years before this historic beginning. This story is least known.

Several serious attempts had been made to produce modern iron in India in the early nineteenth century, but they failed. Individuals who undertook these ventures were mostly lured by official support, the actual scale of which is unknown. The earliest attempt was made by a former Company servant named J.M. Heath, who was given by the East India Company an exclusive right to produce iron through methods 'similar to those in Europe'. He 'undertook production at Porto Novo, a small seaport south of Madras at Palampetti near Salem, and at Beypore on the Malabar Coast. Charcoal was used as fuel and bullocks were employed to power what machinery was used'.[13] This experiment failed. What was more important, George Jessop of Jessop and Company at Calcutta protested in 1825 against this grant of exclusive right to Heath. He identified himself as a partner in one of the largest iron foundries in Great Britain and submitted,

> The works belonging to myself and partners in Calcutta were purchased by my brother, Mr. H. Jessop, for the purpose of forming a connection with the Butterly Iron Works at home. It is therefore not to be expected that under such circumstances I should wish to manufacture in this country an article the sale of which I am interested in promoting in my own country and which formed so great a revenue and gives employment to thousands.[14]

For various reasons, Jessop did not initiate the project, and half a century lapsed before a more serious effort was made in Bengal to produce steel under modern lines. With the development of railways and other public works and the rise of modern industry in India after 1850, India became a large importer of iron and steel, and plans were made to consider the possibilities of import substitution. 'The Bengal Iron Works Company, the first plant designed along modern lines, was formed in 1874 to take advantage of this growing demand and the high prices of imports.'[15] Its first blast furnace and foundry came into operation in 1877, when it was overtaken by a financial and operational crisis. 'It was discovered that the plant had been designed on too small a scale and along outmoded lines . . . (and) the company shut down in 1879.'[16] The Government of India purchased the defunct company at a nominal price and operated it as a public enterprise for some time with limited success. In 1889, doubts and pressures from government and private entrepreneurs in London compelled the Government of India to finally close it down as a public enterprise.

It was at this juncture that the thought for the manufacture of steel under contemporary technological methods was initiated in Cossipore. The Bessemer steel process, introduced in Europe in 1856 in response to the demand of Napoleon III for steel capable of withstanding new explosive shells, had made a breakthrough in defence production. 'The Bessemer open hearth, and Gilchrist processes of steel manufacture lowered the costs of production and led to the general use of steel, in place of iron, for heavy artillery in the last quarter of the nineteenth century.'[17] Much of the failure in the previous attempts to make durable steel in India were rooted in technology. Naturally, years before the erection of the extensive iron and steel foundry in Cossipore in 1892, serious considerations were given to the appropriate technology for the manufacture of steel. As early as July 1887, Captain R.H. Mahon, then officiating Superintendent of the FSF, Cossipore wrote to the Inspector-General of Ordnance, Bengal,

> To manufacture either the cast or forged steel projectiles in India for so many different guns and to set up the large amount of machinery necessary will cause an immense outlay and it will finally commit the Government to a method of manufacture which may after all be superseded by the cheaper method introduced by Messrs Nordenfeldt and Noble. . . . I know of no reason why such a test should not be carried out in India. . . .[18]

The method that Mahon was referring to was described in contemporary reports as Mitis castings. As E.A. Cowper, the late President of the Society of Mechanical Engineers in England, explained, the surface of the Mitis casting was remarkably smooth owing to the special composition of the moulds which required little cleaning on the emery wheel. 'The Mitis casting c[ould] be made in *Chill moulds* without becoming brittle or bad in any way and *never* require[d] annealing.'[19] The advantage of casting 'chill' in shell work was considered to be very great, making a large reduction in cost by saving labour, time, and material required for sand moulding. The superiority of the Mitis iron as a casting metal was due to its greater purity.

The Mitis system was also useful in work other than shell-making. It was claimed that it would furnish a method of doing away, not only with forging almost altogether, but also of greatly reducing the labour of working up the rough forging with machine and other tools. Captain Mahon continued,

> I have examined a number of castings sent out to this country by Major Wace, and one or two I have worked in machines. The castings are wonderfully clean and homogeneous and are cast exceedingly light; this was especially noticeable in the case of a ploughshare which in its thickest part scarcely exceeds 3/8 inch and withal very strong. I find them work up without difficulty, though a trifle harder than ordinary castings.[20]

Mahon was not the only person to have such faith in the discovery of the Mitis process of steel manufacture in India. Colonel H.J.F. Shea (1886–91), the Superintendent of the Gun Carriage Factory at Fatehgarh, was also impressed by it. In his letter to the Deputy Inspector-General of Ordnance, Bengal Circle, in July 1887 he identified a good number of items likely to be made under this system—carriages for horse, field and heavy batteries, frameworks of wagons, axle-trees, naves, shift irons, limber hooks, trail eyes, handles and hinges of ammunition, boxes, nuts, bolts, screws, rivets, stirrup irons, foot-rests, etc. These items then 'should be tested in England for (the) Government of India *even if authorities at Woolwich do not intend to entertain the idea of promptly proving the merits of Mitis castings* . . .'.[21] The Director-General of Ordnance in India was aware of the apparently most valuable discovery and its possible benefits, but as the system had not generally been adopted in Europe, he advised the Government of India in August 1887 to 'proceed with caution before introducing it into India for the manufacture of various Ordnance stores'.[22] Major R. Wace, who had been the Superintendent (1882–90) of the FSF but was deputed on special duty tc England during this time, was more optimistic and emphatic. In his letter to the India Office, London, in September 1887, he was eloquent:

> I recommend that a supply of rough forging sufficient to meet all the requirements for the next two years be sent to Cossipore, as they are now supplied to Woolwich; and at the same time that steps be taken to give the manufacture of these shells, on the system which has been lately adopted by Mr Nordenfelt [*sic*], a fair trial. To this end Mr Nordenfelt might be asked to supply the rough castings for 500 common and 500 shrapnel shell for the 12-pounder B.L. Gun, and these might then be submitted (to) a comparative examination and trial, either at home or in India, with the rough forging of the same nature, as supplied to Woolwich. The shell could be finished at Cossipore, if comparative trial was carried out in India. If the result should be in favour of Mr Nordenfelt's system, I would urge that no time should be lost in introducing the system of casting on the Mitis principle into Ordnance Factories in India, and that arrangements be entered into with Mr Nordenfelt in regard to the patent rights, and the introduction of the manufacture into India.[23]

The Government of India finally echoed the voice of the Director-General of Ordnance in its letter to the Secretary of State on 20 September 1887. It stated that a sufficient quantity of scrap iron suitable for the production of Mitis castings would be available in India and suggested proper testing of the various articles manufactured by this process before their final adoption in India, where 'the climatic changes [we]re far greater than England'.[24] There must have been some trials and experiments on this method of production, and some pressure groups might have operated for and against its adoption. They were finally dismissed in February 1888 when the Mitis castings submitted for trial at the Royal Arsenal had turned out full of blow holes, and not fit to be sent out.[25] But the significance of the development was that Cossipore became the locus of a heated technological debate even before the beginning of its steel production.

It was by the end of 1890 that the Director-General of Ordnance submitted the final proposal to the Government of India for the manufacture of steel in the FSF. The general introduction of breech-loading ordnance in India had necessitated the provision of steel projectiles, thereby providing for the manufacture of steel shells as used for breech-loading guns. As the Director-General elaborated in his letter dated 2 December 1890,

> A considerable part of the plant for the manufacture of these shells already exists in Cossipore, but in order to utilise it to the best advantage, the manufacturing power of the factory must now be supplemented by the appliances and establishments required for the production of steel, both cast and forged. *If this is not done, the only shell factory in India will no longer be self-contained.*
>
> The supply of ammunition for guns is also not the only consideration. Steel enters very largely into the construction of the carriages for field artillery and of all mountings for heavy guns, and owing to there being at the present time no source of supply in India, the smallest repair in cast steel cannot be executed without preferring a demand on England for the necessary material. . . . I am of the opinion that the present requirements will be met by its introduction into the Cossipore Factory, in which the question for the manufacture for steel shell for guns which are actually under supply, has to be considered first of all, and the source of supply of the necessary components determined. If the

manufacture of cast and forged steel should prove successful, of which there is no reasonable doubt, most valuable assistance then could be given to the gun carriage factories in the supply of steel components and material, and there is no limit probably to the eventual requirements of the Ordnance Department as regards the supply of different articles of war materials, machinery and miscellaneous castings. . . .

For the production of cast steel, I propose that a 3-ton furnace on the Siemens-Martin open-hearth principle be erected. This is the system most generally adopted in England and is considered to give the best results, while being easier of manipulation, and less liable to derangement than others. . . .[26]

Certain findings on the product cost of the proposed manufacture were also interesting. The cost of cast steel made at Cossipore was estimated at £22.10s per ton as against £25.13s in the Royal Laboratory, exclusive of depreciation charges. The cost of the forged steel projectiles was estimated at Cossipore on the first out-turn with imported bar steel as £53 per ton, against £56 and £58 per ton in England for the unfinished components of 12-pounder shrapnel and common shells. The Director-General noted that the main reason for the reduced cost was cheap labour obtainable here. Most interestingly, he added,

But it is no secret that a few firms in the trade, having had hitherto almost a monopoly in the supply of steel forgings, have demanded a high price which has been further maintained by the inability of the trade generally to compete for the comparatively small orders such as shells with their ever varying forms and specifications.[27]

It is important to note that the Director-General of Ordnance had received a memorandum dated 2 October 1890 and submitted by the Superintendent, India Store Department, which questioned the Cossipore venture on grounds of production and prices, thereby suggesting a delay in the manufacture. He was furious. He found 'no valid objection' in the memorandum and added in his letter to the Military Department on 31 December 1890,

At present . . . the Ordnance Department is absolutely and deplorably dependent on England for all articles of steel. The pivot of a platform cannot be replaced without an order on England. It is impossible that such a state of things can last long if the policy of independence for

India is to be maintained, and I beg once more to urge that the manufacture of steel shall be undertaken with as little delay as possible.[28]

The Government of India finally approved the project for the manufacture of steel in Cossipore, and Lord Lansdown, the Governor-General and Viceroy of India, in his letter to the Secretary of State on 25 February 1891 asked for a sanction of Rs. 1,25,000 for the purpose.[29] Of this amount, the cost of importing and setting up the new plant was about Rs. 1,20,000 (or £9,000 @ 1s 6d per rupee). There would be a recurring extra charge of Rs. 5,700 per annum to meet the additional establishment after the commencement of manufacture. The final approval came from the Secretary of the State in May 1891:

In view of the advantage of India being, in time of War, independent of this country in regard to an efficient supply of artillery ammunition, and in view of the opinion expressed by the Director-General of Ordnance in India that steel can be made in India at such rates that the cost of projectiles and other articles made from it will compare favourably with the cost to your Government of similar articles manufactured in the Government factories in England, I have resolved to sanction your several proposal.[30]

Captain Mahon, already deputed in England, was authorized to supervise over the supply of the necessary plant of machinery from that country for Cossipore. Mr. Lomas, Foreman of the Foundry, and Mr. Kidd, Foreman of the Shell Factory, along with the four men of the Carnatic Artificer Company working in the Cossipore Factory, were sent to England for training and instruction. The original plan was to bring from England even the shed to contain the plant[31] but it was subsequently decided that the structure would be built at Cossipore.[32]

NEW TECHNICAL DEMANDS AND PERSONNEL IN COSSIPORE FACTORY IN THE 1890s

It appears that the establishment at Cossipore was slowly changing during the thirty years after the Mutiny. A few new posts in the technical and general sections were created to cater to the new demands, but the measure was half-hearted and at least one important post remained vacant for some time. The official indifference to Cossipore continued for a long time, and personnel and production could not fulfil the expectation. However, more vigorous changes were to come

Signs of change and signs of reservation were simultaneously visible in Cossipore in the 1880s. Changes were rooted in the redirection of production, and with the proliferation of technical work an altogether new level of training and qualification was required from its personnel. However, the government did not change its policy immediately. The two trends can be discerned from the very developments of 1881.

Controversy began in Cossipore on the question of the reappointment of a 'Manager' on the death of Mr. Arthur in July 1881. There had been an establishment of European Mechanics in the Factory since 1872, and though gun production had been stopped in 1874, the services of the Head Mechanical Engineer (known as Manager) were retained with a salary of Rs. 600 per month. In 1881, at the time of filling the vacancy, the situation became a bit confused. The government's reaction was negative, as evident from the letter of the Governor-General to the Secretary of State on 25 September 1882:

> On the death, on 17 July 1881, of Mr Arthur, the late Manager of the Factory, the question of appointing a successor to his post arose, but the necessity of a Manager to this Factory, a point which had for some time previously been under consideration, appeared still more doubtful since the manufacture of guns at the factory has been discontinued.[33]

The government thought that it was the opportune moment to abolish the post.[34] There was even a suggestion that the so-called Manager had been in fact only the Head Mechanical Engineer.[35] Accordingly, Mr. McGlashan, the Foreman of the Turning and Boring Room, was appointed Foreman and Head Mechanical Engineer on a pay scale of Rs. 400–500 to substitute a Manager on Rs. 600 a month.[36] There was a search for a successor to McGlashan, and the Inspector-General of Ordnance, Bengal, in his letter to the Military Department highlighted the urgency of finding an adequately trained person.

> . . . I am informed that there is a man by name Mr John Fortune, at present employed in Torpedo Shop, Royal Laboratory, Woolwich (under

Foreman Mr George Anderson), who is a first class mechanic of varied experience and in every respect thoroughly well fitted to perform the duties of Foreman in the Turning and Boring Room of the Factory at Cossipore. . . .

. . . there is no possibility of obtaining the services of any mechanic in India who is competent to undertake the manufacture of Ordnance fittings, the conversion of Ordnance on Pallisar system etc. without a lengthened practical training at the Royal Arsenal, Woolwich. . . .

In consequence of the paucity of the Europeans attached to the Factory at Cossipore, if one more of them now felt sick, it might become necessary to close a portion of the Factory, for it would be dangerous to have large plants of machinery running without efficient supervision.[37]

While the personnel requirements of the FSF, Cossipore were so mounting and urgent, the government attitude was still hesitant and a strange conclusion was reached by Colonel W.T. Chitty, the Controller of Military Accounts, Bengal on the confusing situation.

In 1872, the establishment (a) of European Mechanics . . . was sanctioned (b) for this Factory and the services of Mr Arthur engaged as Manager, on a salary of Rs. 600 p.m.

This establishment was maintained up to 1874, when it was decided (c) to discontinue the manufacture of guns in India. That decision led to the abolition (d) of one of the two appointments of Foremen, but did not affect the appointment and pay of the Manager, which continued as before.

The pay of the Foreman, whose services were retained in the Factory, was subsequently increased (e) from Rs. 300 to Rs. 375 by annual increments of Rs. 15. This rate of pay was also allowed (f) on re-engagement to one of the Assistant Foremen, thus virtually reviving the appointment, which had been abolished in 1874, but reducing the number of Assistant Foremen. . . .

The proposal (g) of the Inspector-General of Ordnance to abolish the post of Manager . . . and reorganise the establishment of European Mechanics . . . (h) may be sanctioned as the proposal involves an average monthly saving of Rs. 231-4-0.[38]

The search for properly trained personnel for Cossipore continued despite financial constraints in the 1880s.The Secretary of State in April 1885 informed the Government of India on the engagement of one John George Lomas as a Moulder for FSF,

Cossipore on a salary of 'Rs. 320 per mensem, rising by annual increments of Rs. 20 to Rs. 400, with quarters and medical attendance'.[39] It is notable that the original salary offered was in the scale of Rs. 220–300 and it was unattractive to Mr. Lomas who had been employed in the Shell Foundry, Royal Laboratory.[40] An increase was thereby granted. In 1888, the Director-General of Ordnance recommended the engagement of a trained man from the Royal Laboratory, Woolwich well acquainted with the manufacture of the latest patterns of B.L. shells, to fill the post of Assistant Foreman of FSF.[41]

A more vigorous era was initiated in the FSF in production in the 1890s, following a re-direction in the 1880s. Two new shops for the manufacture of the metal fuses and tubes were added in 1890, and a major venture for the production of steel was undertaken by Major Mahon (1891–9), the Superintendent, in 1892. Accordingly, an extensive Iron and Steel Foundry was erected to the north of the Shell Factory.

> In 1896, this portion of the Factory was further supplemented by the erection of a rolling mill from which a large variety of steel bars of all manner of sections is turned out for the Indian Marine, the Public Works Department (especially railways), the Military Works Services and for the use of Ordnance Factories and Arsenals.[42]

Captain Mahon had resigned from the Ordnance Department in February 1888 'to go through a course of instruction to qualify himself for a post where special technical knowledge was required' and he was 'reappointed' in 1890.[43] Indeed, 'the whole period of his absence from the Ordnance Department in India was spent in study, and in acquiring information as to new process of manufacture'.[44] Naturally he gave a great boost to the recruitment of personnel with high technical expertise by contemporary standard. As Captain Mahon wrote to the Inspector-General of Ordnance, Bengal on 28 August 1893,

> . . . the work done in the Gun Section of this Factory is to all intents and purposes the same in quality, and will, when the new shop is at work, be approximately equal in quality to that performed by the Field Gun Section at Woolwich. There are or will be 53 horse and field batteries depending on the shop, besides a vast number of other orders

pertaining to the orders of North-Western Frontier siege trains and coast defences.

The nature of the work renders it absolutely necessary to employ in England the most highly trained workman obtainable; except perhaps the torpedo shops, no department required more skilled men. Even under these favourable conditions as to work, a foreman in England at this class of work rarely supervises more than 80 men.

In this country, the supervising foreman has to make every workman, and when made, has to watch him very carefully to maintain the accuracy in the work. . . . As a matter of fact, the foreman in this country is obliged at present to supervise the work of 200 men; and when the new shop is finished, and the new machinery already sanctioned is at work, this will be increased to 300 men, a number which is not possible for one man to overlook in a satisfactory manner.[45]

Under these circumstances, the qualifications expected for the post of the Assistant Foreman, Gun Section, were made more rigorous and grounds for such an appointment were stated to be more serious.[46] What was more important, C.W.H. Penwarden, the new incumbent for the post, was requested to mention the date 'on which he completed the courses of instruction he underwent in England prior to embarking for India to take up his appointment'.[47]

Similarly, new qualifications were stipulated for the appointment of an Assistant Foreman for the Forge Work in Cossipore in 1894. The Superintendent submitted,

(The duty) consists mainly of a smithy in which the principal work is the manufacture of forgings for B.L. Guns fittings including sights of all kinds and of a steel forge. . . .

. . . the applicant should be not only a good blacksmith in the ordinary sense, but also accustomed to the manipulation of steam hammers and able to turn out from them the more complicated forgings . . . as well as bars of sections.

. . . In fact, as men brought up in the Royal Factories are usually specialists while men from outside have frequently good general knowledge, the latter are *more* desirable. . . .[48]

Marshall Lane was appointed Assistant Foreman for the said post in 1894, and this time selection was not confined to the men of Royal Factories and was extended to men belonging to private firms.[49] Similarly, R.B. Kidd was appointed as the second Assistant Foreman for the FSF in 1895. He had served in the Drawing Office of the Laboratory Department at Woolwich.[50]

At the height of the appointments of qualified personnel in Cossipore was the case of Walter George McMillan who had joined the Factory as a chemist at the age of 27 in 1888 and worked for five years with a salary of Rs. 250, rising to Rs. 300 by five annual increments of Rs. 10. He was a student of King's College, Cambridge, a Daniell Scholar for original research and a Demonstrator of Metallurgy in the same College and the author of *A Treatise on Electro-Metallurgy*. At the end of the five-year agreement in 1893, as he was to about leave India, he wrote,

> . . . working without skilled assistance, I have been alone responsible for the whole of the Chemical and Mechanical testing of the Factory. . . . The introduction of a Department for the manufacture of steel by the Siemen's open hearth process, in the beginning of 1892, has added greatly to the volume and responsibility of my work, more especially as the undertaking was a new experiment in India, and suitable raw materials and 'refractories' had to be sought for locally. . . .[51]

Captain R.Mahon, the Superintendent, appreciated McMillan's work greatly in his testimonial dated 8 February 1893. He added that in his 'endeavours to promote to manufacture of iron from indigenous material, Mr. McMillan's reports on coals, limestones and ores obtainable h(ad) been of great value . . .'.[52] McMillan left because the salary was considered insufficient. He had asked for a 'salary of Rs. 750 per mensem to rise by annual increments of Rs. 50 per mensem to Rs. 1250 with a further increase to Rs. 1500 per mensem' but the Director-General of Ordnance found this 'prohibitive'.[53]

H.B. Weeks had been engaged as a Chemist and Metallurgist in the FSF on a monthly salary of Rs. 500 in McMillan's place months before the expiry of the latter's term. Indeed, the Government of India was eager even to pay more, if required.[54] Weeks, who had been employed as Assistant Metallurgist in the Royal Laboratory, had to go through a rigorous process of training for several months before joining the new post. As the

Secretary of State informed the Government of India on 9 February 1893,

> (Weeks) will undergo a course of instruction in tannery at the Yorkshire College, Leeds from Easter to Midsummer 1893, and will afterwards have a course of training at Waltham Abbey and Woolwich, in the chemical testing of gunpowder, cordite, gun-cotton, and nitro-glycerine, and of the ingredients employed in their manufacture; also, in the chemical and mechanical tests applied to bricks, lime and building materials generally. The course will occupy about three months, and on its completion arrangements will be made for him to sail to India in November 1893. . . .[55]

The introduction of the process of manufacture of steel in 1892 was a major venture of FSF, and by the middle of the year the Superintendent submitted a proposal for an increase of permanent staff. He was now very anxious for the appointment of an Assistant Foreman for the Forge Work, which would relieve the conductor then in-charge of the shop. The new man must be from England, his knowledge 'up to date in the forging of steel bars from ingots'.[56] It was not easy to get a suitable man always. In March 1893, for example, one Alexander Grant was selected for the post, but was found not to possess the due qualifications which required a 'blacksmith by trade brought up to and accustomed to steam hammer work, especially in production of small article by stamping'.[57] Ultimately, the Director-General of Ordnance Factories recommended that Harry Joseph Overall, who had been a good workman and well acquainted with smith's work and die forging, complete his training. He should 'undergo a short course of forging bar from the ingot and a three months course of die sinking'.[58] In January 1894 the War Office reported that he had 'satisfactorily completed the course'.[59] What was more important, part of the Superintendent's report for the increase of establishment included a general account of labour and production in FSF around June 1892.

> At the present time the extra establishment of this factory amounts daily to over 1000 men. Work in the Fuze Factory, Steel Foundry, pressure in Shell Factory and double work in the mill, with the attendant increase of general labour, is likely during the coming year to increase

the number employed to at least 1500. For the proper control of this large body of men and the supervision of the several entrances and exits, I submit that the services of a warrant officer can well be employed. . . . I may also add that much steam hammer forging work will follow on the preparation of steel bar from the ingots cast from steel furnace, and a forge man accustomed to this sort of work is desirable.[60]

For the Rolling Mill established in FSF in 1896 more personnel were needed. A plant to the value of £4,658 had been installed in the Factory to supply for ammunition manufactured in India, for which an import substitution of Rs. 31,760 was expected.[61] Accordingly Henry Gurnett was appointed Assistant Foreman for the Rolling Mill to supervise the manufacture of these items, and Charles Jacob, Assistant Foreman to work the plant for the manufacture of vent-sealing tubes.[62] Both of them were slightly over-aged, the former being 39 and the latter 32, but were recruited as younger men were not available. The Factory had to work 16 hours a day to meet the urgent requirement of brass strips, but manufacture stopped after 1898 as it 'could not be taken up economically'.[63] In August 1898, the Director-General of Ordnance reported to the Military Department that the Steel Bar Rolling Mill at Cossipore had 'worked most satisfactorily under Mr. Handley's management with the assistance of Mr. Liddle' and the Secretary of State approved their re-engagement for three years, on salaries of Rs. 550 and Rs. 340 a month, respectively in 1899.[64] It was important to note that the availability of good men from England was exceedingly difficult at that time as men were eager not to go out and when they were willing, they demanded high rate of wages.[65] The qualifications required of prospective candidates were also very demanding. For example, when it was decided in May 1898 that a 'Caster' should be appointed to assist Mr. Gurnett, the 'Roller' in the FSF, Cossipore, the Director-General of Ordnance wrote to the Military Department specifying several requirements.

. . . a mechanic (for the new appointment of a Caster) with the following qualifications may be engaged and sent out from England as early as possible:

1. He should be thoroughly competent to take subordinate charge of

the ingot metal casting for cartridge metal, consisting of brass, copper and cupro-nickel. The metal is melted in gas-fired furnaces, and consequently a man having some knowledge of this system should have the preference.

2. He should have a general acquaintance with the work of the rolling metal so as to be able to the practical experience he will gain at Cossipore to take charge of the mill in the absence of the Foreman and the 1st Assistant Foreman.
3. He should be fairly educated and competent to take charge of usual shop accounts. . . (He should also agree) to provide for the performance of night duty.[66]

By 1900, further changes were initiated in Cossipore. It was decided to move the steel and iron furnaces, the steel bar mill and the cartridge metal rolling mill to Ishapore, to be named the 'Steel and Cartridge Metal Rolling Mill' under the control of FSF, Cossipore. By the early years of the new century, there came a change of far-reaching importance. The introduction of quick-firing guns in England led to a consideration of the question of undertaking their manufacture in India. A plant was set up to turn some 90 to 100 guns in Cossipore under the Superintendence of Major Renny (1899–1905). The FSF was thereafter re-named the Gun and Shell Factory (GSF), Cossipore, and its Ishapore unit was called the Gun and Shell Factory, Ishapore. Major C.T. Bell (1905–12) was in-charge of the entire two-unit factory as Superintendent. The establishment and personnel were further extended, as reflected in the authentic list supplied for 1910.

A 'LIMITED' OR AN 'ABORTED' TRANSFER OF TECHNOLOGY?

One of the vexed, issues in the history of technology is the question of technology transfer. While this question can be approached with reference to any historical regime, colonial or post-colonial, in the context of any country, technology transfer in India under colonial conditions has become a major subject of deliberation. Such a transfer, where it at all took place, is stated to have been very limited and marked by twin imperfections: motivation factors and diffusion effects. These two limitations frequently marred most of the so-called technology transfers, civil or military, in colonial India. Historians are justified to highlight the difference between India and Japan in technology transfer in the nineteenth century, especially as regards institutional and overall technological change (diffusion effects) that such transfer could generate. However, what is not often emphasized, is that the limited motivation and diffusion of technology transfer may significantly vary even under colonialism because of the very nature of certain kinds of technology. Technology of defence production in India under colonial rule, as exemplified by our Cossipore study, is clearly one such case. But was the Cossipore story of the 1890s a 'limited' or an 'aborted' case of technology transfer?

Admittedly, technology was a double-edged affair in Cossipore as far as ordnance production and policy was concerned. While the very limitation in its application had an adverse bearing on production and was the direct consequence of the policy pursued at the higher levels since 1858, the highly competitive world of the technology of warfare forced partial modification of the policy in the late nineteenth century, thereby limiting colonial dependence on this matter. Though apparently a paradox, such a situation is better explained by the neo-Marxist tradition of commentary on the drive of the 'developmental state' more to military technology than to civil one.[67] British science and technology were highly weighed towards the military ever since the late nineteenth century, retarding industrial but encouraging military research and development in the process.[68] Limited prospects of technology transfer in defence production in Cossipore in the 1890s should be seen from these perspectives. It is true that the potential experiments were temporary, success limited, deployment of expert personnel from England was slow, sometimes half-hearted and often not adequately remunerative. Nonetheless, the historical significance of the attempts of technological transfer should not be underestimated. The basic dichotomy was not fully resolved before Independence, and occasional attempts at rejuvenation were marked by failures and obstructions. The more important limitation, however, was the virtual absence of a policy to recruit Indians as technical personnel at the higher levels of the defence establishment in

India, thus minimizing and virtually negating the diffusion effects. At the same time, the availability of qualified Englishmen was thwarted by many factors including their mobility, demands in Europe and unattractive financial prospects in India. Viewed from the signs and considerations of India's long journey in the technological route, the 1890s experience in technology transfer in Cossipore should be taken as a limited but spectacular advance that was finally aborted under colonial conditions.

NOTES

1. Michael Adas, *Machines as the Measure of Man: Science, Technology and Ideologies of Western Dominance*, Delhi, 1990, p. 367.
2. Major-General K.C. Mehra, *A History of the Army Ordnance Corps 1775-1974: A Saga of Two Centuries*, Allahabad, 1980, p. 127.
3. Michael Howard, *The Franco-Prussian War: the German Invasion of France, 1870-1871*, London, 1961; rpt. 1981, p. 5.
4. Correlli Barnett, *Britain and Her Army, 1509-1970: A Military, Political and Social Survey*, Harmondsworth, 1970, p. 342.
5. Ibid., p. 328.
6. Martin van Craweld, *Technology and War: From 2000 B.C. to the Present*, London, 1991, p. 297.
7. J.F.C. Fuller, *Armament and History: The Influence of Armament on History From the Dawn of Classical Warfare to the End of the Second World War*, New York, 1945, rpt. 1998, p. 31.
8. From the Inspector-General of Ordnance and Magazines to the Military Department, 20 June 1860, para 2 in PMD, August 1860, no. 4, National Archives of India (hereafter NAI).
9. From Northbrook to the Secretary of State for India, 4 August 1873 in *Military Despatches to the Secretary of State for 1873*, Letter no. 161 of 1873, NAI.
10. See *Appendices to the Report of the Special Ordnance Commission Appointed by the Order of the Government of India to Enquire into the System of Ordnance Supply in India*, Calcutta, 1875, p. 328.
11. W.A. Johnson, *The Steel Industry of India*, Cambridge, Mass, 1966.
12. Morris D. Morris, 'The Growth of Large-scale Industry to 1947', in Dharma Kumar, ed., *The Cambridge Economic History of India*, Volume II, *c.1757–c. 1970*, Cambridge, 1982, p. 591.
13. Ibid., p. 584.
14. See George Jessop of Jessop and Company to Holt Mackenzie, Secretary to the Government's Territorial Department, May 1825,

quoted in Arun Bandopadhyay, *The Story of Jessop*, Calcutta, 1988, p. 26.
15. Morris, op. cit., p. 585.
16. Ibid.
17. Richard A. Preston and Sidney F. Wise, *Men in Arms: A History of Warfare and its Interrelationship with Western Society*, New York, 1956; rpt. 1979, p. 243.
18. From Captain Mahon, Officiating Supdt., FSF, Cossipore to IG of Ord., Bengal, 5 July 1887 in the Manufacture of Steel in India, L/MIL/7/13792, Oriental and India Office Library (hereafter OIOL).
19. Ibid.
20. Ibid.
21. Colonel Shea to the Deputy IG of Ord., Bengal Circle, 23 July 1887, Ibid. Emphases added.
22. From R.F. Lowis, Director-General of Ordnance in India to the Secretary to GOI, Military Department, 18 August 1887, para 4, ibid.
23. From R. Wace to the Military Secretary, India Office, London, 21 September 1887, see ibid.
24. See extract from a Despatch of GOI to the Secretary of State, 20 September 1887, no. 163, para 14, see ibid.
25. From the Ordnance Consulting Officer in India to the India Office, London, 24 February 1888, ibid.
26. From Maj. Gen. A Walker, Director-General of Ordnance in India to the GOI in the Military department, 2 December 1890, paras 3–5, ibid. Emphases added.
27. Ibid.
28. From the DG of Ord., to the Military Department, 31 December 1890, ibid.
29. From the GG of India to the Secretary of State, 25 February 1891, para 4, see ibid.
30. From the Secretary of State to the GG of India, 14 May 1891, Military Letter no. 70, para 2, see ibid.
31. See extract of the letter from the GOI to the Secretary of State, 25 November 1891, para 29, see ibid.
32. See extract of the letter from the GOI to the Secretary of State, 20 January 1892, para 26, see ibid.
33. From GG to Secretary of State 25 September 1882, Military Department, no. 232 of 1882, in Revision of Establishment in Cossipore around 1882–83, L/MIL/7/6240, OIOL.
34. From the Military department to the IG of Ord., Bengal, 12 August 1881, ibid.
35. From the IG of Ord., Bengal, to the Military Department, 1 January 1882, no. 97-I: 'The individual hitherto called Manager has been in

fact the Head Mechanical Engineer, and such there must be to work the establishment efficiently; but there is no need to call him the Manager, a name which was apparently given to the Head Mechanical Engineer of the Foundry and Shell Factory to add to the dignity and pay of the late incumbent, Mr. Arthur.' Ibid.

36. See note 33, para 3.
37. From IG of Ord., Bengal to the Military Department, 25 July 1881, paras 5, 6 and 7 in L/MIL/7/6240, OIOL.
38. From the Controller of Military Accounts, Bengal to Military Department, 6 April 1882, no. 305, paras 2–5, see ibid.
39. From the Secretary of State to the GG, 2 April 1885, Military Department, no. 78, para 1, in L/MIL/7/6242, OIOL.
40. From F. Lyon, Superintendent, Royal Artillery to the Director of Artillery, 26 January 1885, ibid. 'Seeing that the Rupee is not worth more than 1s 8d, the salary Rs. 220 rising to Rs. 300 appears to be insufficient remuneration for the services required. Mr. Lomas would agree to a payment of Rs. 320 rising annually to Rs. 400.'
41. From the GOI to the Secretary of State, 22 June 1888, no. 100, para 8, in L/MIL/7/6254, OIOL.
42. See *A Brief History of the Government Gun & Shell Factory at Cossipore* (Anon, n.d.), p. 10.
43. See Military Despatch from the Secretary of State to GOI, 7 June 1894, no. 69, para 2, in L/MIL/7/1668, OIOL. Strictly speaking, it was not a re-appointment but a new appointment. However, since the GOI had given the privilege of 'continuous service in India', the Secretary of State did not like to disturb the order.
44. See extract from a letter of GOI to the Secretary of State, 22 March 1893, no. 44, para 18, see ibid.
45. Captain Mahon to IG of Ord., Bengal, 28 August 1893, no. 1247A, paras 1–3, in L/MIL/7/6291, OIOL.
46. 'Qualifications – A well-educated young man, age from 24-28, brought up in the Royal Gun Factories, and accustomed to recent work on breech-fittings. . . .'
 'Illness or absence on leave of the foreman which must be occasionally granted will cause serious inconvenience . . . (as the work is now) more than one European can properly supervise.' From Mahon to IG of Ord., Calcutta, 10 April 1893, no. 68A, see ibid.
47. From the GG to the Secretary of State, 19 December 1894, no. 215, ibid.
48. From the Superintendent, FSF, Cossipore to IG of Ord., 5 September 1894, no. 1253A, in L/MIL/7/6300, OIOL. Emphasis in the original.
49. See extract from a Despatch to the Secretary of State, 31 October 1895, no. 195, para 16, ibid.
50. From the Secretary of State to the GG of India, 29 August 1895,

no. 86 (Military), para 8 as well as the extract from the Government of India's (hereafter GOI) Despatch to the Secretary of State, 28 November 1894 in L/MIL/7/6302, OIOL.

51. See McMillan's letter dated Calcutta, 1893, n.d., in L/MIL/7/6252, OIOL.
52. See R. Mahon's testimonial of McMillan, dated Calcutta, 8 February 1893, ibid.
53. From the Director-General of Ordnance from India (hereafter DG of Ord.) to the Military Department, 13 July 1892 in L/MIL/7/6271, OIOL.
54. See extract from a Despatch of GOI to the Secretary of State, 2 August 1892, para 11, in ibid. '. . . we request that the Chemist engaged may be specially trained before being sent out. Should, however, there being any difficulty in securing the services of a reliable man on the salary proposed, viz., Rs. 500 per mensem, owing to the reduced value of the rupee, we request that a higher salary be offered.'
55. From the Secretary of State to the GOI, 9 February 1893, no. 20, para 1, ibid. He ultimately left England on 10 November 1893 'to take up the post of Chemist and Metallurgist, Cossipore Factory' (Military Despatch to GOI, no. 133, 23 November 1893, para 16, ibid.)
56. From Lt. Colonel R. Wace, the Officiating Inspector-General of Ordnance to the Director-General of Ordnance, 6 June 1892 in L/MIL/7/6275, OIOL.
57. Military Despatch form the GOI to the Secretary of State 16 February 1893, no. 26, see ibid.
58. Military Despatch to GOI, 18 January 1894, no. 6, para 8, ibid.
59. Ibid.
60. From IG of Ord., to the DG of Ord, 6 June 1892, ibid.
61. See extract from the letter of DG of Ord. to the Military Department, 22 June 1894 in L/MIL/7/6315, OIOL.
62. See extract from a letter from GOI to the Secretary of State, 18 December 1895, para 22, ibid.
63. See Financial Despatch from India, no. 290 of 1898, Enclosure I, quoted in L/MIL/7/6315, OIOL.
64. From DG of Ord. to the Military department 10 August 1898, para 2 and from the Secretary of State to the GG of India, 2 February 1899, para 2 in L/MIL/7/6316, OIOL.
65. See Memorandum by the DG of Stores, 8 October 1896, ibid.
66. From the DG of Ord. to the Military Department, 11 May 1898, para 3, in L/MIL/7/6319, OIOL.
67. For a detailed review of this neo-Marxist approach, see David Edgerton, *Science, Technology and the British Industrial 'Decline' 1870-1970*, London, 1996, pp. 5–10. See also David Edgerton,

England and the Aeroplane: An Essay on a Militant and Technological Nation, London, 1991 for the argument of a particular warlike strategy in technological change, and David Coates, *The Question of UK Decline: The Economy, State and Society,* London, 1994, for political and ideological arguments to account for such changes.

68. See J.D. Bernal, *The Social Function of Science,* London, 1939 for a classic account.

CHAPTER 10

Communicating Science: The Bose 'Style' in Perspective

MADHUMITA MAZUMDAR

Styles of scientific communication have in recent times become an important area of study for social historians of science in the West.[1] Though not entirely uncharted terrain in studies of Indian science, there is seldom much analytical interest in understanding the centrality of scientific communication in the making not merely of scientific facts, but of scientific careers, scientific personas, scientific disciplines, and scientific regimes in general. Such communication, it is argued, involves not just the formal discursive practices involved in the discipline, but also the less documented public utterances aimed at non-specialist audiences for specific pedagogical or political purposes. In each case, communication is understood as an exchange between discourse and audience. The underlying assumption is that all discourse is strongly imprinted with cultural codes and marked by concern for audiences, tailored to their expectations, and forged in terms of their effectiveness.

In recent times, biographical studies of the Indian scientist Jagadish Chandra Bose (1858–1937) have begun to comment on the singular style of scientific communication he adopted both in his insistence on the public lecture demonstration and on his use of a distinctively literary idiom in his presentations. Going directly against the grain of Baconian norms, Bose, it is argued, adopted a style that could be described as both 'theatrical' and 'rhetorical'. While this seemed to have gone down well with his lay audiences both in the West and in India, one of Bose's most recent biographers argues that for the scientific community in general Bose's 'style' was an aspect of the tragic flaw in his genius. It marked him as an 'outsider' and made his legacy a deeply ambivalent one.[2]

The problem with this perspective is that it tends to reduce Bose's choice of certain styles of scientific communication as an aberration in his persona and as an embarrassing aspect of his career, best forgotten. Although it does try to make some sense of the unusualness of this style in terms of the 'different' cultural conditions within which it took shape, it tends to slide towards the argument that whatever was not 'conventional' about Bose's science in terms of an accepted Western standard was indicative of whatever was inherently 'Eastern' in his style and persona. Thus what were deemed to be the singular markers of Bose's style of scientific communication—his use of the literary idiom, his invocation of the dramatic mode, and the ideological casting of his research institute as a ritualized space or 'temple'—were held to be illustrations of his susceptibility to the enduring cultural influences of the East.[3]

Such an explanation of his style seems at best inadequate and at worst hasty, largely because of its tendency to steer clear from historicizing the issue in relation to the evolving styles of scientific practice in the West and in relation to the internal logic of Bose's scientific career. What is perhaps even more problematic about this treatment is that there is a certain unquestioned assumption about the existence of a single, homogeneous Western convention, which we are told was either deliberately or whimsically flouted by Bose.

Starting at this point of critique, my paper seeks to take a closer look into some of the defining features of Bose's style of scientific communication and tries to understand its purportedly anti-Western/Baconian strain not as an eccentric response to accepted convention, but as part of a series of contingent strategies deployed to legitimize the claims of his practice. I argue that the important question to be asked is why Bose chose to adopt specific features of his style at specific times in his career. This perspective in other words looks beyond the realm of Bose's whim and eccentricity and locates the issue of his 'style' within a history of the 'effects' it was meant to produce. It takes its theoretical position from histories of early modern science which have sought to conceptualize the variety of public communicative strategies used by scientists as deliberately crafted 'performances'.[4] These performances, it is argued, were meant for both lay and expert audiences and aimed at conveying not merely scientific truths but more importantly, the experiential aspects of the truth-making exercise as a whole. Such public exercises though often looked upon with a degree of ambivalence were seen to constitute a complex cultural strategy that scientists evolved in order to solicit patronage and claim legitimacy. Studies such as these often conclude that contrary to popular belief, the Baconian norms of scientific communication had a distinctly uneasy career in the Western scientific community, and until well into the late nineteenth-century scientists were free to deploy their highly individualized styles of communication without the fear of institutional censure.[5]

Arguing from this perspective, I take the view that to pit Bose's style against a standardized ahistorical Western norm is academically untenable. To engage with the issue of his style, it is necessary to both relate it to the diverse histories of scientific communication in the West as well as to recognize the specific context of scientific practice in colonial India.

The first three sections of the essay look into the evolving elements of Bose's style through a series of defining moments in his career. The fourth and concluding section tries to understand the larger rationale governing his choice of certain styles in terms of both the internal logic of his research interests as well as the larger social and political context of his practice. Although this essay derives its basic argument from standard historical and biographical accounts of Bose's life and career, it refrains from engaging these accounts in any great detail. Focusing specifically on the public/performative aspects of Bose's style and the effects it produced, it attempts to draw attention to his mode of scientific communication as a part of a complex history of the career and habitat of modern science in colonial India.

THE MAKING OF THE BOSE STYLE

In order to map the complex trajectory of Bose's career, historians on most occasions refer both to the creative impact of his social and intellectual inheritance and to the peculiar circumstances of his entry into a scientific career. Heir to the ideas of the nineteenth-century 'Renaissance', Bose, we are told, grew up in cultural milieu where liberal, secular, and rational attitudes coexisted with equal measure of respect for scriptural or canonical authority.[6] Jagadish Chandra Bose was born in 1858 into a Brahmo family of Bikrampur in East Bengal. While his father adhered to a 'monotheistic, anti-idolatrous caste-denying brahmoism', his mother never gave up her Hindu orthodoxy. The family was said to have remained strongly influenced in their lifestyles by a whole range of Brahmanic traditions.[7] Bose's father, Bhagawan Chandra, was deemed to be man of varied interests and tastes. Apart from being an administrator and entrepreneur, he was also an amateur physicist and biologist and promoter of technical education. He also had a deep involvement in Vedantic theology. Notwithstanding his numerous commitments, Bhagawan Chandra took a special interest in his son's education. It was he who introduced Bose to botany, physics, and the wondrous world of technology. These early encounters with science were to leave an indelible imprint on Bose's mind as he entered Calcutta's St. Xavier's College in 1870. It was there that he developed his long-lasting relationship with Father Eugene Lafont (1837–1908), a Jesuit priest, the best known and perhaps the only experimental physicist in Calcutta of his time. It was Father Lafont who encouraged Bose to study physics in combination with Latin and Sanskrit.[8]

Bose graduated from the University of Calcutta with a Bachelor's degree in Arts. Shortly afterwards, he proceeded to study medicine in England. Within three months of his joining the degree course in medicine, he abandoned it and left London for Cambridge where he entered Christ's College to read the natural sciences. In 1884, he obtained a bachelor's degree in Arts with a second class in the natural science *tripos*. In the year before, he had obtained a bachelor's degree in science from the University of London. Armed with his two English degrees and his 'Cambridge connections', Bose returned to India. In Calcutta he was appointed (the first Indian) Professor of Physics in Presidency College in 1885. Despite the significance of this achievement, the controversy surrounding the appointment revealed the extent to which Indians continued to find themselves vulnerable before the politics of racial prejudice. Alfred Croft, then Director of Public Instruction for Bengal, typified such attitudes with his consistent claim that Indians did not possess the requisite temperament for the 'exact sciences'. He registered his disagreement with Bose's appointment and withdrew from this position only after the intervention of Viceroy Ripon. Bose's appointment, however, remained mired in debate as a result of the discriminatory structure of salary and allowances assigned to him.[9]

For Bose, the years between 1885 and 1894 constituted a kind of interregnum. Soon after his appointment in Presidency College, he began to teach practical classes at the Indian Association for the Cultivation of Science. Here his work was primarily in the realm of electricity and magnetism.[10] His biographers say there is no evidence to suggest that he undertook any serious research during this period. In 1888, he discontinued his practical classes at the Association. The official centenary history of the latter, referring back to its Report of the Eleventh Annual Meeting in April 1888, attributes this to Bose being preoccupied with experimental studies of electric waves at Presidency College.[11]

Bose's own account places the beginning of his systematic researches soon after his thirty-sixth birthday. His motivations, he said, were to pursue knowledge and achieve personal glory. His resolve was to become a scientist rather than remain a science teacher.[12] It was a difficult choice, for he neither had the resources nor the infrastructure to carry on research on the scale he had envisioned. He could look to no mentor or patron for support. Father Eugene Lafont, his old teacher, was active at St. Xavier's College and continued to lecture at the Indian Association till 1893, but Bose's research interests had by then moved into areas where Lafont could not be of any help. What he did learn from Lafont, apart from possible research areas, was a certain norm of scientific communication: the 'lecture demonstration' which made it imperative that the results of experimental knowledge be taken out of the private domain of the laboratory or observatory into the emerging public sphere. Lafont's insistence on the public lecture demonstrations provides an early clue to understanding Bose's own enthusiasm for this mode of communication at the very outset of his career as a professional scientist in Bengal.[13]

In the years between 1895 and 1898 Bose devoted himself to the study of the optical properties of electric waves or electric rays as he chose to call them. In May 1895 he read before the Asiatic Society of Bengal a paper 'On the Polarisation of Rays by Double Refracting Crystals'. In October of the same year, he dispatched to London an article 'On the Determination of Indices of Refraction of Sulphur for the Electric Ray' for publication by the Royal Society.

Bose, it was clear, had worked out an elaborate research programme between 1894 and 1898. According to Subrata Dasgupta, a reader of his papers could not but be struck by the enormous scope of his research agenda. Most of the work was conceived and planned in his laboratory, 'a small room about twenty feet square with an attached bathroom'.[14] The object of the present inquiry, he stated in his very first paper, was to 'find natural substances which would polarize the transmitted electric ray'.[15] Bose was clearly intent on establishing himself as an authority on the optical properties of electromagnetic waves. The objective of his first research programme was to devise experiments to produce a variety of optical phenomena of interest—refraction, reflection, or polarization. Bose, in other words was seeking to devise means for observing the nature of electromagnetic waves in a controlled fashion.[16] The first task he therefore set for himself was the construction of suitable apparatus. His tireless effort in this direction led him to design apparatus known as the 'oscillator'. This was meant to produce electromagnetic waves of the order of millimetres. Here he was said to have supposedly improved upon the device of the great Oliver Lodge himself in the sheer effectiveness of the transmitter. In terms of the originality, Bose's early experiments on electric waves lay in the production and demonstration of smaller length waves than had been previously produced and more particularly in the compactness and effectiveness of his total apparatus.[17]

In 1895 Bose gave the first major public demonstration of his new experiment at the Town Hall in Calcutta. He transmitted electric waves that passed through three walls, the large frame of Alexander Mackenzie, the Lieutenant-Governor of Bengal, and activated a receiver which in turn caused a bell to ring, a pistol to fire, and a small amount of gun-powder to explode.[18] The success of the experiment was acknowledged by the city's scientific and political elite and became a subject of social comment.

Yet for Bose, the excitement that the event generated did little to alleviate the resentment he felt at the isolation that conditions of work imposed on him in Calcutta. He desired to travel to England and present his findings to a larger and more informed audience. In the early months of 1896 he wrote to the Government, 'I have hitherto worked in complete isolation with every imperfect appliance. . . . In order to plan an efficient laboratory it would be of inestimable advantage to visit the laboratories of Britain and the Continent.' The authorities agreed, and in the summer of that year Bose set sail for England.[19]

In England, the connections he had built during his stint as a student served him well. He was invited to the sixty-sixth meeting of the British Association for Advancement of Science in Liverpool (1896) where he took the opportunity to demonstrate his apparatus for the study of electric waves. He followed this with a copy of the paper he wrote on this to the University of London, which conferred upon him the degree of Doctor of Science. More honours followed. On 29 January 1897 Bose was invited to deliver at the Royal Institution a lecture on 'Electro-Magnetic Radiation and Polarization of the Electric Ray'. The occasion was chaired by Crichton-Browne, Treasurer and Vice-President of the Royal Institution. It was attended by an audience of about 550 people.[20] The Royal Institution lecture is said to mark the beginnings of what would later be identified as the characteristic Boseian style. Commenting on the emerging features of this style, Subrata Dasgupta writes,

> As befitting a public lecture, Bose's tone was expository, relaxed and first person singular. . . . He showed three different versions of his transmitter, that produced waves 'which are approximately, ¼ inch, ½ inch and 1 inch in length' and displayed his spiral spring coherer. He performed experiments. He showed how various substances selectively absorbed the radiation and thus appeared coloured. . . . Experiments followed one after another. . . . Electromagnetic radiation, Bose

concluded, near the end of the lecture, behaves much as does visible light. 'The two phenomena were identical'. . . Bose did not wish his audience to overlook the larger canvas. His language, as he completed his discourse became oratorical. He appealed to imagery and metaphor. He asked his audience to 'imagine a large electric organ' having keys 'capable of producing different ether notes' the notes will rise gradually in frequency. But even as the ethereal sea . . . is thus being agitated the human witness is left unmoved since he cannot see, feel or hear the agitation. At some point, however he will receive a 'sensation of warmth'—the infrared radiation; he will then detect a 'red glimpse of light', colours will be seen for a brief time—this the realm of visible light and then as the keys of the 'electric organ' produce ethereal notes of even higher frequency 'our organs of perception fail us completely. There is again total darkness'. It is the region of 'invisible light' which was being explored, Bose reminded his audience. And, tacitly they were asked to realize that his work formed part of that exploration.[21]

For Dasgupta it becomes clear that the unusual features of Bose's discourse were manifest both in the 'oratorical' style he deployed and in the dramatic appeal he made to his audiences to 'imagine' a phenomenon instead of directly observing it. He was asking them to travel with him into the region of 'invisible light' and imaginatively or perhaps intuitively experience a phenomenon that could prove that electromagnetic radiation behaved in much the same manner as light. As a concluding remark on this unusual strain in Bose's appeal, Dasgupta writes, he had 'strayed beyond the highly conventional language of scientific discourse'.[22] His concluding sentences were meant to have provided his audience and readers with a 'glimpse of the man behind the impersonal mask of universalized western science, the one indication that this was the work of one belonging to a cultural tradition not rooted in Europe'.[23]

For Dasgupta, this was the distinctiveness that marked Bose's style of communication all through his subsequent career. As his interest moved steadily and perhaps controversially into the realm of plant physiology, the style of presenting his scientific findings became progressively unconventional. This was evident in the ways in which Bose presented his results in written documents as well as in the lecture demonstrations he organized for his audiences in the West. From his choice of apparatus he designed for his experiments, to the deliberate presentation of his findings in monograph form, and the ideological element he invested in his overall scientific pursuit, Bose, it was evident, was following a deliberately different line.

The directions of this 'difference' were particularly evident in 1906 when the first detailed descriptions of the experimental methods and results of his findings in plant physiology were brought out in two major monographs entitled, *Plant Response* (1906) and *Comparative Electro-Physiology* (1907). The choice of the monograph form was regarded as unusual but Dasgupta contends that it was a last recourse. In his account of this phase of Bose's struggle, he argues that in accordance with convention the scientist had initially sent his first findings to the Royal Society, expecting it to accept and publish them after the formalities of vetting were completed. To his utter disappointment, however, the Society remained unmoved by these findings and even displayed open hostility by refusing to publish any of it in its *Transactions*.[24]

Dasgupta concludes that the effect of this perceived indifference was in a sense responsible for changing the way in which Bose would henceforward communicate his results to audiences in the West. He decided to do away with the convention of communication through the scientific paper and for the next four years wrote two massive books both under the imprint of Longmans Green. The first book was titled *Response in the Living and Non-Living*.[25] The consequences of this were to have a significant bearing on Bose's public persona both in Bengal and in the West.

This was evident when *Plant Response*, suffered a curious twist of fate. The nationalist press greeted it as the greatest *swadeshi* event of the year but to his scientific peers in the West, as Dasgupta observes, Bose presented himself an 'outsider' once more.[26] The hard work that he put behind it had come to naught. This was deeply disappointing not only to Bose himself, but to all nationalists who had placed their hopes on it. Bose had put in immense amounts of effort to ensure the acceptability of his thesis. The result was a tome of unprecedented proportions. It consisted of nine parts with a total of fifty-two chapters spread over 780 pages with descriptions of 315 experiments and

278 illustrations.[27] All this was meant to establish with certainty the two central claims of the Boseian thesis: that response to stimulus was the universal, shared character of all plants sensitive and ordinary, and that there was no line demarcating plant from animal substance. Woven into the text of *Plant Response* were descriptions of all the new instruments Bose had devised in the course of his investigations. He had coined Sanskrit names for them. There were, for instance, two varieties of the *soshungraph* for recording suction of water, the *kunchangraph* which derived its name from the Sanskrit root *kunchan* meaning contraction and was used to measure contraction in certain plant organs and finally there was the *morograph* once again from the Sanskrit term *mri* or death an instrument that was meant to record the 'death point' of a plant as a function of temperature. Apart from this there were the two other critical instruments that Bose would use in all his experiments, the crescograph and the balanced crescograph, which were used to magnify and record plant growth.[28]

Notwithstanding the meticulous research that lay behind the book, Bose's professional audiences in the West received it with a degree of scepticism. Most remained indifferent, while some expressed wonder at the 'fertility of resource' and the 'ingenuity of his experiments'.[29]

Bose responded to the welter of criticism against this work with the publication of another volume, *Comparative Electro-Physiology*, in 1907. By this time he had also begun to feel the necessity of travelling to Europe in order to make his case with greater conviction. The Secretary of State, after much deliberation, approved of Bose's deputation for a year. Bose took advantage of this opportunity to visit Britain once more and renewed association with the Royal Institution. He sought and was granted admission once more as a worker. This brief stint at the prestigious institution however did not yield the desired results. The community of botanists to whom Bose addressed his findings were unwilling to allow him access into what was projected as their exclusive domain. In all senses, therefore, despite the great preparations, this visit to England and America proved to be disappointing once more. The paper read in Dublin was the only one to be formally accepted by the 'botanical community' by way of normal channels of scientific communication.[30]

In the ten years between 1903 and 1913 Bose did not submit a single paper to the Royal Society. He concentrated on the invention of a completely new apparatus by which plant responses could be observed and recorded with greater precision. The effort brought into existence the new 'Oscillating Recorder', whose magnifying power was increased by a hundred thousand times. The new series of apparatus was designed to confirm and verify the results obtained previously and to demonstrate the transmission of nervous excitation in plants against the generally accepted view that there was nothing in the plant comparable to the nervous system in animals.[31] Results of one of these experiments with the Oscillating Recorder were after much hesitation communicated to the Royal Society. The Society without much delay decided to publish it in the *Philosophical Transactions* in 1913. This official intimation of acceptance was for Bose the beginning of a new spurt of creativity. The deputation to Europe that followed soon after served, in his words, to 'lift the gloom that had engulfed his life for fourteen long years'.[32]

A new phase now began in Bose's career. His biographers have often focused a great deal of attention on his tours to Europe and the general impression he left on the minds of his Western audiences, but seldom have they seen these tours in terms of the impact they had on the evolution of Bose's public persona.[33] This point assumes importance because it seems as if in the years following his formal entry into the discipline of plant physiology, he found it imperative to explain to his audiences the reasons, motivations and philosophical underpinnings of his research interests. It almost seemed as if by doing so he would bring his audiences to an understanding of not merely the new 'truths' he had discovered but the totality of the intellectual/spiritual experience that made possible the discovery of those truths.

Looking closely at the purport of this new spurt of public activity it seems that the burden on Bose to legitimize his practice to his peers in the West as well as to his Indian patrons compelled him to choose a mode of communication that involved an active and participatory public.[34] It now seems possible that the key to understanding the unusual nature of his style of scientific communication lies in understanding the nature and context of this public-orientedness.[35] Driven by a programme of research that lacked the required infrastructure and epistemological support, Bose was forced to look beyond the laboratory to a forum where his experiments could be witnessed and comprehended by both lay and expert audiences. The material context of his practice, in other words, forced him adopt a mode of communication that could arouse a particular kind of public response. The lecture demonstration, his preferred mode, was tailored to meet this end. He was to put this mode to greater and more intense effect in the years to come when his experimental field allowed him to carry his laboratory with him.

Commenting on the significance of the public lecture demonstration in early modern Europe, social historians of science argue that such presentations were always different from the original experiments and had great dramatic potential. The demonstration in essence was the rehearsed reiteration of an already achieved effect, in which the desired outcome was known in advance. It was accomplished through manipulation of material apparatus, together with a specific kind of discursive packaging that conveyed its specific meaning for the audience. It also required the establishment of certain kinds of relations between demonstrator and audience, such that the audience knew what to expect and how to respond.[36]

This significance of Bose's increasing enthusiasm for this style of communication was amply demonstrated during his Fourth Deputation in Europe in 1914 when it became clear that notwithstanding the ambivalent responses his research invited in the established institutional domains of the Western scientific community, the public lecture would be the forum wherein both he and his work would be validated.

THE STAGING OF 'EASTERN' SCIENCE

Ever since his first success abroad, each of Bose's visits to Europe became a subject of a great deal of public interest in Bengal. The scientist perhaps owed the magnitude of this adulation to Tagore, who along with Sister Nivedita had by the turn of the century become one of the most discerning interlocutors of his science. Tagore's eulogistic rendition of the scientist's first trip in 1897, in a way set the tone for public appraisals of each of the succeeding ones. All of these came to be represented as odysseys of adventure, discovery, and heroism. His struggles to achieve recognition for himself and his science were often depicted in the allegorical format of the lonely voyager's journey into strange and distant lands, his encounters with prejudice and hostility, and his eventual redemption through a formidable display of courage, forbearance, and moral rectitude.[37]

Bose's personal sense of disillusionment following the third deputation to Europe seemed to have dampened public spirit for a while, but the announcement of yet another visit soon after brought about a resurgence of hope and anticipation among many of his admirers.[38] In Bengal, it was the *Modern Review* which took the initiative to arouse and sustain public interest in what was to be his Fourth Deputation to Europe. It carried the most extensive coverage of this trip and included such novel features as the extended *tour diary*.[39] Basiswar Sen, one of Bose's students who accompanied him on this tour, was assigned the task of preparing press notes for this fairly widely-read fortnightly. He sent in some of the most detailed accounts of the visit and the kind of response it generated in the 'lecturing empires' of London, New York, Washington, and Chicago. Sen's tour diaries entitled, *Round the World with My Master* were evidently inspired by the spirit of Passepartout, companion to the great Phileas Fogg, whose fictional eighty-day journey round the world had come to acquire a kind of universal appeal among youngsters everywhere. Sen's tour diaries were in fact to become the prototype of most other press reports on the Fourth Deputation.[40]

Sen's enthusiasm about this odyssey with his Master was evident in the extravagance of detail he recorded about each phase of the visit. In the very first of his pieces, written shortly after they had begun to sail, he began by a grateful acknowledgement of the fact that he had had 'the supreme privilege' of being accepted as 'disciple' by the 'Master'. 'Days before the journey', he wrote, 'the Master would come up to him and present him with sealed envelopes containing the results he foresaw of the new experiments he was about to perform'. To his utter bewilderment, Sen would find that everything would come out exactly the way the Master had anticipated. This was enough to convince him of the success of the tour ahead.[41]

The interesting aspect of this particular trip was that Bose decided to carry his plants specimens along with him from India. For his young companion, this was the most daunting responsibility thrust upon him. The 'vital psychological and practical' preparation for the trip, he wrote, 'involved the rather difficult task of carrying the plants from India' and caring for them right through the entire period of travel and stay.[42] Sen was entrusted the enormous responsibility of making all the necessary preparation for the long journey by sea on the *S.S. Egypt* from the port of Calcutta. His companions on board were 'two Mimosas, fairly vigorous and two telegraph plants, one a weakling and the other lame and halting'. '. . . This was the strength of the expedition', he wrote, 'that went out to meet the hostile shock of the world.' 'But I had no misgivings of the ultimate victory—I knew my Master and the flag he carried.'[43]

The tour was indeed a success, as reports from England and the United States began pouring into newsrooms in Calcutta. Sen's tour diaries to the *Modern Review* were undoubtedly the longest and most detailed accounts of the response that greeted Bose on this occasion. The solemnity, grandeur, and spectacle that surrounded each of Bose's public lectures, seemed to have acquired almost epic proportions in the eyes of this young disciple. Bose's reception at the Royal Institution evidently left the most vivid and indelible impression on his mind. The editor of the *Modern Review* took the cue and printed this part of the diary with a pictorial display of all the greats of Western science who had lectured there. With Raleigh and Tyndall, was a photograph of Bose resplendent against the imposing backdrop of the portals of the Royal Institution. In Sen's own rendition,

> Such is the great tradition of the Royal Institution that no recognition could be greater than to be asked to lecture before this Institute. . . . To be chosen to deliver the Friday Evening Discourse is regarded as an unique honour. The lecturer enters the historic hall and stands at the identical place occupied by Davy and Faraday and commences his

discourse without any introduction, for one who is asked to deliver the Friday Evening Discourse is well known to the scientific world and therefore requires no introduction. He begins his lecture abruptly without making any reference to the President or the audience. For it is understood that his message is to the world and not to any group or people. The audience at Albemarle Street is crowded by the aristocracy of intellect and culture. The street itself becomes almost impassable on account of the waiting carriages and motors. In order to avoid serious dislocation of traffic it is absolutely necessary that a lecture should begin at the stroke of nine and end with the stroke of ten. This tradition however was broken only once when the enthusiastic audience at the Royal institution demanded with one voice a longer duration of my Master's discourse.[44]

Conventions were broken on other occasions too, when scientists who had earlier been indifferent or even hostile to Bose's research went to visit the laboratory he had set up at Maida Vale in London. And this in general seemed to be the response everywhere. Sudhindra Bose, lecturer of Political Science at the University of Iowa, sent in some of the most interesting descriptive accounts of Bose's lecture tour to the United States soon after. The significance of Sudhindra Bose's account lay in its vivid descriptions of Bose's experiments, personality, and style. Sudhindra Bose's dispatches were as effusive as Basiswar Sen's, although unlike Sen, his account seemed to have been informed by the desire to present Jagadish Bose as an exemplary 'scientist of the East'.[45]

Upon his arrival in the United States, and even in the weeks before, Bose, according to this account, had become the subject of several magazine articles, newspaper editorials, cartoons, and poems. While some of these looked upon his science as novel and inspiring, there were others that were clearly amused. His visit to New York, for example inspired a creative soul in the *New York Times* to render his science into verse. Entitled 'The Song of the Sensitive Plant', this creative rendition of the theory of plant response ran as follows:

> Be kind to the hypochondriacal plant!
> It's nervous and lady-like qualms,
> Its delicate frailty you surely must grant,
> For it faints at all songs excepting psalms.

Speak low near a maiden fern! You will note,
It trembles with fright if you shout,
It flourishes best when you've got a sore throat'
And is pleased by the very devout.

Be firm with a rubber plant! Put it away,
When your friends come to make a call,
It is dreadful to find how a secret will stray,
When you thought no one knew it at all.

Be gay near Begonias! Their gorgeous array,
Betokens a sensitive blush,
To secure best results, a story risqué,
Will bring the desirable flush.

Then be good to the plants! For a great botanist,
Says their sensitiveness is intense,
They are shocked if a girl should chance be kissed,
And will die at moral offence![46]

Notwithstanding the note of condescension that underlay the humour, such responses were perhaps important as they spoke of the unprecedented levels of public interest that attended Bose's visit. The institutions where he was invited to speak, were indeed some of the best in that country. The long list included such prestigious names as the New York Academy of Sciences, the American Association for the Advancement of Science, the Brooklyn Institute for the Arts and Sciences, the Philosophical Society of Philadelphia, and the Universities of Harvard, Columbia, Iowa, Illinois, Chicago, Michigan, and Wisconsin.[47] One of the largest and most appreciative audiences that were to have greeted him was at the Cosmos Club in Washington City. Here the meeting was said to have commenced in the evening and carried on till late at night. Long before the scheduled time the big lecture hall was filled to capacity with the city's academic elite. In Sudhindra Bose's words,

> Prominent men and women were seen perched upon window sills or even seated on the floor. Dr. Graham Bell, the inventor of the telephone, came twenty minutes before time but the crowd at the hall was so large that he could not get within half a block of the hall. The enthusiasm of the indomitable inventor however was not chilled. In the following morning he was said to have called together a group of the noted

savants of Washington at his home in honour of the distinguished Hindu scientist.[48]

To Sudhindra Bose, there was something more than Jagdish's scientific experiments that made him elicit this kind of public adulation. It was the singularity of his personality and the style of his scientific presentations that struck all as being so distinctive.[49]

The general topic of Bose's lectures during this leg of his tour of the United States was 'plant autographs and their revelations'. The theme of plant 'autographs' in fact was for Bose one of the most creative evocations of the phenomenon of plant response. His biographer, Subrata Dasgupta, observed that shortly before the trip he had expounded on the phenomenon in a lecture to young students at the Presidency College.[50] In this he was said to have claimed that just as hand-writing could reveal the physiological and mental conditions of human beings, so could plant autographs reveal plant responses to various stimuli. Bose had worked upon this principle to devise a special kind of 'writer' for his plants. This in effect was an electromagnetic contraption that could induce the vibratory movement of the 'writer'. The 'writer' was then made to move at right angles to the surface of a sheet of paper attached to it and record the response of the plants in the form of a graphical representation. The device of the 'writer' was for Bose the most compelling source of evidence for his new experiment.

Commenting on the significance of the scientists' use of such elaborate modes of experimentation and demonstration, Steven Shapin in his observations of the career of the early modern English scientist, Robert Boyle, has shown how critical these were in the production of scientific fact. Scientific facts, he argued, acquired the status of facts when scientists were able to make use of three key technologies either separately, or in combination: a material technology, a literary technology, and a social technology. Material technologies are the various instruments used in an experiment both within the confined space of the laboratory and in larger lecture theatres. Literary and social technologies refer to all other aids that went into the enormously challenging exercise of converting 'passive audiences' into 'involved eye-witnesses'. The relevance of this observation bears out when the techniques that Bose used to persuade his audiences are placed under scrutiny. Each of Bose's lectures was illustrated with intricately crafted lantern slides, diagrams, and other visual aids. He was said to have used all these to demonstrate to his initially sceptical audiences that plants felt pain much in the same way as animals, and that the stimulus of motion to plants was of the same nervous character as in animals. All plants, he averred, were sensitive and in some of them there were tissues which beat spontaneously like the heartbeat of animals.[51] Moving on to an account of more dramatic aspects of the scientist's presentations, Sudhindra Bose wrote,

> Professor Bose is not an orator nor does he care to be one. He is simply a clear, forceful and convincing speaker. He walks slowly to the edge of the platform stands very still with his left hand behind him and looks at the audience for a full half minute. 'Ladies and Gentlemen', he says and plunges at once into his subject. In spite of the formal 'ladies and gentlemen' he is very informal—he does not orate, he talks. Robert Burns made poetry out of his works and days, Jagadish Bose finds a poem, a drama and an epic in his scientific researches. He is intoxicated with the fascination of his work. He speaks therefore out of the fullness of his heart.[52]

The aura of his presence was supposed to have been enhanced too by the extraordinary cast of his features. Sudhindra Bose described them with unreserved admiration.

> There is something peculiarly *attractive about* his personal appearance. His thick wavy hair that is tinged with grey has a tendency to project itself on either side of his massive forehead in poetic fashion. There is a bit of pride about his burning black eyes, which look life squarely in the face and challenge it. His face is the face of man sure of himself, the face of a high-breed, intelligent, confident, successful but not altogether satisfied man. Professor Bose has a strong physique and a slow resolute stride. He has a deep chest and broad shoulders. Yet he is not an athlete, everything about him suggests the student. Dr. Bose is pre-eminently a scientist. He is however not a scientist of that type that possesses a gelid [*sic*] intellect encased in an insulated covering. Profoundly intellectual that he is Dr. Bose is more than a thinking machine. He has a throbbing, feeling heart, he is human and very much so.[53]

The significance of this account perhaps lies in the fact that it sought to portray an image of Bose as a counterpoint to prevailing notions of the scientist in the West. This was the figure that in popular imagination inhabited the isolated world of the laboratory, thought in abstractions, and spoke a language that few understood. The contrasting image of Jagadish Bose that emerged from these accounts was of a scientist who was more 'human' and hence more accessible. He was in every sense a figure of infinite charm and warmth, and one endowed with the ability to make his audiences respond to his experiments as much with the heart as with the head. In fact, what seemed to be a common point of admiration for Bose amongst his Western audiences was the characteristic spirit that informed his researches. The *Scientific American*, commenting on the successes of Bose's researches, wrote that one is struck by its 'unique character where the synthetical methods of the East co-operated with the analytical methods of the West in a single mind'.[54]

This recognition of the singularity of his researches in a sense became for Bose the crucial point of contact from wherein he could address both specialists and lay audiences in the years to come. Critical to the elaboration of this idiom was Bose's metaphysics. Commenting on the significance of this on the overall spirit of Bose's research, Ashis Nandy wrote, the point to note about the second phase of Bose's researches was that from this time on his ideological concerns remained focused on 'the oneness of existence'.[55] The consequent attempt to ideologically tame and transcend individualism so as to instil in man the awareness of the ultimate unity between *brahman* and *atman* (essential reality of the self) often acted as the primary source of his support and originality. The consciousness of both the scientist and the poet he had once argued went out in the search of the inexpressible one. The difference lay in that the poet ignored the means that the scientist did not.[56]

This reference to the similarities between the creative impulse of the poet and the scientist, which Bose made in the concluding remarks of a lecture at the Mymensingh session of the Bengal Literary Academy or the *Bangiya Sahitya Parishat* in 1911, was articulated with greater enthusiasm to Western audiences during his fourth visit to Europe. The series of experiments that he took with him during this visit allowed him to elaborate in the most dramatic ways his new persona as the scientist of the East. The lecture at the Royal Institution in 1914 was resonant with the new note that he had struck.[57]

> I have shown you this evening the autographic records of the stress and strain in both the living and the non-living. How similar are the two sets, writing so similar indeed that you cannot tell them one from the other? They show you the waxing and waning pulsations of life and climax due to the stimulants, the gradual decline in fatigue the rapid setting in of death rigour and the toxic effect of poison. It was I who came on this mute witness of life and saw an all pervading unity that binds together all things—it was then that for the first time I understood the message on the banks of the Ganges thirty years ago—they who behold the one in all the changing manifoldness of the universe unto them belongs eternal truth, none else none else.[58]

With this speech it was clear that Bose had come to a powerful legitimating argument for his enterprise. He had broken out of the condition of being a 'mute spectator of life' and taken upon himself the task of being its messenger. It had now become imperative that his audience receive that message both cognitively and emotionally. The public lecture demonstrations of these years testify to this constant striving to get this message across.

The key component in Bose's style of scientific communication during these years was in effect a neatly evolved 'performance'[59] aimed not merely at the presentation of proof of his experimental findings but at mobilizing a series of symbolic effects in order to generate an active, participatory audience. The lecture demonstrations involved in his Fourth Deputation abroad had in fact already acquired the attributes of what in terms of such stylistic manoeuvres could be defined as a specialized 'cultural performance'.[60] In the first place, these demonstrated Bose's desire to communicate with his audience on multi-sensory levels by invoking both visual imagery and dramatic posture. The aim was not to merely tell them something but to lead them into experiencing it. From the first series of lectures on the electromagnetic waves to the later ones on plant response Bose strove to create an active, participatory audience for his science. But it was also clear that by the turn of the century he was keen to do a little more. Bose was by this time also eager to present a political statement through his work. Protagonists of *swadeshi* had already identified the larger significance of his work and Bose was now prepared to elaborate in his distinctive style.[61] On his return to India he devoted his energies towards creating that enduring 'frame' for his scientific 'performances' that would symbolically take his work out of the ambit of routine reality and place it in a realm of transcendent truths.[62] The research institute he was set to establish in 1917 in a sense marked the mature face of the strategy of communication he evolved over the years in order to consolidate the audience-base for his endeavour. From its early aims to draw his audiences into a cognitive and emotional understanding of his science, the mature phase of the Bose style aimed at creating an ideological constituency for it.

THE BASU VIGYAN MANDIR: CONSUMMATION OF A STYLE

It was evidently on a note of personal triumph that Bose set out to present to the Indian public the Bose Research Institute or the Basu Vigyan Mandir in November 1917. On this occasion, his idealized notions of science and its practice came to be most powerfully articulated, not merely through the formal procedures of language but through the far more visually evocative structure of the Institute itself. When it was finally complete in November 1917, he chose to dedicate it to the nation not as the expression of a sense of personal gratitude to his countrymen, but as a tribute to a spirit of scholarship that had in his mind inspired many of the nation's greatest thinkers and sages. He saw to it that the Institute represented that spirit not only in its commitment to the advance of scientific knowledge, but in its efforts to create those conditions in which this 'ancient' ideal of scholarship could flourish and permeate all institutions of learning in the country.

According to Ashis Nandy, Bose, inspired by the aesthetic sensibilities of his friend and associate Sister Nivedita, served as the main architect for the building of the proposed Institute.[63] For his designs, he was said to have borrowed heavily from Ajanta and Ellora. The Institute was meant to be a centre where students would flock 'to search the ultimate unity which permeated the universal order and cut across animal, plant and inanimate lives'. Called the Temple of Science or *Vigyan Mandir*, its most distinctive feature was the special *vedi* or platform in the garden on which scholars and students could meditate. Also in the garden were deer, peacocks, and birds of various hues. As emblems of the Temple, Bose selected sculptured representations of the Sun God (a symbolic pointer to the identification between *bhakti* or devotion and knowledge), a *vajra* or thunderbolt (the weapon of the Gods and the traditional symbol for legitimate fury), and the *ardha amalaka*, the Buddhist symbol of total renunciation.[64]

The largest of the many chambers of the Temple was the lecture hall, which could seat 1,500 listeners. On the rostrum facing the seats in the hall as well as on the ceiling, was a sculptured relief in gold, silver, and bronze of the sun-god rising in his chariot to embrace the cosmic rhythm of light and darkness. The body of the lecture hall was stark and bereft of ornamentation, but behind the rostrum and above the screen hung a most remarkable piece of allegorical painting by one of Abanindranath's most famous pupils, Nandalal Bose. It depicted, as Bose himself suggested, 'the idol of knowledge floating down the Ganges with the eternal woman beside him'—a representation of *shakti* inspiring *purusha*.[65]

All this was meant to create an ambience for scholars for whom certain moral qualifications were far more essential than academic attainments. 'A student using the Institute for his personal ambitions would have no place in it, for he would lower the ideals of others.'[66]

The principal objective of the institute was to create an atmosphere of

> complete detachment from the outside world with its numerous distractions. . . . Here the aim would be to encourage the student to learn to identify completely with nature and immerse himself in his quest. . . . It was not the excitement of publicity but a sequestered life that was the condition of success in great scientific achievement.[67]

In a speech delivered in Calcutta shortly before the final touches were being put to the plans for the institute, Bose proclaimed that there would soon rise a 'temple' where the teacher, cut-off from worldly distractions, would go on with his ceaseless pursuit of truth and, on dying, hand over his work to his disciples. 'Nothing would seem laborious in his inquiry, never was he to lose sight of his quest. For his was the *sanyasin* spirit.'[68]

The shorthand Bose used to describe it was the 'Eastern spirit' in which knowledge was pursued neither as a social attribute nor as a route to material advantage, but as a way that led to a deeper understanding of the self and of God. Knowledge understood in these terms had been cultivated in the spirit of a *sanyasi* or renouncer. One of the earliest occasions in which he spoke about it was in a speech he delivered before the Bengal Literary Conference in Mymensingh in 1911.[69] Here the 'Eastern spirit' was invoked in order to dissent against the norms of Western scientific practice, its fetish for specialization and its narrow utilitarian concerns. In his own words,

> You are aware that the prevailing tendency in the West at the moment is upon the excessive sub-division of learning. The result of this specialization is rather to accentuate the distinctiveness of the various sciences so that for a while the great unity of all tends to be obscured. Such a caste system in scholarship helps at first in the gathering and classification of material, but if followed too exclusively it ends by limiting the comprehensiveness of truth. . . . . The Eastern aim has rather the opposite, namely that in the multiplicity of phenomena, we should never miss their underlying unity. After generations of this quest, the idea [comes] to us almost spontaneously and we apprehend no insuperable obstacles in the path of its attainment.[70]

Crucial to the Eastern quest for unity was the cultivation of a strict code of discipline and morality and scholarship untainted by the ambition of material gain.

> The greatest wealth of external appliances was of no avail whether there was not a concentrated pursuit utterly detached from personal gain. Those whose minds wander hither and thither, those who hunger for public applause instead of truth itself, by them the quest is not won. To those on the other hand who do long for knowledge itself, the want of favourable conditions does not seem the principal obstacle. . . . In the first place we have to realise that knowledge for the sake of knowledge is our aim and that the world's common standards of

utility have no place in it. The enquirer must follow where he is led, holding the quiet faith that things that appear today to be of no use, may be of the highest interest tomorrow. . . . It is necessary then the enquirer and his disciples should work on ceaselessly undeterred by years of failure and undistracted by the thunder of public applause. We may one day come to realise that India in the past has shared her knowledge with the world and we may ask ourselves, is that destiny now ended for us? Are we to remain debtors only? Perhaps when we have felt this new Nalanda may arise ![71]

The speech at the Bengal Literary Conference set the tone and agenda of Bose's subsequent addresses on 'true' requirements for scientific research in India. For him the necessary conditions for success in scientific research related less to the attainment of material conditions than to the moral attributes of the researcher. This was what the Eastern spirit of scholarship, exemplified in the attainments of Nalanda, had taught and this was what had to be revived if India was to take her rightful place in the scientific world. The Institute he was about to dedicate to the nation was meant to be the clearest statement of his commitment to this ideal.

The opening of the Bose Research Institute on 30 November 1917 was a great cultural event in Calcutta. Going by the names of the principal donors, it was evident that the idea of the Institute had already acquired the support of the city's rich and famous. One of the greatest enthusiasts of the project was Rabindranath Tagore, Bose's old friend, philosopher, and guide. Bonded by mutual admiration and empathy, the two had come to share a common vision of national regeneration. Though unable to be present on the day of the inauguration, Tagore went on to compose a memorable song for the occasion, a tribute to Bose himself and the spirit which inspired his research, the song was one of the finest expressions of Bose's vision of the Institute. The following translation from Tagore's original composition in Bengali was published in the *Modern Review* as part of its extensive and in-depth coverage of the event. In the poet Manomohan Ghose's rendition, it ran as follows:

'Tis to the Mother's temple ye are come
Her sacred inner courtyard; light ye then
Her precincts, ye who are favoured sons,

Make here your dwelling; and with omen fair,
The conch-shell horn auspicious sound

Say, 'Victory to this man of man,
This kingly sage, school'd in austerities,
And victory! Still shout victory, victory!'
Come with the mother's blessings ye, whose minds,
Unshakeable!

Throne on the thunderbolt,
Come, all that struggle upward and aspire,
O glorify this dear country come!
Come all ye, who meditating, on one thought,
Come all ye, who have renounced,
Come all ye, whose lot insufferable is woe,
Come ye, whose earned wealth is unconquered strength,
Come, brotherhood of freedom in the soul
Come ye, who know, come ye who would destroy,
The long shame of Bharatland
Come O thou blessedness, thou glory come,
Thou fragrance of unfading righteousness
Come burning sun, blazing amidst the sky of deeds,
In strength of virtues' heroism and righteous acts, live thou,

. . . . . .

Say, 'Victory' to this peerless man of men,
This kingly sage, school'd in austerities,
And 'Victory still shout victory, victory!'[72]

Tagore's composition set the tone and ambience of the inauguration of the Vigyan Mandir. The association of the institute with the sacred space of a temple, the infusion of research with the values of righteousness, virtue, and heroism, and the patriotic vision of national glory through scholarly attainment were to be the central themes of the founder's dedication. The ethos of the Vigyan Mandir came to acquire its particular rhetoric from the aesthetic idealism that came to inform contemporary nationalist attempts to recover the identity of other forms of 'Indian' creativity. Here Bose was evidently indebted to Sister Nivedita with whom his association was a long and productive one.[73] Apart from her being one of the most serious and enthusiastic interlocutors of Bose's science, Sister Nivedita was a friend of Bose and his wife, one who enjoyed a special presence in their lives. The ideas and experiences that the two shared with her was to have a series of significant cultural consequences. Sister Nivedita's ideas on Indian art and aesthetics, for instance, would have a profound influence on Bose's own sensibilities. The decision to cast the Vigyan Mandir both architecturally and ideologically in the mould of the ancient universities of India was influenced in large measure by a visit to these places in the company of Nivedita and Tagore.

Patrick Geddes observed the significance of these visits for Bose in these words,

> Nor is it to be wondered at that among the universities of Taxila and again among the Nalanda he should feel that it was not only their old university spirit thrilling within him but the common spirit of all universities. These visits peculiarly awoke and strengthened in him the perception that his life-work was to be more than one of personal purpose and scientific character—more than the organization of a physical laboratory and what he must henceforth aim at, work for should be nothing less than the recreation of some yet fuller centre of intellectual quest and diffusion like those of old. . . . In this old pride of India as she was and hopes of her as she may be, on one hand, no less than his peculiarly full and wide participation in Western science on the other, we see at once the two uniting forces which found expression in the founding of the Bose Research Institute.[74]

For her vision of Indian art Nivedita drew upon the views of Swami Vivekananda as they were first expressed in the Paris Universal Exhibition of 1900. The Swami's exposition of Indian art, writes Tapati Guha-Thakurta in her study of the ideology of nationalist art, 'had entered the main current of the Orientalist interpretations by rejecting the model of Hellenistic influence and exalting early Buddhist art of India as the greatest in the world'. For Vivekananda as well as for Sister Nivedita, she argues, 'projecting the independent evolution of Indian art through an exclusive inner domain aesthetic idealism became an important theme of nationalist assertion'.[75] This specific search for an 'Indian' art evidently informed by an Orientalist spirit animated Bose's endeavour in the realm of science. Here his points of reference were drawn from both Sister Nivedita and Abanindranath Tagore. The similarities between the two were unmistakable.

Abanindranath Tagore, one of pioneers of the Bengal School of Painting, was the most easily recognizable counterpart of Jagadish Bose in the realm of arts. The interpretation of an Indian art tradition that he provided in a mood of assertiveness was informed by similar Orientalist assumptions of a fundamental divide between East and West. His approach to tradition too was firmly rooted in nationalism. He believed that preserving his identity as an Indian artist entailed the retention of an organic link with past traditions, even as he continued to move toward new directions. And as Bose chose the metaphor of the 'eternal flame of thought' that had been handed down to each generation, Abanindranath used the metaphor of the 'perennial flow of a river' which could never sever its bond with its original source, nor ever cease to surge forward.[76] The similarity of rhetoric was also evident in their perception of their practice as religion. Abanindranath's definition of art and the artist much like Bose's definition of science and the scientist drew heavily upon the mystic terminology of *dhyana* (meditation) and *sadhana* (austere dedication).

> It is only through a lot of disciplined worship, devotion and edification that one can become an artist. . . . True, that the brush or the compass are important instruments (towards achieving this end; but these are useless unless one can learn, at the same time the deeper secrets and formulae (mantras) of their utilisation. (*Shilpe Bhakta Mantra*)[77]

These precepts had a direct resonance in Bose's formulation of the ideal conditions necessary for scientific practice.

In carrying out research, Bose asserted, there were other difficulties beside the want of well-equipped laboratories.

> We often forget that the real laboratory is one's own mind. The room and the instruments only externalize that. Every experiment has first to be carried out in that inner region and to keep the mental vision clear, great struggles have to be undergone.[78]

Bose evidently drew a large measure of these values from Rabindranath, who gave them new life through his writings on the 'ashram ideal' and its institutionalization within the nationalist schemes of education during the *swadeshi* years. His numerous essays on the goals of education pointed to the same need to revive the spirit of indigenous scholarship with values of detachment, self-abnegation, and discipline. 'Struggle', 'renunciation' and 'restraint' through the rigours of *brahmacharya* were themes common to most of his writings during this time. A modern education could be given a national character if it was pursued with commitment to an indigenous ethic of scholarship.[79]

Whether or not Bose actually adhered to the ethic of renunciation in his professional or private life is open to debate. But his much-publicized struggles were indicative of his predicament as a scholar while his choice of the icon of the 'renouncer' or *sanyasi* as moral exemplar was proof of his commitment to this ideal. It reflected his concerns about both the requirements of scientific research in the country as well as larger and perhaps more fundamental preparation for the struggle against colonial rule. Asceticism as an ideal was as essential to scientific scholarship as it was to the cultivation of a larger spirit of resistance. The Vigyan Mandir was meant to convey this deeper nationalist impulse. It signified Bose's own perceptions of his style of scientific practice as well as the ideal he held out for the Indian scientist.

CONCLUSION

The Vigyan Mandir in many ways marked the consummation of the Boseian style. For what essentially linked the stylistic elements invoked in 1897 at the Royal Institution with the Fourth Deputation abroad in 1914 and the establishment of the Vigyan Mandir in 1917 was the invocation of a conscious act of 'performance'. Bose's style when conceived in these terms brings out the complexities of the communicative strategy he evolved to forge a cognitive and emotional connectivity with his audiences as participants, observers, eye-witnesses, and, above all, as allies.

In 1897 Bose had taken upon himself the near impossible task of making visible to his audiences what was essentially invisible to ordinary sensory experience—the optical properties of electric waves. The apparatus he had with him was evidently inadequate and he was compelled to persuade his audience to use their 'imagination'. The language at this point was unusually literary, a first sign of his defiance of convention and the glimmerings of own distinctive style.

Yet if the term 'convention' alludes to any standard developed in the West, a closer look at the evolving modes of scientific communication in Britain and Europe shows that the use of literary tropes in scientific discourse was common practice to the end of the nineteenth century.[80] Although looked upon with increasing reproach by the new standards of professionalization, Bose, it could be argued, was well within Western convention in his choice of literary and dramaturgical discourse in his public lecture demonstrations. Studies of the complex nature of the relationship between the lecture-demonstration and the playhouse show that already in the seventeenth century, Francis Bacon had warned of as the dangers of the 'idols of theatre' in connection with experimental science. Established systems of philosophy, he argued, represented 'worlds of their own creation after an unreal and scenic fashion'.[81] Surrendering to the lure of narrative, simplicity, and dramatic spectacle that characterized theatre, traditional philosophy subordinated critical curiosity to wonder. Although Bacon's aim was to free experimental enquiry from this taint, 'the idol' it has been shown was not as easily dislodged as he wished. In the eighteenth century, much scientific activity moved into settings characteristic of what Jurgen Habermas has called the emergent 'public sphere', including coffee shops, public houses, recreational resorts, and assembly rooms. In many of these settings experimental demonstrations continued to invoke the model of the playhouse with its ambivalent moral implications. Critics were often keen to label such scientists and their interlocutors as vulgar projectors or wonder-mongers. Controversy surfaced repeatedly about the displays of electrical effects as lecturers were accused of making natural phenomena appear miraculous. It seemed as if the effort to display experimental knowledge publicly was inherently compromised by the lure of dramatic spectacle.[82]

In Britain, the issues surrounding the theatrical form of public science took on a strongly political tone after the outbreak of the French Revolution. In a climate of intense political polarization, public science was subjected to unprecedented strains that continued to shape it in the following century. On the one hand, scientific showmanship reached across to broader audiences among the working classes with a programme that included phrenology and mesmerism. On the other hand, a more specialized experimental science emerged within the newly formed institutions, exploiting technically advanced instrumentation.[83] This specialist science was nonetheless still oriented to a public audience—the upper and middle-classes that were viewed as the potential patrons of research. Humphrey Davy's stage-managed and rhetorically crafted displays at the Royal Institution in London invoked a new model of theatre. Davy's successor at the Royal Institution, Michael Faraday, however, maintained a firm demarcation between the places of public work and public demonstration. He made creative use of transitions between the two spaces to consolidate his experimental findings and to communicate them. David Gooding's studies have shown that the movement from the laboratory to the theatre strengthened Faraday's own understanding of electromagnetism and helped persuade other physicists that he was correct. Gooding shows Faraday working to magnify phenomena while masking the labour involved in their production, so as to make nature speak directly to his audiences. Gooding remarks,

> To become accepted as part of the scientist's experience [a natural phenomenon] must be transferred from the personal realm to the public domain, where it can be reproduced and witnessed by all. . . . Faraday was good at moving his discoveries from the personal domain of the contingent to the public forum of the demonstrable and self-evident.[84]

From this perspective, Bose's predilection for the public lecture, the staged experiments and the evidently theatrical mode of his presentations appear less unconventional. The internal logic of his research interests, the unavailability of advanced technical instrumentation, and the imperative of securing patronage from both the colonial government and his immediate social context drove him towards a mode of communication that demanded the active involvement of his audiences. In his investigations into the optical properties of electromagnetic waves and his experiments with plant response, he publicly replicated and presented his findings to a range of sceptical audiences. The impulse was not merely to announce his findings but to draw in his audiences into the entire experience of conducting an experiment and recording its results.

The experiment had to be performed on more than one level. It had to communicate knowledge to audiences who were to be eyewitnesses to phenomena inaccessible to the normal range of sensory experience. Bose's instruments were part of the panoply of literary and material technologies deployed to ensure effective communication of not merely a scientific fact but a scientific experience.

It is in the context of this impulse to communicate the experience of doing science (rather than convey its purely factual content) that one can understand the significance of positing the Vigyan Mandir as a 'ritualized space' devised for the pursuit of knowledge and also for the contemplation of truth. Going by the care and effort devoted to its architectural design it becomes evident that the entire attempt was to 'frame' the experience of research in the laboratory, and also within a larger culturally coded domain of spiritual contemplation, the temple. Since the real world of scientific practice was rarely experienced as a coherently ordered totality, the research institute constructed through the stylized temple was meant to provide the experience of a substitute reality in which the scientist could pursue knowledge in a spirit of self-imposed solitude.

The Vigyan Mandir, in other words, sought to project an idealized totality within which the scientist was meant to pursue his research. For Bose, to understand nature one had to commune with nature. The research institute had to create the optimum conditions for this communion. The physical plan of the institute was such that the laboratory would imperceptibly lead into the garden—thus both in natural surroundings and in the laboratory, plants would write their autographs and in turn their autobiographies. In such an environment the student, free of all distractions, would observe the 'panorama of life' unfold and come to realize how 'community outweighs apparent dissimilarity'.[86] Bose's elaboration of the image of the solitary student in communion with nature was a deliberate act to draw public attention to the singularity of the scientist's pursuit. The solitude of the scientist drew him closer to the realms of truth.

The posture of solitude, in other words, was also an act of performance directed towards eliciting a particular kind of public response.[87] In the context of the moral idealism of *swadeshi*, this anti-materialist ethic embodied in the ideals of the Vigyan Mandir had a particular emotional appeal. Seclusion, asceticism and a moral pursuit of truth were ideals held up for the development of national character. The experience of doing science within the confines of the Mandir was projected to be the way to achieve that political/pedagogical goal.

Bose's active search for a public may in the final analysis be regretted, but in the constraining circumstances of his scientific career his attempts are worthy of commendation. The production of scientific knowledge in poorly equipped laboratories and in the face of uncertain financial resources may have been a self-defeating enterprise had Bose not actively courted an audience. The effort to look beyond the confines of the laboratory and create a public sphere of science had of necessity to be part of a scientist's brief. The singularity of Bose's style of communication must be seen and appreciated in this context rather than through an explanatory framework that chooses to regard it as a sign of the scientist's eccentricity or as an expression of some abiding Eastern disposition that failed to read the conventions of the West.

NOTES

1. One of the most useful surveys of the various approaches to the study of scientific communication is available in Jan Golinski, *Making Natural Knowledge: Constructivism and the History of Science*, Cambridge, 1998, pp. 47–66, 79–91, 103–19, 133–45.
2. This perspective on Bose's style is elaborated in Subrata Dasgupta, *Jagadish Chandra Bose and the Indian Response to Western Science*, Delhi, 1999, pp. 55–9.
3. Ibid.
4. Golinski, *Making Natural Knowledge,* pp. 79–91.
5. Ibid., pp. 86–91.
6. Ashis Nandy, *Creativity and Authenticity in Two Indian Scientists*, Delhi, 1980.
7. Ibid.
8. Ibid.

9. Dasgupta, *Jagadish Chandra Bose.*
10. Ibid.
11. Ibid.
12. Ibid., p. 45.
13. For a detailed study of Father Lafont's scientific activities, see Arun Kumar Biswas, *Father Eugene Lafont of St. Xavier's College and the Contemporary Science Movement*, Calcutta, 2001.
14. J.C. Bose, Speech to the students of Presidency College, Calcutta, 1928, reprinted as 'The Uphill Way', in D. Sen and A.K. Chakrabarty, eds., *J.C. Bose Speaks*, Calcutta, 1986, pp. 38–43.
15. Ibid.
16. Dasgupta, *Jagadish Chandra Bose*, p. 50.
17. Ibid., p. 51.
18. Not merely a scientific but a great public event, widely reported and circulated as part of Bose lore. See Anil Chandra Ghosh, *Vigyane Bangali*, Dhaka, 1931, pp. 44–5; J.C. Bose, 'Telegraphy Without Wires', in *Abyakta*, Calcutta, 1921, pp. 57–61.
19. Dasgupta, *An Indian Response*, p. 56.
20. Ibid.
21. Ibid., pp. 5–8.
22. Ibid., p. 58.
23. Ibid., p. 59.
24. Though initial researches on *Plant Response* had began much earlier (around 1900), this was the first time that Bose decided to put together all his experiments and findings in a coherent thesis.
25. Dasgupta, *Jagadish Chandra Bose*, pp. 192–3.
26. Ramananda Chatterjee, editor of *Prabasi*, hailed it as the greatest *swadeshi* event of that year. 'Not the breaking of bangles, or the burning of Manchester cloth, nor even the establishment of National Universities, the greatest *swadeshi* event of the year is the publication of Acharya Jagadish Chandra Bose's *Plant Response*. It is unfortunate that none of the major papers have cared to talk to about it.' *Prabasi*, *Bhadra,* Calcutta, 1913. Contrary to Chatterjee's allegations, however, *The Statesman* did carry a feature on the book on 1 July 1906. In it said, '[I]t is seldom that a work so revolutionary in its attitude and scope as this new publication of Professor Bose makes its appearance in the world of science. For the great bulk of scientific works in all periods consist and must per force consist in the testing and verification of observations and predictions made by earlier investigators. The birth of epochs in Science as in history is rare. This being so, Calcutta all the more to be congratulated upon the issue from her university of a book whose claims are of the present order.' Quoted in Dibakar Sen, ed., *Patrabali, Jagadish Chandra Bose*, Calcutta, 1994, p. 134.
27. J.C. Bose, *Plant Response*, Calcutta, 1906.

28. Bose had initially solicited Tagore's help to coin Sanskrit names for his new apparatus. On the latter's inability to help him, Bose fell upon his own resources, but soon realized that the effort was futile. In his own words, 'initially I thought I would call the crescograph *briddhiman*, but it didn't happen that way. When I first decided to name my instruments in Sanskrit I didn't realise how difficult the task of propagating *swadeshi* could be'. First there were the British newspapers which greeted the new names with derision. They were grotesque and absurd. It was only the *Boston Post* which was appreciative. In its editorial it said that the scientist who devised his instruments had the right to name them. And if hitherto all scientific apparatus derived their names from Latin or Greek roots, why couldn't an ancient language like Sanskrit provide another source? The *Boston Post's* approbation notwithstanding, for Bose it was clear that his Western audiences would never get those names right. Referring to an incident where a famous American scholar from a famous American University urged Bose to speak on his instrument the 'KANCHAN-MAN', Bose wrote ruefully that this was proof of the futility of the entire exercise. It was easier to make Hirnaykashipu chant the name of God than make the Westerner pronounce a name in Sanskrit. 'Our *Hari*' he added, 'will forever be condemned to be Harry!' See Dibakar Sen, *Patrabali*, p. 55.
29. Dasgupta, *Jagadish Chandra Bose,* p. 118.
30. Ibid., p. 210.
31. Ibid., p. 211.
32. From the 'Voice of Life', Bose's dedicatory address at the inauguration of the Bose Research Institute in 1917, reprinted in several publications including Amal Home, ed., *Acharya Jagadish Chandra Bose Birth Centenary, 1858-1958*, Calcutta, 1958 pp. 56–67 and D. Sen and A.K. Chakrabarty, eds., *J.C. Bose Speaks*, Calcutta, 1986 pp. 7–20.
33. To most European admirers Bose had, in the 1920s, evolved from mere 'scientist' to venerated 'Hindu Savant'. Reports in French newspapers, for instance (often reprinted in the *Modern Review*—hereafter *MR*) carried this title. *MR*, February 1923, pp. 230–2; *MR* July 1926, p. 87, etc.
34. The material context of Bose's science and the chronic financial and infrastructural constraints have been documented and though not always clearly stated, it becomes evident that the public lecture demonstration was a means by which Bose could address potential patrons. An active and participatory audience in this context meant not merely those who could cognitively access his science, but some who could support it financially.
35. This is a point which has been made repeatedly by historians of early modern science who have illustrated, with the help of a range of case-

studies, the intellectual, financial, and social constraints that compelled scientists to move their place of production from the laboratory to the public sphere.

36. For a discussion of the specific nature of the public lecture demonstration, see H.M. Collins, *Changing Order: Replication and Induction in Scientific Practice*, Beverley Hills and London, 1985.
37. See for instance the compositions of Tagore, *Vigyan Lakshmir Priya Paschim Mandire, Acharya Jagadish, Joy hok taba Jay* collected in Pulinbehari Sen, ed., *Chithipatra, Santiniketan*.
38. Ramananda Chatterjee, editor of *Modern Review*, was foremost among them. He was a student of Bose at Presidency College. Recalling his days as student there, he wrote, 'When Jagadish Chandra Bose became Professor of Physical Science at Presidency College, Calcutta in 1885, I was a student of that college. My classmates and I had the privilege to attend his lectures and see his experiments—his lectures inspired many of his students with enthusiasm for science. So far as I am concerned, he helped me with some valuable contributions in Bengali and English and with affectionate encouragement and guidance.' Ramananda Chatterjee, 'Kasthipathar', *Prabasi*, 1932, cited in Shanta Devi, *Bharat Mukti-Sadhak Ramananda O Ardha Satabdir Bangla*, Calcutta, 1950, pp. 19–20.
39. Ramananda took special interest in this tour as was evident in the kind of space reserved for its coverage, never missing an opportunity to garner and print every bit of information pertaining to the trip, from tour diaries to photo features, from newspaper clippings and notes to elaborate editorial comment.
40. The idea of a tour diary kept by a student was perhaps encouraged by Chatterjee in the hope that it would provide an informed account of an event loaded with political significance.
41. Basiswar Sen's, 'Round the World with My Master' was published serially in *MR* between 1915 and 1916.
42. Ibid., *MR*, April 1915, pp. 81–5.
43. Ibid., p. 82.
44. Basiswar Sen, 'Round the World with My Master II' *MR*, March 1916, pp. 356–7.
45. Sudhindra Bose, 'Professor Jagadish Chandra Bose in America', *MR*, May 1915, pp. 559–64.
46. Ibid., p. 560. Notwithstanding this apparent trivialization of Bose's science, there was enough reason for Bose to believe that his American audiences were impressed by his experiments on plant response. He is said to have conveyed this to a senior civil servant at the India Office. 'The Americans regard my methods to be of the highest importance in practical and theoretical science. They say that since the establishment of the (Washington) Academy of Sciences they have never had another

lecture of equal importance.' (See Dasgupta, *Jagadish Chandra Bose*, p. 226.)
47. Ibid., p. 559.
48. Ibid., p. 562.
49. Ibid., p. 564.
50. Dasgupta, *Jagadish Chandra Bose*, p. 211.
51. Steven Shapin, *Scientific Revolution*, Chicago, 1996, pp. 106–17.
52. Sudhindra Bose, 'Prof. Bose in America', *MR*, May 1915, p. 561.
53. Ibid., p. 562.
54. Cited in 'Notes', *MR*, July 1915, pp. 1–2. The image of 'Scientist of the East' stuck with Bose through his life. While admirers like Ramananda Chatterjee resented this dubious distinction for its racial overtones, audiences in the West waxed eloquent on the new spirit he brought to bear upon the practice of Western science. A report from the *Fortnightly Review* reprinted in the *MR*, December 1928, pp. 564–5 said, 'In Bose is seen an invincible perhaps immortal quality which has given permanence to Indian civilization such as no other nation has approached. In Sir Jagadish, the culture of thirty centuries has blossomed into a scientific brain of an order we cannot duplicate in the West. We find in him a spiritual sense, difficult to define, intangible, yet evident permanently of the East—the quality out of which all great faiths have grown. . . .'
55. Nandy, *Alternative Sciences*, pp. 58–9.
56. Ibid.
57. Speech delivered at the Mymensingh session of the Bengal Literary Conference in 1911. This remains the first cogent statement of Bose's singular engagement with his practice. He used his salutation as President of this session of the Congress to articulate his views on the pursuit of scientific knowledge as a spiritual/aesthetic experience akin to the writing of poetry. For a collection of his popular scientific pieces written in the poetic diction, see *Abyakta* (The Unmanifest), Calcutta, 1921.
58. Nandy, *Alternative Sciences*, p. 59.
59. For a brief but useful discussion of 'performance' theories see Catherine Bell, *Ritual: Perspectives and Dimensions*, New York, 1997, pp. 72–6, 159–64. For a more original perspective see Victor Turner, *The Ritual Process: Structure and Anti-Structure*, London, 1969; *From Ritual to Theatre: The Human Seriousness of Play*, New York, 1982.
60. Bell, *Ritual*, pp. 160–1.
61. See reference above to Ramananda Chatterjee's description of the publication of *Plant Response* as the greatest *swadeshi* event of the year 1906.
62. For a perspective on the 'framing' of performances see Bell, *Ritual*, pp. 161–2.

63. Ashis Nandy, *Alternative Sciences*, p. 80.
64. Ibid.
65. Calcutta Municipal Gazette, December 1928, p. 164. Commenting on the theme Nandy writes, 'obviously this was compatible with the traditional Indian concept of *purusha*, the ultimate male as the passive creator and *prakriti* as the active participant in the processes of creation'. Nandy, *Alternative Sciences*, p. 80.
66. Cited in *Application from Dr. J. Bose, Education Proceedings, Simla Records* 3, p. 16.
67. Ibid.
68. Quoted in Anil Chandra Ghosh, *Vigyane Bangali*, Dhaka, 1931, pp. 26–30.
69. Substance of the Presidential address delivered in Bengali by Prof. J.C. Bose, to the Bengal Literary Conference at Mymensingh, 14 April 1911, *MR*, May 1911, pp. 515–25.
70. Ibid., p. 516.
71. Ibid., p. 517.
72. Published in Pulin Behari Sen, ed., *Chithipatra*, vol. 6, pp. 65–6, tr. Manmohan Ghosh.
73. Bose is said to have first met Sister Nivedita near the end of 1898. She was horrified to find those about him who took every opportunity to prevent Bose from pursuing his work. She was amazed, she wrote, to see this great scientific man working so entirely alone. Soon after she got to know him, she became in effect his editorial assistant, working with him on most of his publications till 1911. See Dasgupta, *Jagadish Chandra Bose*, p. 76.
74. Patrick Geddes, *The Life of and Works of Sir J.C. Bose*, London, 1920, p. 135.
75. Tapati Guha-Thakurta, *Making of a New 'Indian' Art: Artists, Aesthetics and Nationalism in Bengal, c. 1850-1920*, Cambridge, 1992, p. 173.
76. Ibid., p. 201.
77. Ibid., p. 202. This reconstruction of a traditional Indian aesthetic, argues Tapati Guha-Thakurta, coincided with the Romantic theories of art and artistic creation. They now acquired an Indian pedigree with a base in ancient Sanskrit treatises. The recurrent theme of the primacy of the mind over sight shaped the artist's entire approach to a painting in contrasted values of feeling versus form, emotion *versus* skill.
78. There were echoes of this Romantic theory of creative expression in Bose's understanding of the practices of science. The most eloquent defence of this Romantic view was made at the Presidential Address to the Mymensingh Literary Conference in 1911. Two years after this seminal address, similar ideals were expressed in an article written for

the *Bengal Educational Journal* (1913). In it Bose said, 'It is knowledge itself and not the accessories of knowledge, fame, comfort and power that should be all in all to him (the scientist). The seeker of knowledge must fix his eyes on the distant goal of truth itself.' Reprinted in D. Sen and A.K. Chakrabarty, eds., *J.C. Bose Speaks*, pp. 1–4.

79. Tagore's essays on this subject, include 'Shikhsha-Samasya', *Rabindra Rachanavali* (hereafter *RR*), Birth Centenary Edition, vol. 11, 1961, pp. 559–72; 'Tapoban', *RR*, vol. 11, pp. 589–96; 'Asramer Shiksha', *RR*, vol. 11, pp. 711–15.
80. See the discussion of 'Darwin's Metaphors', in Golinski, *Making Natural Knowledge,* pp. 124–5.
81. Ibid., pp. 92–3.
82. Ibid.
83. Ibid.
84. David Gooding, 'In Nature's School: Faraday as an Experimentalist', in D. Gooding and A.J.L. Frank, eds., *Faraday Rediscovered: Essays on the Life and Work of Michael Faraday, 1791–1867*, London, 1985, pp. 105–35.
86. Dasgupta, *Jagadish Chandra Bose*, pp. 230–1.
87. In the words of Steven Shapin, '[S]olitude was often an intensely public pose . . . it made no sense without that public audience and its meaning depended on a publicly understood stock of evaluations.' 'The Mind Is Its Own Place: Science and Solitude in Seventeenth Century England', *Science in Context*, no. 4, 1990, pp. 191–218.

SELECT BIBLIOGRAPHY

Bell, Catherine, *Ritual: Perspectives and Dimensions,* New York, 1997.

———*Biographical Memoirs of the Fellows of the Indian National Science Academy*, Delhi, 1976.

Bose, D.M., S.N. Sen and B.V. Subarayappa, eds., *A Concise History of Science in India*, Delhi, 1971.

Bose, Jagadish, Presidential Address at the Bengal Literary Conference, Mymensigh, 14 April 1911, rpt. in *Modern Review*, May 1911.

———'The Voice of Life', *Modern Review*, December 1917.

———*Abyakta*, Calcutta, 1921.

———'Life and its Mechanism', *Modern Review*, December 1924.

———'The Unvoiced Life', *Modern Review*, December 1925.

Bose, Sudhindra, 'Professor Jagadish Chandra Bose in America', *The Modern Review*, May 1915.

Chowdhury, Indira, *Frail Hero, Virile History: Gender and the Politics of Culture in Colonial Bengal*, Delhi, 1997.

Colby, R.C., G.N. Cantor, J.R.R. Christie and M.S. Hodge, *Companion to the History of Modern Science,* London, 1990.

Collins, H.M., *Changing Order: Replication and Induction in Scientific Practice*, Beverley Hills and London, 1985.

Das, R.K., 'Sir J.C. Bose in Europe', *Modern Review*, October 1926.

Dasgupta, Subrata, *Jagadish Chandra Bose: An Indian Response to Western Science*, Delhi, 1999.

Geddes, Patrick, *An Indian Pioneer in Science: The Life and Work of Sir Jagadis C. Bose*, London, 1920.

———'The Bose Research Institute Revisited', *Modern Review,* December 1922.

Ghosh, Anil Chandra, *Vigyane Bangali*, Dhaka, 1931.

Golinski, Jan, *Making Natural Knowledge: Constructivism and the History of Science*, Cambridge, 1998.

Gooding, David, 'In Nature's School: Faraday as an Experimentalist' in D. Gooding and Frank A.J.L. James, eds., *Faraday Rediscovered: Essays on the Life and Work of Michael Faraday 1791–1867*, London, 1985.

Guha-Thakurta, Tapati, *The Making of a New 'Indian' Art: Artists, Aesthetics and Nationalism in Bengal c. 1850–1920*, Cambridge, 1992.

Home, Amal, ed., *Acharya Jagadish Chandra Bose Centenary Volume*, Calcutta, 1958.

Magitot, Dr., 'The Hindu Savant', *The Modern Review*, February 1923.

Mukherjee, Visvapriya, *Jagadish Chandra Bose*, Delhi, 1983.

Mukhopdhyay, Ashim Kumar, *Colonialism, Nationalism, Scientism: A Study of Acharya Jagadish Chandra Bose*, Calcutta, 1995.

Nandy, Ashis, *Alternative Sciences: Creativity and Authenticity in Two Indian Scientists*, Delhi, 1995.

Prakash, Gyan, *Another Reason: Science and the Imagination of Modern India*, Delhi, 1999.

Sarkar, Sumit, *The Swadeshi Movement in Bengal, 1903-1908*, Delhi, 1973.

Sen, Basiswar, 'Round the World with My Master I and II', *The Modern Review*, April 1915 and March 1916.

———'The Bose Research Institute', *Modern Review*, December 1933.

Sen, D. and A.K. Chakrabarty, eds., *J.C. Bose Speaks,* Calcutta, 1986.

Turner, Victor, *The Ritual Process: Structure and Anti-Structure*, London, 1969.

——— *From Ritual to Theatre: The Human Seriousness of Play*, New York, 1982.

CHAPTER 11

The World of an Engineer: M. Visvesvaraya and Irrigation Engineering in Twentieth-century India

RAJ SEKHAR BASU

In India, technologies of hydraulic manipulation have been practised over the centuries. These had evolved as responses to the material developments and the changing needs of society. The continuous search for appropriate technologies have given rise to highly skilled as well as scientific understanding of both land and water use.[1] More recently, it has been argued that hydraulic practices in the Indian subcontinent spanned three distinct and overlapping phases. In the earliest times, tanks, inundation canals and temporary bunds had constituted the operational mechanism of water harvesting.[2] Nonetheless, the shift from traditional to modern irrigation in the days of the Raj had been made possible through the efforts of the scientific and bureaucratic establishment seeking to disregard people's knowledge as unscientific and backward.[3] Satyajit Singh has pointed out that the entire domain of irrigation practices during the colonial period was dominated by a civil engineering bias, which had a contempt for all other methods of science and technology.[4] In early nineteenth-century India, colonial policies initiated a sharp break in technique and hydraulic principle through the introduction of perennial canal irrigation. In a period of little more than a hundred years, the Royal Engineers constructed a series of barrages and weirs across riverbeds, thereby diverting the flow of the river water into intricate and extensive canal systems. Such irrigation works operated through a system of shutters, making it possible to regulate flows during lean and rainy seasons. Rohan D'Souza describes this transition from tradition to a canal-dominated system as the advent of modern irrigation in South Asia.[5] The years between 1892 and 1947 witnessed an increase from 20 to 28 million in terms of acreage irrigated by wells and tanks. The colonial bureaucracy continuously advocated the need for an administrative system capable of mobilizing large amounts of money and manpower, which could all be utilized for the construction of large-scale water works, reservoirs and centres of power production. Big dams and proliferating system of canal networks also seemed to have caught the imagination of Indian technologists working under British superiors in the irrigation and water supply departments. Such an ambience enabled a group of British and Indian technologists to outmanoeuvre the achievements of the great emperors of the past.[6]

In the colonial era, the new profession of hydraulic engineering was nobody's monopoly.[7] M. Visvesvaraya was perhaps the most important personality, who influenced the builders of irrigation projects. He was both an engineer and economist. He tried to draw the attention of people from the purely political wrongs to the fundamental economic wrongs. As one of his biographers has observed, 'He is no standard bearer of lost causes but born to give the sluggish masses a good shaking.'[8] It was natural for man of such proclivities to devise irrigation schemes that would turn the arid tracts into basins of vegetation. The colonial irrigation schemes aiming to utilize the water flow of the Godavari and the Kistna rivers largely influenced Visvesvaraya's ideas on irrigation engineering, water supply, and drainage—all of which later found the fullest expression in the construction of the Krishnarajasagar reservoir in Mysore. The civil engineering concepts guiding the Godavari system had led Visvesvaraya to believe that scouring sluices and regulators could be used into facilitate the flow of water into the main irrigation canals that made their way into the interior. The masonry work revolving around the construction of walls and revetments seem to have greatly influenced him. At the same time, the plans of Sir Arthur Cotton and Sir Atwell Lake to construct an anicut at

Bezwada, about 60 miles from the mouth of the Godavari River, influenced his idea of the economic potential of large-scale irrigation works in India.[9]

In this paper, I explore the wide range of strategies and tactics employed by Visvesvaraya during the first three decades of the twentieth century. I argue that his achievements as a civil engineer should not be seen only as the influence of Western hydraulic engineering, guided by the motive of marketability of water resources. His involvement with large-scale irrigation works spread over different parts of the country reveals the fact that Visvesvaraya was a technologist who attempted to combine the modern with the traditional and differentiate the pan-Indian from the regional. Like his English predecessors, he believed in the financing of public works through loans and finances from governmental and non-governmental sources, but was not totally oblivious to the realities of the Indian rural scenario. He showed flexibility and moderation in his approach while expounding his ideas on the 'block system of irrigation'. He was able to convince the peasantry of the Bombay Presidency that this was a sound project, which would distribute the benefits of irrigation over a large number of villages and also concentrate irrigation in each village within specified zones determined by topographical and ecological factors. While developing such ideas, Visvesvaraya did emphasize on the applicability of traditional skills and methods.[10]

In his later days, irrigation figured as a major subject in Visvesvaraya's discourse on planned economy. It is this aspect that has made him a controversial figure. Critics argue that Visvesvaraya's long association with the feudal ruling groups and the bourgeoise accounted for his myopic views of rural economy and development. His irrigation schemes came to be looked upon as attempts aimed at the privatization and commercialization of natural resources. Such criticism holds some validity, as these schemes proved to be beneficial to an emerging rich peasant class. As with the case of canal irrigation, the construction of weirs also led to a strengthening of the nexus between the state mechanisms, landholding groups, and the class of rural creditors. Seen in this light, the achievements of Visvesvaraya might be laudable from the point of view of engineering sciences, but hardly the same from the context of equality and justice.

Mokshagundam Visvesvaraya was born in the small village of Muddenahalli in the Kolar district of the Mysore State on 15 September 1861. Visvesvaraya was much indebted to his maternal uncle for his education and subsequent career. Born into a family rich in scholarship, but poor in terms of material wealth, Visvesvaraya's life as a student was filled with struggles. In 1875, he moved to Bangalore for higher studies and joined first the Wesleyan Mission High School and later the Central College. When Visvesvaraya passed his B.A. examination in 1880 his Principal certified him as a 'Capital Mathematician and a very good English Scholar'. Subsequently, he secured a fellowship to study engineering in the College of Science, Poona. There he stood first among the successful candidates in the examination held in 1883. This success fetched him the James Berkeley Prize and also an appointment as Assistant Engineer in the Bombay Public Works Department. He began his official career in 1884, at the age of twenty-three. In 1908, he rose to the position of Superintending Engineer, the highest post open to Indians at that time. In fact, during these years he worked in several places, planning and constructing irrigation works.[11]

In the early days of his career as Assistant Engineer, Visvesvaraya incurred the displeasure of his superior, the Executive Engineer. The trouble began over the execution of an engineering project during the heavy monsoon season. Visvesvaraya thought that the execution of the project would be a wasteful expenditure. The Executive Engineer disagreed and felt that his subordinate lacked energy and was disobeying his orders. Faced with this challenge, Visvesvaraya completed the work in time, whereupon the same Executive Engineer advised him to appear for a practical examination, much before the usual time required for that purpose. Visvesvaraya appeared for the examination and was successful, and soon promoted to the first grade.

Subsequently, he was posted in Poona and was given the charge of maintaining buildings and roads. In 1893, he was sent to Sukkur in Sind to execute water supply and drainage schemes. This tenure enabled him to make a place for himself. On the completion of his mission, he earned encomiums from Lord Sandhurst, the Governor of Bombay as the 'most able engineer' the project could ever have had.[12]

However, Visvesvaraya's career took off only when he was reposted in 1899 as the Executive Engineer of the Poona Irrigation District. At this time, he was faced with the challenge of winning public support for a system of regulated supply of irrigation water without the prevalent wastage. The concept of rationing of water on a ten-day basis was bitterly opposed by the peasants, who were inspired by politicians like Lokmanya Bal Gangadhar Tilak. The Marathi newspaper *Kesari* run by Tilak launched attacks against him. In order to diffuse the tension, Visvesvaraya convened a conference at the Fergusson College to explain his scheme of rationing of irrigation water. He pointed out that the large-scale extension of irrigation in provinces other than Bombay had resulted from the demand for irrigation for grain and other staple crops. Such crops, it was pointed out, were ordinarily grown in the Deccan on rainfall, and other than the years of scarcity did not require irrigation facilities. However, Visvesvaraya stated that the system of revenue management devised in Bombay by the Government of India was based on conditions prevailing in the large irrigated tracts of the United Provinces of Agra and Oudh, Punjab, and elsewhere. The application of the same system for the sake of uninformity and administrative convenience was believed to have been the main reason behind the retarded development of irrigation in the Bombay Presidency.[13]

Visvesvarya argued that there was a fairly steady demand for irrigation for a few selected classes of crops in the Bombay Presidency. In fact, crops like sugar cane, which required capital manure and enterprise could not be grown over a large continuous area like wheat or rice, year after year, and without resting the land or practising some kind of crop rotation. Thus the cultivation of sugar cane and other high-class crops in the Deccan needed a system of rules and management specially adopted for their development. Visvesvaraya did not think that the small tanks in the open plains of the Deccan would be of much use. They would not fill except in years of exceptional rainfall and even if full would not hold water long during the fair season. He proposed to 'have storage tanks along the course of a canal to be filled, as required, from the canal'. He proposed to link this scheme with that of his block system of irrigation.

The cardinal principle in the management of irrigation works in the Bombay Presidency, according to Visvesvaraya, was 'protective', the aim was to protect the standing crop in the area under command against serious losses during seasons of deficient rainfall. He argued that the object to be kept in view in the management of irrigation was to get the best return every year, and not to treat the works as mere reserves for the protection of ordinary crops in seasons unfavourable for agricultural operations. The block system of irrigation proposed to distribute the benefits of an irrigation work over a large number of villages and to concentrate the irrigation in each village within blocks of specified limits and on selected soils. The total area of the blocks in each village had to be large enough to enable everyone 'able to grow an irrigated crop to have a share, but not too large to constitute a surfeit, or lead cultivators to neglect the advantages of water supply in good seasons'.[14]

The block system proposed that one-third of the area in each block was to have sugar cane or other perennial crops at a time, and in the remaining two-thirds either vegetables or a monsoon or *rabi* crop could be grown up to the end of February. After February, water would be given for only one-third of the area, that is, for the area on which a perennial crop was grown. Thus a sort of triennial crop rotation was proposed in the case of each block. Visvesvaraya stated that water rates had to be charged on the basis of a fixed rate per acre, irrespective of the crop sown. There would also be a general limitation that only one-third of the area would receive water throughout the year. The rate was fixed at between Rs. 8 and 15 per acre, according to circumstances. There was also a proposal that excess water in the lower reaches of the canal would be diverted at the end of the *rabi* season for the cultivation of crops outside the blocks. It was believed that such an arrangement would stimulate irrigation from wells.[15]

Visvesvaraya placed his ideas before the cultivators of the Poona district at a conference. These had been expounded earlier in his depositions before the Irrigation Commission, and he invited the village communities to choose their areas and regulate the distribution of water equally to each block.[16] The cultivators approved the system, but favoured the idea that the irrigation works should be managed by the government rather than by themselves. The cultivators adapted to the rules issued by the Irrigation Department, but there are reasons to argue that grave maladies were inherent in Visvesvaraya's proposed scheme of block irrigation system.

Visvesvaraya's ideas on the block irrigation system need not be evaluated from the standpoint of an engineer interested in providing sufficient water for irrigation during the agricultural season to the peasant communities in India. His ideas seemed to have been heavily influenced by the logic of economics and management, as well as technological methods that emphasized on the preservation of water.[17] The block system of irrigation was essentially a system that had been borrowed from the American experience. It operated on the premise that the canal officer would have a thorough knowledge of the area and the time within which it could be watered through discharge. There were proposals to introduce a flume and a self-recording water register in all outlets, which could not be inspected more than once a day. This system, however, had several disadvantages. First, the failure on the part of the cultivators to utilize water for the whole area of the blocks in any one rotation could result in a short supply to the crops during the next rotation, as their negligence had led to wastage of water. Second, the cultivator had to submit water applications for each crop and until he obtained permission he would not be given the right to use water. In fact, a cultivator could prepare land for irrigation and then find that there was no water because others had forestalled his efforts to obtain a supply from the watercourse. In any case, this system, with its emphasis on a host of bureaucratic tangles, could never be a step towards equitable distribution of resources to rural communities as whole. It actually led to unequal distribution and greater economic inequalities. Visvesvaraya's scheme undoubtedly suited the interests of the rich peasantry, who used their superior monetary resources and their closer proximities with the bureaucracy to deny the other sections of the peasantry of their legitimate rights over the usage of water.[18]

Visvesvaraya seemed to have been oblivious to the demands and aspirations of a vast section of the rural society. He failed to realize that the availability and supply of water was as much an issue for the poor peasants as it was for the rich. He believed that capital investment was perhaps the only guarantee for a regular supply of water, and the poorer cultivators who could not obtain capital were left with no chance of irrigating their fields in a year of plague or famine.[19] In other words, the block system of irrigation had very few regulations to prevent growing inequalities in the rural society. It remained essentially an arrangement for safeguarding the interests of the rich farming lobby, wholly engaged with the production of commercial crops.[20]

Visvesvaraya's ideas on wastage of irrigation water were based on the statistics provided by the Irrigation Department of the Bombay Presidency. He argued that in Bombay only about 33 per cent of the area estimated to be annually irrigable was actually irrigated, while in other provinces, except Bengal, the corresponding percentage was 80 or more. It was pointed out that in the Bombay Presidency, perennial irrigation was wasteful. The area of perennial crops was about a tenth of the total irrigable area, and was often not more than a thirtieth of the area estimated as irrigable. This meant that water flowing in the channels over 30 sq. miles was used to water an area of 1 sq. mile.[21] This peculiarity had long been a subject of enquiry: as early as 1893, engineer R.B. Buckley had observed,

> The right conclusion to be drawn from the extremely low duty of the canals generally was that very large quantities of water were lost or wasted between the head sluice and the fields. This is a result which is not uncommon in Bombay where the water is often led for considerable distance from the source of supply along shallow channels to irrigate fields scattered widely over the district; this is the chief explanation of the very low duties which are obtained in that province. . . . Irrigation under such conditions must be wasteful of water.[22]

In Visvesvaraya's own reports, these ideas were replicated. He stated that much water was wasted at the site of distribution. He observed, 'Each irrigator works for himself and often times against his neighbour. The waste causes loss to government of water, which might produce revenue, and is harmful to the cultivators as it leads to the over-saturation of the soil. It affects the climate making it colder and damper.'[23] Visvesvaraya therefore, argued that if water was given by rotation and by measurement, such waste could easily be checked. The proposed block system would act as an automatic check against waste and misuse. This idea of saving water through careful regulation was a product of sheer hard-core economic thinking. It was stated that the preservation of water would obviously lead to an extension of the area of valuable crops grown from year to year. With such an assured area, the government was in a position to reduce the rates and yet derive large returns. There was an expectation that revenue would be considerably larger than what was obtained in a situation when no such conditions of water regulation and management existed.[24]

Visvesvaraya also proposed that the blocks be revised once in six years and the system could eventually become permanent like the *thal* system in Nasik and Khandesh.[25] He felt that if it satisfied all parties and interests, the government would be assured of a sufficient return covering the interest on the capital invested. On the basis of the statistics provided by the colonial bureaucracy, there was an assumption that the flow in the rivers and channels was principally during the monsoon and *rabi* seasons. It was calculated that while works with a precarious supply were able to earn Rs. 5 to 8 per acre and with a good hot weather supply as much as Rs. 16 per acre, the revenue derived from water rates on new irrigation (first class) works did not frequently cover the expenses of maintenance and organization. Visvesvaraya pointed out that in the second-class works, where the distribution of water was done by the irrigators themselves, there was a constant yield of revenue. Thus, these and the smaller village works had to be put into a thoroughly efficient mode to increase the prosperity of the country.

A thorough study of the block system for the Nira Canal[26] at Poona also reveals several interesting details. The formation of blocks and the allotment of the block areas necessitated the appointment of two officers, one from the Public Works and the other from the Revenue Department, in the rank of Assistant Engineer or a Senior Sub-Engineer and a Mamlatdar. This committee had to be assisted by a *panch*, comprising three persons from the village and two private gentlemen of standing, and interested in agriculture within the taluka. This committee had to work for three or four months in the first year of settlement.[27]

Visvesvaraya favoured several proposals related to the functioning of the block system for the Nira canal. He stated that the village would have to be treated as a unit, and the area to which the cultivators would be entitled had to be fixed once in six years (or during the full term of a land revenue settlement, as the case may have been). This fixity of area was thought to be a distinct gain to the village as a whole. The whole village could be called upon to lay out the allotted area in blocks of not less than 60 and not more than 300 acres each. It was argued that if the cultivators succeeded in distributing the area among themselves and submitted statements of distribution signed by all the applicants for shares, the distribution of the village would be settled without the intervention of the government, though subject to the approval of the Collector and the Canal Officer.[28] Visvesvaraya strongly stated that the land falling under the command of the canal had to pay a protective tax known as owners' rate in some parts of India. He observed, 'At present water is given on yearly leases and the cultivators have no claim for water any more than Government has to compel a landowner to irrigate his land, or to make him pay a water rate for the privilege of using canal water whenever he wants it.'[29] The distribution of areas was made on the spot by the committee and the *panch*, who took into consideration the just claims of people who had shown enterprise in developing irrigation in the past.

However, difficulties arose from the decision to allot irrigable lands on the basis of the functioning of the block system. In the first place, a man who has lands within the block will usually be entitled to irrigate a small portion of it. On most occasions, he did not favour the idea of handing over the remaining area to others. Second, all the farmers who had held irrigated lands during the three years preceding the introduction of the system would not always be in a position to obtain land within the block. Whatever may have been the case, two classes of cultivators had to be dealt with. While there were individuals who were unwilling to let their lands get into the hands of others, there was a second category who were unwilling to apply capital and manure for the improvement of lands within the blocks, so that they qualified for the use of canal water.[30] Such differences were not taken seriously and there was an expectation that the inclusion of land in a block would lead to an instant increase in its value so that the owner would benefit from the transfer of his land on lease or in exchange for land outside the blocks. It needs to be explicated that though there might have been a willingness on the part of the cultivators of any particular area to allow others to irrigate their lands on payment of reasonable rent or *khand*, the situation was no less discriminating that what it actually seemed to be. There was every possibility of rich peasants within the blocks demanding unreasonable terms. This situation seemed to run to the advantage of the capitalists, who without owning lands were greatly involved in large-scale sugar cane cultivation.[31]

Visvesvaraya felt that inclusion of any land in a block would lead to an enhancement of its value and the owner would benefit by giving it on lease or exchanging (combined with a small rent) it for land outside the blocks. As irrigation would promise better returns, the farmer would find it more profitable to lease it to others, who would be entitled to irrigate it rather than to grow dry crops on it themselves. In other words, it was assumed that the general willingness of the cultivators of any particular area to allow others to irrigate their lands on payment of a reasonable rent or *khand* was a factor in determining the positions of the blocks.[32]

There were also proposals to meet the demands of the people who could not obtain lands within the blocks. The one way of doing it was to select a block area 25 or 50 per cent larger than the actual area required, and then to realign them to the smaller limit after every applicant had found space. The object was to prevent owners within the blocks from demanding exorbitant or unreasonable terms. However, such a plan might have benefited a section of the peasantry, but not the rural classes as a whole. There was very little opportunity for ordinary peasants to exchange their lands in return for a small rent or to acquire lands by paying rent for a term of six years. The situation was heavily tilted towards the fortunes of the capitalists who organized sugar cane cultivation on lands taken on lease from others. At the same time, the authority that had been vested in the *panch* could lead to greater divides within the rural society. In the name of excluding unwilling people and including willing people, the legal claims of the ordinary peasantry could very often be overlooked. The committee assisted by the *panch* would in almost every case decide all matters relating to allotment by keeping the richer section of peasantry in mind.[33]

Significantly, the block system of irrigation approved by the Indian Commission of Irrigation, encountered opposition from a wide section of the society. It encountered opposition from the revenue officials, European and Indian, from the District Collector down to the village Munsiff. The vast majority of the rural classes were opposed to it, fearing a loss over their rights on land. The scheme was only supported by the rich peasantry of Bombay Presidency, which had expectations of profits from their alliances with the sugar barons.[34] In 1908, the Bombay government stated that the block system was a 'complete success' and its development had been due entirely to the genius of M. Visvesvaraya, one of the ablest officer of the Public Works Department. Yet the fact remains that despite the novelties in the scheme, the government failed to notice the glaring rural distress that could arise. In a sense, rural development as envisaged by Visvesvaraya could lead to a situation where a direct correlation would be established between prosperity and distress.

Visvesvaraya's block scheme was followed by his efforts to introduce certain new concepts of civil engineering in the case of the Nira canal and the Lake Fife, both of which were situated close to Poona. The Nira canal had been constructed to provide water for irrigation in the eastern part of the Poona district where rainfall was scanty and uncertain. The works comprised of a storage reservoir about 30 miles south of Poona and a canal on the left bank of the Nira River, whose total length was over 101 miles. The canal head works were located at Vir, 10 miles east of the Poona–Satara Road. The storage reservoir was formed by damming the Yalvandi River, a tributary of the Nira, just above the junction of the two rivers at Bhatghar. The object of the reservoir was to store water in the monsoon in order to feed the canal in the fair season when river supply was deficient or had failed altogether. The discharge of the river was usually abundant till the end of the monsoon, but diminished rapidly between October and December, when it had to be supplemented by the storage reservoir. In fact, from the month of January to the beginning of the following season, the canal was almost entirely dependent on the storage for its supply.[35] The reservoir was formed by a dam of masonry and concrete 3,020 ft long. The dam was 127 ft in maximum height above the lowest foundation and 103 ft above the riverbed. There were two waste weirs or waste sluices, one at either end of the dam. The sluice openings were 10 ft wide, the piers 3 ft thick, and the gates impounded water to a height of 8 ft above the sill. The gates used were Reinold's patent roller gates, which were suspended by chains.[36] The gates were actuated by counter-weight floats working in a chamber in the body of the weir wall which enabled them to open automatically when water rose to full supply level and close as soon as water fell below level. With a rise of water level 8 ft above sill, there were calculations that there would be a discharge of maximum flood of 56,457 cu.ft per second.

The principal food crops in the area served by the canal were raised in the *rabi* season. Experience showed that there would be a demand for water from 20,000 to 80,000 acres of *rabi* crops after a normal monsoon. After a deficient monsoon, the demand for water would be very large, but the water supply would not ordinarily suffice for more than 60,000 acres. By 1900–1, sugar cane cultivation which had temporarily fallen off since 1895–6 owing to the low price of 'gul' or 'raw sugar' was regaining its earlier position. There was an opportunity for the expansion of sugar cane cultivation, but as the *rabi* crops required water, not more than 4,000 acres of sugar cane could be watered at a time. In 1899–1900, the assessed revenue from sugar cane amounted to Rs. 86,774 and the actual revenue for *rabi* crops was Rs. 43,199. The balance of Rs. 18,556 represented water rates from all other crops and other miscellaneous items of revenue.

Visvesvaraya stated that if the canal was worked on existing lines, the ordinary growth of net revenue would amount to 2½ per cent on the outlay within 1911–12. But if irrigation was developed with an eye to increase the revenue, the irrigation work had to be made pliable for the payment of interest charges on the capital outlay. However, this could not be done without increasing the sugar cane area and withdrawing water supply from food grain crops.

Visvesvaraya was more emphatic in his report related to the arrangement of automatic gates for Lake Fife. The automatic gates planned by Visvesvaraya were employed on a large working weir at the Bhatghar reservoir. The partial balancing of one gate by chains attached to the next functioned smoothly and when the gates opened one moved down and the other moved up, making it possible to reduce the weight and number of the counter-weights. The weight of the heavy gate opened the water way and the weight of the counter-weight closed. Visvesvaraya suggested that as one gate had to be heavier than the other, all gates had to be as large as possible and of the same size. However, the main feature of the proposals comprised the construction of cisterns for the counter-weights. The cisterns were viewed as separate structures on the downstream side of the weir and for blasting out a channel in the rock. The different cisterns were to be connected by means of a pipe leading to the sluice valve. Such an arrangement would lead to discharge below the level of the counter-weight, when the gates would be opened. Viewed from the point of hydraulic engineering, the arrangement was admirable for its simplicity and the attention to the protection of the pipeline connecting the cisterns. There were plans to construct the pipes in trenches, where rocks would be filled in with concrete. In other words, the security of the cisterns was given a great deal of importance.[37]

A detailed study of Visvesvaraya's project for automatic sluice gates for Lake Fife waste weir reveals several interesting points. The gates patented by Reinold were to work in pairs. They were to be suspended by pulleys and the members of each pair were to partially balance the other. The falling gate was to be the heavier of the two and would weigh about 10,000 lbs and the rising gate 4,000 lbs. Each pair of gates were to be closed by hauling up the heavier gate by means of a chain attached at the other end to a line of chains and wire ropes, moving on pulleys or rollers. The counter-weight was to be formed by a hollow watertight cylinder 15 ft in diameter and 6 ft in height. The action of the counter-weight on the gates was to be dependent on the function of an inlet channel from the reservoir and an inlet pipe that was to be laid at the bottom of the cistern. The outlet pipe from each cistern was to be throttled by a sluice, which was to be regulated in such a manner that flow from the pipe was always less than that into the cistern through the inlet. When the water would rise to the full supply level in the reservoir, it would begin to flow into the cistern through the masonry inlet, gradually immersing the counter-weight. The latter was to float, the tension on the chains attached to the falling gates would be slackened, and the gates thus released would go down by their own weight. At the same time, when the water in the reservoir fell below the inlet level, the supply to the cistern was to be cut-off and it was to be gradually emptied by the outlet pipe. The counter-weight would then regain its weight and as it would fall the heavier one of each pair of gates were to be pulled up and the lighter one was to fall by its own weight. These would lead to the closing of the sluices.[38]

These schemes proposed by Visvesvaraya were not totally influenced by the logic of hydraulic engineering. While Visvesvaraya pointed to the increase of discharge capacity and increase of storage, the financial implications of such projects were also taken into consideration. He observed, 'The true test of the value of an irrigation work is its capacity for providing water for irrigation in a year of scanty rainfall. Whenever rainfall is favourable, there is also a corresponding falling off in the demand for water for irrigation. The stability of irrigation depends on the absence of interruption or contraction of irrigation in a year of drought.'[39] It was calculated that the use of automatic gates would lead to a great increase in storage. The total storage was estimated to be 362.57 m.cu.ft, equivalent to a continuous supply of a little over 17 cu.ft per second for eight months. Since there would be no appreciable increase in evaporation and leakage, the additional storage was proposed to be utilized for sugar cane cultivation. The net revenue to be earned from sugar cane cultivation was calculated at Rs. 27,200, which figure represented a return of nearly 7.10 per cent on the estimated works outlay, or 5.6 per cent on the total capital outlay of Rs. 489,301.[40] What he failed to perceive was that the projected rate of irrigation charges (fixed at Rs. 40 per acre) was well beyond the reach of the ordinary peasant of western India. In fact, though the use of new techniques was a novelty, the efforts to meet a part of the project costs proved to be ruinous for a vast section of the peasantry. In a sense, the entire idea of commercialization of water resources as envisaged by Visvesvaraya had all the potentials of a situation where land alienation and rural indebtedness would be part of a reality.

In 1904, Visvesvaraya was elected member of the Institute of Civil Engineers, London. He also attended the Irrigation Conference held at Simla and submitted four papers. In the same year, he was appointed Sanitary Engineer to the Bombay government and was the first Indian to hold that office. In 1908, at the request of Lord Morley, Secretary of State, he was sent on deputation to Aden to prepare schemes there for water supply and drainage. At the instance of Gokhale, Visvesvaraya was also involved with the Poona Municipal Sewage Scheme.[41]

Visvesvarya resigned from the service of the Bombay government in 1908. The Government of Bombay granted him a pension in view of his exceptionally meritorious services. In his long stay of nearly fourteen years in Poona, as a government servant Visvesvaraya had opportunities of coming in close contact with leading personalities such as Mahadev Govind Ranade, Gopal Krishna Gokhale, and Bal Gangadhar Tilak. He did not take part publicly in politics, but he did contribute to the journal of the Poona Sarbajanik Sabha. With the support of Mahadev Govind Ranade, Visvesvaraya secured the historic building, 'Hira Baug', for starting a modern club where club members could meet, discuss, read, and resolve their differences. This club was established in 1891. Ranade and Gokhale served as members of its managing committee and Visvesvaraya was its Joint Secretary till 1894.[42]

After his resignation from the Bombay government Visvesvaraya decided to spend a couple of years in travel abroad with a view to study infrastructural developments that could be applied in India. He visited Italy, Sweden, Denmark, Holland, England, Russia, Canada, and America. He visited farms and factories, irrigation and drainage works. Those visits widened his perceptions on development. However, his plan to spend about two years abroad was interrupted by an urgent invitation from Hyderabad. Nawab Ali Nawaz Jung, himself an eminent engineer, invited Visvesvaraya to take up the office of the Special Consulting Engineer and save the city from the flood waters of the Musi River. Visvesvaraya accepted the offer, but stated that he could only take it up following his European tour. When he did not receive a confirmation he wrote back to the Nizam's government rescinding his consent. Soon negotiations were resumed and Visvesvaraya demanded a salary which in those days was given only to Britishmen.[43] Visvesvaraya took up his new post in April 1909. After collecting the necessary documents, he prepared a scheme to protect the city from floods and another for the drainage of the city.

The flood of 28 September 1908 had been the most destructive witnessed in Hyderabad for over three quarters of century. The river rose 15 to 20 ft high in the inhabited area, inundating houses and property for an average width of half a mile.[44] Four bridges were over topped and their parapet walls carried away. The Purana Pul had its approaches cut open but suffered very little damage otherwise. The Musallam Bridge and the Afzal Bridge suffered the most. The flood also uprooted the masonry flank wall of the weir across the entrance to the north arm of the river round the Imli-bun Island. It washed away the surface soil over a portion of the island and obliterated the head wall and a portion of the irrigation canal at the anicut. This led to the demolition of houses on the north and south banks of the river.[45]

The heaviest rain had occurred at Shamsabad in the neighbourhood of two of the largest reservoirs in the basin, namely the Palmukul and the Parti. They had a combined storage of 392 m.cu.ft up to full supply and probably 800 m.cu.ft including the surplus on the day of the breach. The bursting of these reservoirs made a very serious addition to the water flow and proved to be the last straw on the back of the riparian camel.[46]

The great flood at Hyderabad occurred not only because of

the quantum of rainfall, but also due to the fact that a number of tanks had burst simultaneously. Tanks pouring large and uncertain quantities of water into the river was a new liability to be reckoned with in the solution of the problem of flood prevention. Visvesvaraya suggested several policy changes. He favoured a proposal to enlarge the river channel so that, with a depth nowhere exceeding 25 ft, it would carry a flood equivalent to a runoff of ¼ inch per hour (or 1,39,000 cusecs) from the entire catchment. There were also proposals to construct two storage reservoirs for storing the excess discharge in a flood, one on each branch of the river, at distances of over 6 and 3 miles respectively above the city. The capacity of one of them (Easi reservoir) for storing floodwater would be 8,379 m.cu.ft and that of the other (Musi reservoir) 5,360 m.cu.ft—in total, 13,739 m.cu.ft. This was equivalent to a runoff of nearly 6.9 inches from 862 sq. miles, the entire catchment above Hyderabad. It was felt that as the uncontrolled area below the dam sites might give an extra half-inch runoff per hour on occasions, while the reservoir sluices were discharging the full quantity of 1,39,000 cusecs, a margin of about 23,000 cusecs had to be allowed in the capacity of the river channel. The maximum capacity to be provided in the river channel through the city was taken to be 1,62,000 cusecs. The storage space favoured was equal to about 25 per cent of the monsoon rainfall. This was an extremely wide margin of safety and was felt to be important for reservoirs intended to protect any congested city from floods. This entire idea of storage was borrowed from the flood prevention schemes that had been devised for the Passaic Valley near New York.[47]

Visvesvaraya argued that the prevention of floods by means of a controlling reservoir was possible in the case of Hyderabad because of the limited catchment and the existence of suitable reservoir sites in the valley. Influenced by Sir Vernon Harcourt, the author of *Rivers and Canals*, Visvesvaraya stated that the reservoirs were not likely to become economically applicable for dealing with floods until their cost were reduced by favourable local conditions. This implied that the reservoirs had to be employed for supplying irrigation canals and water to the towns.

Visvesvaraya believed that larger the storage reservoir, the cheaper it would be to store water.

The Musi River Valley was thought to be ideally suited for the development of irrigation. It was more likely to be used for irrigation purposes than any other river in the Hyderabad Dominions. There were 17 *sarf-i-khas* and *jagir* anicuts on the river below Hyderabad and below these again, two government canals, the upper and lower Rajakalwa, watered a large part of Nalgonda district. These ancient channels crossed the watershed of the Musi and discharged their surplus waters into the Krishna River.[48]

Visvesvaraya pleaded that a complete irrigation scheme required reservoirs for storing the water and canals for its distribution. In the previous schemes the provision had, however, been made only for the former. By opening the reservoir sluices water could be made to run into the river and be utilized on the numerous anicuts as well as the lower Rajakalwa canal. This last canal had a carrying capacity of 395 cu.ft per second at the head. Since the existing works were greatly in need of storage, there was no immediate necessity for providing new irrigation canals. Thus it was suggested that for the next decade or so an additional sum of Rs. 10 or 15 lakh had to be expended for utilizing the storage facilities fully.

It was estimated that the storage for which the provision was made in the reservoirs would be sufficient to irrigate an area of 50,000–60,000 acres. This was seen as an improvement, since the largest irrigation system in the state did not irrigate more than 10,000 acres. Visvesvaraya entertained the idea that by undertaking such large protective irrigation works, a measure of famine protection could be made available to the people. These schemes would be of great benefit to famine tracts, which usually suffered from a chronic deficiency of rainfall.

As a civil engineer of great standing, Visvesvaraya also offered several new suggestions to the Nizam of Hyderabad. He suggested a lowering of the high anicut at Imli-bun and widening of the river channel along the right bank of the river, almost immediately above the anicut. He argued that the construction of the anicut had been responsible for frequent overflows and the formation of a new channel and an island. In the high floods, the river waters flowed in the direction of Chandarghat and this had led to the erosion of a portion of the bank. Visvesvaraya emphasized that an enlarged section, which would be given to the river on the south arm, would be more than sufficient in controlling future floods.[49]

A deeper look into Visvesvaraya's scheme for flood abatement and drainage for Hyderabad city brings several issues to the forefront. While it was argued that reservoirs were necessary to protect human lives and property from the flood, there were several other reasons as to why its applicability was desired. Viewed from common sense, it would seem that water would be stored in the reservoirs during high floods and then released gradually, so that the river channel at no time was overburdened. At the same time, improvements of riverbeds and tanks were believed to be necessary for increasing the carrying capacity of the channel to the maximum level. Remodification of the drainage works were suggested because water had already been laid into the city and there were no means of removing it. This led to liquid filth and choking of earthen drains. Thus the new drainage schemes were envisioned with an aim to arrest the pollution of the sub-soil and for reducing the cost of conservancy arrangements. But for Visvesvaraya the economic implications of the new irrigation schemes were no less important. He pointed out that the budgetary outlay on irrigation storage could lead to a return of over 6 per cent per annum. In this connection, he stated that it had to be investigated as to whether *sarf-i-khas* lands and private *jagir* lands could receive water from the new reservoirs. As an ardent advocate favouring commercialization of natural resources, Visvesvaraya favoured rules and an organizational setup that would facilitate the introduction of a water rate in some parts of the state. He strongly argued for a water rate imposed on all classes and that the use of excess irrigation water would obviously lead to greater charges.[50]

In the case of Mysore, Visvesvaraya proposed several new schemes from the point of view of drainage and sanitation. He prohibited the use of pit and well privies and suggested measures for the daily removal of nightsoil from all occupied areas. It was also suggested that household liquid waste from latrines, washing platforms, baths, urinals, and kitchens needed to be collected in impermeable cisterns and channels, which would lead to a gully trap outside the house. In the houses of the poor, all these channels would be open and in the houses of the rich, pipe drains were believed to be more suitable. Visvesvaraya also clearly stated that the existing open drains in the most crowded areas of the city had to be replaced by underground sewers. He emphasized that a comprehensive project of urban development could only be prepared by an Improvement Trust Board under the advice of the Sanitary Board. It was also imperative to have a report on the programme, as it would bring out the reasons for selecting the works.[51]

Visvesvaraya left the Hyderabad service in November 1909. He was soon invited by the then Dewan of Mysore, V.P. Madhava Rao, to accept the office of the Chief Engineer of the Mysore State.[52] Initially, Visvesvaraya was reluctant to accept this offer. He had no intention of taking up routine service under any government. He enquired whether the Mysore government would be interested in initiating large schemes of technical education and industrial development. He wanted his services to be utilized for such purposes. After receiving an assurance he joined the Mysore service as Chief Engineer in November 1909.

In pursuance of the assurance given to him as a condition of his appointment to Mysore service, the Government of Mysore appointed two committees. While one of them was to look into technical education, the other would be exploring the opportunities for industrial development. Visvesvaraya was appointed chairman of both these committees. In his addresses to several bodies between 1910 and 1912, Visvesvaraya outlined his plans of development for Mysore. With the help of statistics, he showed how and in what respects Mysore and India lagged behind countries like Japan, Britain, and the United States.[53] He also pointed out that moral and material efforts were necessary to forge ahead in agriculture and industry, education and administration, planning and execution.

As Chief Engineer, one of Visvesvaraya's first important duties was to take up the task of constructing a dam across the Cauvery River at Kannambadi to provide irrigation and electric power. Based on studies of the Aswan Dam[54] in Egypt, as well as other experiences, he drew up a project to suit local conditions. However, Visvesvaraya faced difficulties in executing the project both from within and outside the state. Some of the senior officials of the Mysore government criticised the large expenditure that had to be incurred on the project. In fact, the Maharaja was persuaded to hold up the project. Subsequently, Visvesvaraya proceeded on short leave. He was later requested by the Maharaja Krishnaraja Wadiyar to proceed with his plans and the Kannambadi reservoir scheme ultimately came to be sanctioned.[55]

This was only a partial victory. The Madras government too was interested in the Cauvery waters for its own Mettur Dam, and apprehended that the new dam in Mysore would cut into its needs. The Madras officials also felt that it could necessitate a redesigning of their own plans. At the intervention of the Viceroy Lord Hardinge, Mysore was ultimately permitted to build a smaller dam.[56] Confident of his success, Visvesvaraya adhered to his original design for a larger dam and won in the end. He timed the completion of the dam and electric installations for 1 July 1915.

In Visvesvaraya's opinion the Cauvery Valley afforded ideal natural facilities for irrigation and power supply. He felt that the Mysore State would be throwing away its opportunities if it failed to utilize them. The gold fields required storage for ensuring their demand for power, which was roughly around 8,900 HP. The mining companies were also asking for storage and if they were denied this facility, they favoured erection of steam or oil plants on their own. Visvesvaraya estimated that if the demand for power supply was met, then an additional revenue of Rs. 4 lakh to 6 lakh could be earned from the third or fourth year after the commencement of the project.

Initially, there was a scheme to erect a small reservoir for meeting the power supply demands. Visvesvaraya argued that this would prove to be a costly venture, if it was attempted on the main river. There was also the prospect of the river drying up from silt deposition. Therefore, he proposed to construct a large reservoir and utilize the remaining storage for irrigation on a scale hitherto not attempted in the state.

In Mysore, over 8,00,000 acres of crops were raised by irrigation every year. Of these only 1,00,000 acres were dependent on the Cauvery supply and were considered more or less safe.

The tanks in the Mysore State were useful both in the good seasons and moderately bad ones. But during periods of extreme drought, they failed and this led to disastrous consequences. Visvesvaraya expressed the opinion that if there was a perennial canal fed from a reservoir filled every year by the unfailing Malnad rainfall, there could be some protection against the famine. At the same time, irrigation would not be confined chiefly to the monsoon months, as was the case with irrigation tanks and river channels. The supply of water should be available throughout the year and would enable cultivators to raise commercial crops.[57]

It was pointed out that the expenditure on this combined irrigation and power supply scheme would probably amount to Rs. 160 lakh, which included Rs. 20 lakh required for generating power. Visvesvaraya proposed that in a space of 8 years, the state would be able to get about Rs. 1 crore worth of produce and about Rs. 8 lakh of net revenue. He felt that since the mining companies were demanding additional power, there was every prospect of an assured revenue almost from the commencement of the project.[58] Subsequently, the return on the irrigation share of the capital was calculated as much as 8 per cent. There was a forecast that the state would be able to produce commodities valued Rs. 1½ crore from irrigation alone.

As in the case with other irrigation schemes, Visvesvaraya's ideas on the Cauvery Reservoir Project were guided by the logic of economic gain. He argued that surveys and investigations of existing irrigation facilities would lead to more efficient collection of revenue, a portion of which could be used by the government for financing such grant schemes. He believed that the extension of irrigation facilities, following the construction of the Cauvery reservoir would make the finance available for the early completion of irrigation projects. In other words, it was felt that the prospect of an irrigation project did not solely depend on state intervention, but was also dependent on the inclination of the ryots to invest in such schemes. Visvesvaraya seemed to be more concerned with the interest of the rich peasantry, since he believed that their profits would lessen the government's financial burden *vis-à-vis* irrigation works.

However, Visvesvaraya's ideas relating to the economic benefits that were likely to arise from the Cauvery River Project came to be strongly opposed by the top officials of the Finance Department of the Mysore State. The Comptroller of Mysore pointed to the financial difficulties of executing the scheme. On the basis of the Dewan's Dasara Address of 1910–11, the Comptroller highlighted several interesting facts. He pointed out that according to the Dewan's statement, the total assets of the state in cash, securities and amounts were to the tune of Rs. 285 lakh. The total liabilities amounted to Rs. 176 lakh, resulting in a net asset of Rs. 109 lakh. It was stated that the minimum balance required at the commencement of a year for working the treasuries was about Rs. 70 lakh. But the growth of permanent recurring expenditure did require an addition of Rs. 5 lakh to the minimum. A deduction of this amount from the net asset of Rs. 109 lakh, made a balance of Rs. 34 lakh. In effect, this was the sum available for the project.

The Comptroller suggested that under existing conditions, the next two or three years' budgets could be balanced to meet the government's financial demands. A balance of Rs. 52 lakh making up a total of Rs. 86 lakh was earmarked as government's budget on irrigation projects. The balance of Rs. 52 lakh was to be found from other sources, i.e. by direct borrowing or by indirect borrowing and by utilizing such portion of funds at the disposal of the government, which were classified as Debt Heads.[59]

Thus it was assumed that the execution of a work of exceptionally profitable character could ensure the financial liability that was to be borne by the government. It was argued that until the work had paid back Rs. 52 lakh to the government as net profit, the government would not have sufficient money on its own to keep its treasuries functioning for purposes of administration. The Comptroller assumed that the usual public works grant would remain at Rs. 25 lakh during the three years under consideration. But there was every possibility that during those years the usual grant of Rs. 25 lakh, which included at least Rs. 10 lakh for turnkey projects, could be reduced to Rs. 2 lakh, if not indeed to a lower figure. Even then it was calculated that the Public Works Department would have Rs. 48 to 50 lakh to spend every year during the period under question.

This figure was exceptional, since it had never been reached before in the history of Mysore's finance.[60]

The Comptroller stated that the idea of unlimited future development often found powerful support in official circles. On the other hand, an advocate of the less expensive scheme had the risk of appearing as a narrow and short-sighted individual caring only for the immediate present and blind to the prospect of future generations. But to keep open the way for future development a vast sum of money had to be invested. This expenditure could lead to an improvement in the material and moral condition of the people, but it was an imperative for the government to examine closely as to how far such development was feasible and whether it was necessary to keep the opportunities open at such enormous cost.[61]

On the basis of his correspondences with the Madras government, the Comptroller argued that there was little hope of getting the unqualified assent of the latter to the 39,000 m.cu.ft reservoir. After much discussion a qualified assent was obtained, but the Madras government would retain some sort of right to watch and investigate the distribution of water. This could lead to an interference in the internal affairs of the Mysore State. Such joint interests were likely to prove embarrassing and unsatisfactory. Thus it was felt that if in a zeal to construct a very large project, the Mysore State was drawn into an arrangement involving supervision and control on the part of the Madras government, it would be better to settle for a scheme which would give rise to lesser disputes.[62] There was a suggestion that a less costly scheme with a slightly larger reservoir would enable Mysore to have a protected supply of 20,000 HP, with a power station situated at Sivasamudram. By removing this power station to Shimsha Falls, this supply could be doubled. Besides this continuous and protected supply, there was also a possibility of generating unlimited power for the greater part of the year.

The Comptroller also stated that even if there was a demand of more than 40,000 HP, the undertaking of the less expensive scheme would not force the state to forego its profit. In the first place, it was not at all clear whether the reservoir of the less expensive scheme was absolutely incapable of being enlarged.

Second, there could be some other sites on the Cauvery for reservoirs, and two middle-sized reservoirs in different places would in some respects be better than one huge reservoir. Third, with the powerhouse at Shimsha, a steam or all plans of Rs. 10 lakh–15 lakh would enable Mysore to protect 20,000 HP. The actual expenses of generation by oil during the two dry months would not enhance the cost of generation per HP per year and the profits would also not be affected. Given the great broad Cauvery for the year, there would hardly be any refusal of the demands of the stable and permanent customers for power. Whatever arrangement was made for protecting the supply beyond 40,000 HP, a wide margin of profit would be left to the state. The Comptroller therefore felt that giagantic lake, of which the reservoir of the Rs. 86 lakh scheme was the first stage, was absolutely unnecessary. In other words, the extra expense in the Rs. 86 lakh project was thought to be judicious if it was set down as the price to be paid by Mysore for retaining the chance of undertaking the second and third stages of the scheme for irrigation purposes.[63]

Subsequently, the Comptroller also raised a few other objections on the basis of observations in the government reports. It was argued that irrigation projects could hardly be remunerative as the storage works in most cases were non-paying. The enormous amount of rock cutting in the second stage to carry the canals over elevated and unfavourable tracts also proved to be too expensive. In this connection, it was also opined that the irrigation portion of the scheme, if at all undertaken, would be of the protective sort but not productive.[64]

In fact, it was also pointed out that the less expensive scheme would be highly remunerative. A question was raised as to why its highly remunerative character should be destroyed by tackling on to it something, or the hope of something, itself unproductive. The only way in which the combination of the electric side could help the irrigation side of the scheme was by relieving the latter of a certain part of the capital cost. The Comptroller observed, 'We would have to build a reservoir for electric purposes costing so much, at any rate. So the cost of the irrigation side of the scheme should be taken as the excess of the cost of the whole scheme over that sum.' In other words, the revenue which could be expected from the irrigation side was not expected to give a fair return on the capital debitable to the irrigation side. It was thus observed, 'the extra expenditure to the extent of Rs. 30 or 40 lakh brings absolutely no return, for whatever irrigation will be practicable from the Rs. 86 lakh scheme will also be practicable from the less expensive scheme'.[65]

The logic of remunerativeness was perceived to be the yardstick of utility. It was clearly stated that if the irrigation side proved to be non-remunerative, it would not be of much use to the people and would hardly be worth undertaking. A serious financial blunder would result from the proclivity to spend money borrowed directly or indirectly on interest on the proposed schemes. Such borrowing would hardly be fair to future generations and it would not be judicious on the part of the present generation to use, on ordinary current expenditure, on unproductive works, the proceeds of the mines which belong to all generations.[66]

Such arguments were expressed to highlight the benefits of the less expensive scheme. In fact, the implementation of this scheme would lead to the extension of irrigation facilities over 25,000 acres, apart from meeting all requirements related to the generation of power. Its construction would also leave the finances of the Mysore State easy and would expand irrigation networks in the other districts of the state. Thus there would be a fair distribution of the resources of the state amongst several useful channels, guaranteeing overall progress and prosperity.[67]

Visvesvaraya placed his own arguments to refute these objections. He began by stating that there was no financial risk whatever, and that the government would be fully justified in investing a much larger part of its financial reserve in such productive schemes. The Comptroller's suggestion regarding the curtailment of the volume of expenditure through the construction of a reservoir of a permanent height of 70 ft was dismissed on technical grounds. Visvesvaraya explicitly stated that it would be unsound to build behind an existing dam for the purpose of raising its height. This had been followed at Aswan, but the circumstances there had been exceptional. There had been an emphasis on cement mortar and this alongside the expenses incurred for masonry had been up to three times higher than the cost of ordinary mortar. Visvesvaraya proposed that the Cauvery reservoir would be higher than Aswan and there would be greater risk in building on unsafe foundation. It was not advisable to incur such risk considering the population and the property situated on both banks of the Cauvery.[68]

Visvesvaraya disagreed with the Comptroller's opinions on the proposed Cauvery Reservoir Project. He stated that a reservoir of 7,000 m.cu.ft, less than a third of the size proposed for the third stage, would hardly be of benefit to the state. He also dismissed the idea that irrigation in the state had largely been guided by idea of protected irrigation. The profitability of the irrigation works in Mysore was highlighted in order to justify the plea for the construction of a large irrigation complex. Visvesvaraya was convinced that with a reservoir added to protect channel irrigation, there would naturally be expectations of higher returns.[69] The Comptroller was also criticized for having misinterpreted the views of the Irrigation Commission. There was an assertion that many of the irrigation works in British India had been paying 10 per cent or more. It was also strongly stated that had the irrigation reservoirs been so unprofitable, the Madras government would never had contemplated the construction of a reservoir on river Cauvery.[70]

In Visvesvaraya's opinion, the value of irrigation was not to be judged solely by the returns anticipated. The real value had to be measured in terms of the increase of the productive power of the country. The Comptroller had overlooked these aspects and had involved himself solely with debt heads, cash balances, and interest charges. Visvesvaraya observed,

> There is no harm in borrowing if we use the loans for creating new property and wealth . . . since the productive works of the Government of India are constructed with borrowed capital, we shall be in good company. This scheme will start with initial advantages rarely vouched safe to similar schemes in British India. We shall have handsome revenue from power from the very start; we shall be building up an estate and the shape of a large irrigation work. This with the power supply which the scheme will render possible, will, if rightly used, lay the foundation of the future prosperity of Mysore.[71]

Visvesvaraya's defence of the Cauvery Reservoir Project was based on several other issues. He did not favour the idea of constructing a large reservoir, all at once. The beginning had to be made by constructing a dam capable of storing 80 ft of water above the riverbed, all most equivalent to a capacity of 11,000 m.cu.ft. The storage would be utilized to protect the existing power supply to the Kolar Gold Fields and for ensuring an additional supply around 5,000 HP. Improvement of water supply to the existing channels below the dam was emphasized in strong terms. There was also a plan to utilize some portion of the stored water for power generation between March and May, when it would not be useful for Madras. Nonetheless, the bulk of it used for the rest of the year would be distinctly beneficial to Madras irrigation. The steady supply of 5,000 cusecs which was being maintained in the river in summer by artificial means was also proposed to be doubled. There was an expectation that when the scheme was fully developed, the total quantity of water which would be abstracted for irrigation would be 37,300 m.cu.ft, less than 40,000 m.cu.ft.[72]

Visvesvaraya also stated that while the quantity passed down at the head of the delta irrigation was known, there was no record what portion of it was actually used for irrigation. In fact, the address of the Dewan of Mysore to the Resident was cited to draw attention to the relative extent of irrigation practised in Madras Presidency and in Mysore. The area in Madras was calculated to be 10,05,000 acres and that in Mysore only 1,08,000 acres. The latter was less than 1/9 of the former and admittedly nearly three-fourths of the discharge obtained in the Madras Presidency was derived through Mysore. The first claim to all surplus water therefore belonged to Mysore.[73]

According to Visvesvaraya, the Madras government was in favour of constructing a reservoir close to the boundary of Mysore. This reservoir was to have a storage capacity of more than 80,000 m.cu.ft, which was double the effective storage contemplated in the Mysore project. There was also a proposal to extend irrigation networks over 3,55,000 acres. Visvesvaraya argued that through such arrangements for regulating the supply of water, the Madras engineers were trying to utilize 19,00,50 m.cu.ft instead of the average of 2,94,084 m.cu.ft. which flowed down into the delta in the past two decades. Interestingly, this left a surplus of 2,35,505 m.cu.ft available for storage in Mysore or Madras. The quantity required by Mysore was calculated to be 1/6 of this surplus.[74]

The proposals of the Chief Engineer for Irrigation, Madras Presidency, were also discussed by Visvesvaraya. C.A. Smith, the Chief Engineer, had entertained doubts on the ability of the Mysore government to control the supply of water for irrigation during seasons of scarcity. Visvesvaraya felt that not more than 40,000 m.cu.ft. would be utilized in any year for irrigation within Mysore territory as long as the existing conditions continued. It was proposed that the discharge needed to be calculated at the head of the one or two canals that would be constructed and necessary measures had to be taken for preservation of the records of discharge.[75] However, Visvesvaraya opposed several recommendations of the Madras government. He felt that the Shivasamudram anicut could not be brought within the measurement of discharge, because there was a large catchment area which brought appreciable supplies during the north-east monsoon in the areas lying below Sivasamudram.[76] Visvesvaraya also disagreed with Colonel Ellis's view that during periods when the discharge at the Cauvery dam was less than 7 ft on the gauge situated close to the head of the delta, there should be no abstraction allowed for Mysore. In Visvesvaraya's perceptions the limit of 7 ft on gauge seemed to be a tough condition. Based on certain set of statistics, it was argued that the average depth corresponding to the maximum fortnightly discharge had rarely exceeded 7 ft in seven months out of eight. In December and January a depth of 7 ft was hardly reached. Visvesvaraya proposed a compromise and presented a list of future limits of discharge for the purpose of impounding in Mysore.[77] Thus with the use of statistics, an attempt was made to arrive at a consensus. Visvesvaraya stated that between March and May, the entire supply of the river belonged to Mysore and no restrictions were necessary for this period. The usual period of storage proposed would be from the middle of June to the end of August. But there would be additional supply from the hot weather storms and the north-east monsoons.

Visvesvaraya also challenged the issue of deficient supply for the years between 1890 and 1910, as claimed by the Madras government. He argued that barring six years there had been no deficit.[78] In fact, without even lowering the 7-ft limit there had been enough surplus in all these years to fill the Kannambadi reservoir. Subsequently, it was also pointed out that in view of a reservoir being constructed in Madras, the regulation at the Cauvery dam could be arranged to a nicety. The average surplus then available would be sufficient for both reservoirs. In fact, if there were any deficit in poor rainfall years, Mysore would have the first claim over the water of the river Cauvery.

Finally, Visvesvaraya stated that if the Madras government entertained any fears that Mysore might use excess water unauthorisedly in times of drought for the protection of dry crops, an assurance could be given that the quantities issued to the canal or canals would be strictly limited to the totals agreed upon. This condition could also be given effect by providing surplus escapes or weirs, if necessary. Subsequently, the Mysore government could agree to any reasonable restrictions, which the Madras government considered necessary, so that nothing beyond the stipulated quantity was withdrawn in a year.

Visvesvaraya also suggested that if these proposals were found to be unacceptable, the Mysore Darbar could agree to a modified form of the suggestions of Colonel Ellis as an alternative. In that case, correct quantities of the supply admitted into the reservoir and then passed down into the river had to be maintained from hour to hour on a day-to-day schedule. At the same time, Mysore could go on impounding for a week or a fortnight at a time, and at the end of each period the quantities impounded could be compared with the surplus at the Cauvery dam. The logic being that the difference, if any, in favour of Madras could be returned to the river within a week. In this connection, it was stressed that the art of river gauging was imperfect, particularly when applied to flood discharges. There was an expectation that with the growth of years, gauging would improve accuracy.[79]

Visvesvaraya made it clear that the suggestion that Mysore could store when there was a surplus flow at Sivasamudram after allowing the equivalent of 7 ft discharge at the Cauvery dam, was not acceptable. The process of impounding had to be always accomplished with reference to the gauge at the Cauvery dam itself. It was assumed that a flood at Kannambadi would take place several days before the corresponding one at the

Cauvery dam. The interval would depend on the intensity of the flood. Therefore, it was proposed to store continuously at Kannambadi and to make an adjustment of the difference at the end of a week or fortnight. Subsequently, to facilitate impounding in Mysore without causing any inconvenience to Madras, it was decided that Mysore could lend up to 10,000 m.cu.ft. from its own reservoir if required. But this process of filling up a reservoir could last till 31 August, provided Mysore was unconditionally allowed to recoup from whatever flow there may be in the river after the stipulated date. As the Chief Engineer of the Mysore Public Works Department, Visvesvaraya, stated that it would be harassing and unusual for one state to control the internal distribution of water within another to the extent proposed by the Madras government. In fact, this to him was probably the first time that such a control was being exercised over any native state in India.[80]

Subsequently, it was opined that the fact that Madras government's proposal for extension of irrigation by means of a reservoir of double the size contemplated in Mysore was itself sufficient proof that the former had no misgivings 'as to the Mysore reservoir filling without trenching on the requirements of the deltaic irrigation'.[81] Almost the whole supply that Madras proposed to impound would be obtained from Mysore and not for the Madras territory. In view of such a situation, the Madras government's objection to storage within Mysore limits was highly improper.

It needs to be understood that technical reasons alone did not influence Visvesvaraya's ideas on the Cauvery Reservoir Scheme. As in the case of the other irrigation projects, economic considerations also seem to have played a great part. Visvesvaraya felt that this scheme afforded ample opportunities to the Mysore State to subsequently increase its incomes from the industrial as well as the agricultural sector. The construction of a reservoir promised a guaranteed income from the mining companies, who would be the major recipients of power supply. In this context, Visvesvaraya who was an avowed supporter of industrialization offered several suggestions to the Dewan's administration in Mysore. He drafted a detailed scheme for the utilization of power being generated at Sivasamudram. The establishment of iron smelting works, fertilizer plans, cotton and silk textiles, and electric locomotives received top priority from him. It was also envisioned that supply of power would boost the prospects of the silk industry based in Bangalore and Mysore. The silk mills established in Bangalore and Mysore or situated along the railway line connecting the two cities could make huge profits and extensive cultivation of mulberry would be possible under the new scheme. Visvesvaraya favoured the appointment of a Director of Industries for co-ordinating the business opportunities that would arise from the implementation of the scheme. He felt that if the government was opposed to such an appointment, a committee of two or three technical officers of government or two or three businessmen from within or outside the state be formed with an energetic secretary. A sum of Rs. 50,000 to 1,00,000 could be placed annually at their disposal for undertaking a study of the development of the market for existing power and for making recommendations for obtaining the largest return from it. It was also proposed that the committee facilitate the process of involving specialists who would make reports on the natural resources and prospects for introducing new industries into the state. Last, the committee could also undertake an industrial survey of the state.[82]

As regards the agricultural sector Visvesvarya had several other ideas. He argued that out of 8 lakh acres of crops cultivated every year in Mysore, about 1,00,000 acres were dependent on Cauvery supply. The Cauvery Reservoir Scheme was a necessity as the tanks in Mysore State dried up during periods of extreme drought. Irrigation from a perennial canal fed from a reservoir field every year by the unfailing Malnad rainfall was favoured, since it protected the state from famines. The reservoir would make water available to cultivators all through the year and would provide them with an opportunity to cultivate commercial crops.[83] The management of the irrigation schemes was to be handed over to a mixed agency of government officers and local committees. It was proposed that this agency could also be entrusted with the responsibilities for carrying out special repairs and for the construction of the less important masonry works. Visvesvaraya favoured a competent class of local contractors for undertaking such work.[84]

In the 1920s, Visvesvaraya was involved with the functioning of the Aoda Dam in Gwalior State. He visited the site of the dam in 1925, in company with the Technical Adviser of the state. The original project for the Aoda Tank had been prepared by S.K. Gurtu and an amount Rs. 16 lakh was sanctioned for it. The work remained in progress till 30 June 1922 and when it was stopped, an expenditure of Rs. 8.5 lakh had already been incurred. There had been two reasons for stopping it; one was a suspicion that the masonry dam constructed in the riverbed had not been sufficiently deep and had been found to leak; the other out of a doubt about the results being commensurate with the ultimate costs.[85]

The Technical Adviser had suggested certain modifications for this project. But the principal points that remained to be settled essentially revolved around issues related to the size of the reservoir and the waste weir. There were also a diversity of opinions on the catchment area, rainfall, and the configuration of the tank site. The direct catchment area drained by the river Sipra into the tank basin was 83 sq. miles and an addition of 50 sq. miles by the diversion of an adjoining stream known as Panahar Nala was contemplated. The Technical Adviser proposed to carry out the work in two stages. In the first stage, the reservoir would be built to suit the Sipra River catchment of 83 sq. miles. In the second stage, the reservoir would be adapted to the entire catchment including the Panahar Nala. For 83 sq. miles, the discharge was calculated at 44,000 cusecs, and for 133 sq. miles, it was 62,400 cusecs. In the former case, the maximum runoff was calculated to be 0.82 inch per hour and in the latter it was 0.73 inch. In the case of the latter, it was assumed that the entire flood from the Panahar Nala would be diverted into the river Sipra. The Technical Adviser made a provision for a waste weir 2,600 ft long with a flood depth of 3 ft for the first stage.[86]

Visvesvaraya favoured a complete detailed scheme with plan estimates and descriptions regarding the Aoda Dam. He felt that a larger scheme would hold 380 m.cu.ft more storage and would only result in an additional cost of Rs. 4.5 lakh. The additional storage would make the dam more suitable for the needs of cultivation. In the second revised scheme the canal would bring under command 68 villages comprising of 15,598 inhabitants and 94,401 bighas of land, of which 46,716 bighas were arable. It was argued that the canal when extended would facilitate irrigation over an area of 1,41,120 bighas.[87]

Visvesvaraya opined that 1 m.cu.ft of storage in this reservoir would cost about Rs. 2,000, which at 5 per cent interest would give a return of Rs. 100 per annum. In fact, the same m.cu.ft water when utilized for irrigation would bring in revenue of Rs. 60 under favourable conditions, the actual direct return being 3 per cent. Thus Visvesvaraya felt that unless water was economically used, the storage would only yield a net income between Rs. 20 and 30 per m.cu.ft, or a return of 1.25 per cent on the capital outlay of Rs. 2,000 involved.[88]

The work on the Aoda Dam was considered to be unfavourable from a financial point of view. It was calculated that direct net revenue of 2 per cent on the capital expended could be regarded as a reasonable return for a protective scheme. With careful management, this could be realizable from the Aoda tank. Visvesvaraya stated that there were several other considerations for which the construction could be supported. In the first place, the tract was extremely dry. In bad seasons, people had to travel several miles to procure drinking water. The ruling family favoured a liberal expenditure for providing water supply to the people of the Sheopur district. Second, there was an expectation that leakage from the tank would flow back into the river and was likely to be used up at the several pick-up weirs lower down. Last, since there had been a partial construction, expansion work though heavy could be justified.[89]

Nonetheless, Visvesvaraya proposed that the cost of the project should be reduced by careful modification of the plans and estimates. Along with these, forecasts of financial results had to be prepared similar to those for protective irrigation projects in British India. These had to be approved by the government along with the revised project before the work was recommended. It was also opined that since the riverbed was not a natural tank site, storage would be expensive. But on account of the arid character of the tract, there was a great demand for water in this region and this made it desirable to adopt measures for the completion of the project. Every effort had to be made to utilize water economically and to expand irrigation rapidly, so as to obtain a direct revenue of about 2 per cent. Finally, to prevent leakage, the construction was proposed of a regular waste weir which could also act as an over flow weir.[90]

Visvesvaraya's efforts to harness the flow of the rivers for purposes of irrigation led him to subscribe to the idea that large dams were the most suitable instrument for river basin development in India. Others in colonial and post-colonial India too strongly believed that projects aiming towards large-scale supply side augmentation had to be encouraged for meeting the growing requirements of water. Thus there were attempts to solve the water problem by storing the huge monsoon runoff in the main rivers, primarily with the help of large-scale structural interventions.[91]

Nonetheless, some questions could be raised in connection with Visvesvaraya's plans to construct large dams for the overall economic improvements of the rural society. In most cases this encouraged agrarian capitalism and sharpened divisions between different social categories. The construction works benefited the rich farmers, migrants and locals. Visvesvaraya was oblivious to the relations between location and accessibility to water on the one hand and socio-economic status on the other, which usually guided the mechanism of the rural society. The extension of irrigation facilities, as most scholars have pointed out, led to 'a shift in the cropping pattern to high input, high output and high value crops'.[92] The small peasant producers had few resources to take advantage of this shift in agricultural operations and management. Their insecure financial base led them to borrow heavily from the rich peasants, who also served as moneylenders. The alienation of land was a crucial issue over a large part of southern and western India. In this context, Visvesvaraya's plans seem to have been imposed from above, rather than from the complexities that might have prevailed at the ground level.[93]

Visvesvaraya also seems to have been indifferent to the human dimensions of large-scale irrigation works. In the case of all such projects, one of the major issues related to the problem of displacement. He was silent on this and hardly ever mentioned the problems of those who were deprived of their ancestral lands, their only source of livelihood.[94] At the same time, certain environmental issues too were ignored. The masonry structures like causeways with foundations very often reached depths where the soil was impervious. This led to a rise in the subsoil water table. Obstructions in subsoil drainage paths led to the recurrence of floods in flood prone areas. Visvesvaraya and others who place the main emphasis on economic and engineering aspects failed to realize that in the case of the rivers swelling up owing to torrential monsoon rains, breaches in the embankments can wash out fertile agricultural tracts and invaluable animal and human lives, thereby erasing the economic benefit of such human interventions.[95] The most serious problem that was also overlooked related to waterlogging and the development of salinity/alkalinity. Visvesvaraya seems to have given less emphasis on the effects of the rapid expansion of various surface irrigation facilities. Overemphasis on canal irrigation obviously led to a rise in ground water. The lack of equilibrium between precipitation and the natural process of evaporation, transpiration, and runoff led to the transformation of productive lands into saline or saline alkaline lands.[96]

In sum it could be argued that environmental and other concerns, which recently have been regarded 'as disagreeable external impositions', hardly played role in project planning. Environmental issues were wholly ignored and preference seems to have been placed on revenue maximization and on engineering aspects of dam construction. Visvesvaraya perhaps belonged to a school which might have helped the view that water planning had to be left totally to engineers. There was very little idea that water planning needed to involve a multidisciplinary body of agricultural experts, peasant organizations, environmental scientists, economists, sociologists and lawyers.[97]

NOTES

1. Satyajit Singh, *Taming the Waters: The Political Economy of Large Dams in India*, Delhi, 1997, p. 7.
2. Rohan D'Souza argues that such technological practices were mostly 'directed towards either impounding precipitation, tapping river inundations or retrieving ground water charge'. He also believes that

the underlying hydraulic principle was to evolve a water harvesting structure that could adopt itself to microclimates, topography and fluvial process. Rohan D'Souza, 'Supply-Side Hydrology in India: The Last Gasp', *Economic and Political Weekly*, 6 September 2003, p. 3786.

3. Lt-Gen Sir Arthur Cotton, the main proponent of canal construction in British India, had a contempt for Indian society and its traditions. He realized the enormous revenue potential in irrigation and favoured a linking of Indian navigation with the Yangtse and the internal navigation of China. In other words, he felt that canal irrigation provided more agricultural production with less labour, leading to high returns in the market. While Cotton favoured the anglicization of the system of governance, he did realize the positive aspects of traditional irrigation systems. In a lecture at the School of Military Engineering in Chatham in 1874 he observed, 'These are multitudes of old native works in various parts of India, but principally in the southern parts of the peninsula where there are numerous river channels, some of considerable extent, and innumerable tanks, containing from 100 m.cu. yards downwards. These are noble works and show both boldness and engineering talent. They have stood for hundreds of years. Many of the weirs are built in the sandy beds of rivers without any rock or sound stratum to found them on, and it was from them we learnt how to secure a foundation in loose sand of unmeasured depth. In fact, what we learnt from them made the difference between financial success and failure, for the Madras river irrigations executed by our engineers have been from the first the greatest financial successes of any engineering works in the world, solely because we learnt from them how to make secure and cheap foundations in loose sand, and the Ganges canal was a financial failure for 25 years solely from a fear to attempt to build a weir in such a situation, which drove the engineers to go up into the hills to lead off the water, where the river bed consisted of pebbles, and it cost them a million to lead the water from there into the plains, the whole of which would have been saved if the water had been taken off at a point clear of the hills, and the loss of this million made the whole difference between financial failure and great success.' However, the colonial state alienated its colonial subjects from their traditional skills and instead appropriated them for its own agenda. This agenda was initially developed around revenue considerations. The management and construction of irrigation works subsequently became the responsibility of the army. Thus what followed was essentially the development of the irrigation works during the times of peace and their wholescale neglect during times of war. See Satyajit Singh, op. cit., pp. 36–8. *Lectures Delivered by Sir Arthur Cotton on Irrigation in India*, Collected and Published by Uddaraja Raman, Vijaywada, 1968, pp. 23–4.
4. Satyajit Singh, op. cit., p. 9.

5. Rohan D'Souza argues that modern irrigation was represented by the construction of several large-scale canal irrigation schemes with permanent headworks such as the Ganges Canal (1854), the Godavari Canal System (1852), and the Krishna Canal System (1855). He opines that all these were large projects which had been made possible by a dramatic hike in cropping intensities, and subsequently all these led to the growth of commercial farming. But there were also problems resulting from salination and waterlogging.
6. Henry C. Hart, *New India's Rivers*, Bombay, 1956, pp. 251–2.
7. Ibid., p. 252.
8. Y.G. Krishnamurti, *Sir M. Visvesvaraya: A Study* (Foreword by Sir Purushottamdas Thakurdas), Bombay, 1941, pp. 6–7.
9. For more details on the Godavari and Kistna (Krishna) river schemes, see Herbert M. Wilson, *Irrigation in India*, Delhi, 2nd edn., 1903, pp. 83–5.
10. For details, H. Rangachar and P. Kodanda Rao, 'A Brief Life Sketch', in *M. Visvesvaraya Birth Centenary Commemoration Volume* by Contemporaries and Admirers, published by Visvesvaraya Centenary Committee. See also Henry C. Hart, 'New India's Rivers', p. 47.
11. Rangachar and Rao, 'A Brief Life Sketch', pp. 295–6.
12. Ibid., p. 297.
13. 'Note on Irrigation in the Block System', Sir M. Visvesvaraya Private Papers, Reel No. 1, Nehru Memorial Museum & Library (hereafter NMML), Delhi, p. 1.
14. Ibid.
15. It was believed that the cultivators would be able to obtain canal water in the monsoon and *rabi* seasons when their labourers and bullocks were busy; in the hot weather when work was slack, they could irrigate their crops from wells. 'Note on Irrigation in the Block System', p. 3.
16. *Times of India* (Bombay Edition), 10 January 1902.
17. Visvesvarya stated that maximum, minimum and average supply of water available from any irrigation work could always be computed on the basis of past experience. In the case of minimum supply, he argued that a fraction should be set apart in the lower reaches of the canal, so that it could be used till the end of the *rabi* season. The balance, usually 2/3 to 4/5 of the total minimum, was taken as dependable supply for every year. Furthermore, the area of the blocks that could be watered by this quantity was to be calculated and then distributed among villages to which perennial irrigation facilities had been extended. The distribution was also done after taking into consideration the extent of perennial irrigation that had been done as well as by considering the possibilities of future extension or curtailment. The block areas allotted to each village rarely exceeded

a quarter of the total cultivable area of the village. In cases where there was more water, it was proposed that it needed to be given for cultivation of crops in the villages outside the blocks. Such a system would enable the cultivators to obtain canal water in the monsoon and *rabi* seasons. 'Note on Irrigation in the Block System', pp. 2–5.

18. The disadvantages of the peasantry resulted from several factors. In the first place, submitting water applications in time, and qualifying for a permission proved to be really difficult. Every crop had to be measured by a staff of measurers and an Assistant Engineer or subordinate was required to check a proportion of the measurements and prepare assessment returns. Then demand statements had to be submitted to the Collector for collection twice a year. To fulfil these regulations was almost impossible for the majority of the peasantry. Incidentally, even if they complied they were exposed to greater economic exactions, especially during a period when the prices of agricultural commodities remained depressed. Ibid., pp. 3–5.
19. Visvesvaraya did express his own views about the problems faced by the poorer people. He argued that even if exceptions were made to support the poor cultivators, there was hardly any improvement in their condition. The concessions frequently led to a repetition of the irregularities that had prevailed in the past. It was also stated that poor cultivators did not need irrigation facilities to the extent needed by the rich cultivators (since they did not cultivate sugar cane for want of capital). Ibid., pp. 3–7.
20. Visvesvaraya expressed the opinion that successful irrigation required organization, cooperation, and capital. Cultivators lost confidence when water was given in one year but refused the following year. It was argued that under the prevailing conditions, there was no means of preventing this. The uncertainty and want of parity between demand and supply was said to have retarded the development of irrigation. Thus it was felt that an understanding between the supplier and purchaser and a long-term contract for the supply would enable the government to charge lower rates and would provide cultivators an opportunity to prepare their lands and arrange the capital in advance. Ibid.
21. Visvesvaraya pointed out that irrigation was carried out in scattered patches, about 5 acres in one place, 2 acres in another, and half an acre in a third place. In the case of sugar cane cultivation, if about 8,000 cu.ft were required for watering an acre and for 8 months with three watering per month, the quantity required would be 1,92,000 cu.ft. According to Visvesvaraya's estimates, the supply of ½ m.cu.ft of water was usually sufficient for one acre and even a less than that would serve the purpose of cultivation. Thus, he observed, 'the loss between the canal head and the field is fully 60 per

cent. If the supply is given by rotation and for the fixed period for which it is absolutely required, the users will have no temptation to extravagance. If the fields are properly levelled and the channels kept clear, as they will be under the block system, water will run over the fields quickly and the waterings will be finished rapidly. The block system cannot fail to lead light waterings and economy of water.' Ibid., p. 4.

22. Ibid., p. 5.
23. Ibid.
24. Ibid.
25. Visvesvaraya argued that the block system was based on the analogy of the *thal* system in the north Deccan. The Superintending Engineer on Special Duty, Beale (cited by Visvesvaraya), was said to have observed, 'The *thal* or area commanded by a *bandkhara* and channel is usually divided into 4 equal parts or *phads*, the whole of which nearly is annually brought under irrigation. Each *phad* is subdivided into an equal number of shares of fields, according to number of owners. In any year, the whole of one *phad* is devoted to rice, another to sugar cane, the third to wheat, gram, peas and others, and the fourth to bajri or kharif, jawari and others. In the next year the same crops are grown, but on different *padhs*, namely, in the following order—sugar cane, wheat, bajri, rice; and in the next two years the crops move round in the same rotation until the fifth year, when they regain their original position. This rotation continued without interruption, the only periods of fallow being the intervals between the crops. This system is termed as quadrennial rotation. In triennial rotation, the bajri is omitted'. Ibid., p. 6.
26. The Nira Canal was a little over 100 miles long and had headworks at Virwadi on the Nira River, about 30 miles south-east of Poona. Lake Whiting, which fed the canal after the cessation of the monsoon flow in the river, was situated about 15 miles above the canal headworks. The dam of the lake was built across a tributary stream just above the confluence. The lake held 5,313 cu.ft when full. In the monsoon, spread over a period of three or four months, the river discharge was sufficient for sustaining a full supply to the canal. Up to November and occasionally till the end of December, the storage was only partly used. In fact, from December to the beginning of the following monsoon, the reservoir was the only source of supply. Ibid., p. 7.
27. Ibid., p. 10.
28. 'Note on Irrigation in the Block System', pp. 10–12.
29. Ibid., p. 14.
30. Ibid.
31. Ibid., p. 15.

32. Ibid.
33. A different sort of opinion was expressed by Visvesvaraya. Ibid., pp. 16–20.
34. H. Rangachar and P. Kodanda Rao, 'A Brief Life Sketch', p. 297.
35. Completion Report Nira Canal, Poona Irrigation District, M. Visvesvaraya Private Papers, Reel No. 1, NMML, Delhi, p. 1.
36. There were 36 sluice gates on the south waste weir, which were opened and closed by chain-worked by hand power. There were 45 gates on the north waste weir, whose action was automatic. Ibid., p. 2.
37. No. CE-27 of 1901 (From the Chief Engineer, Public Works Department, Bombay Presidency to the Superintending Engineer, Central Division, Poona, Dated February 10, 1901) in M. Visvesvaraya Private Papers, Reel No. 1, NMML, Delhi, p. 1.
38. Ibid., p. 4.
39. No. CE-27 of 1901, p. 12.
40. Ibid., p. 13.
41. H. Rangachar and P. Kodanda Rao, 'A Brief Life Sketch', pp. 302–3.
42. Ibid.,
43. Nawab Zain Yar Jung Bahadur, 'Sir M. Visvesvaraya—A Tribute', *Visvesvaraya Birth Centenary Commemorative Volume*, pp. 85–6.
44. The flood occurred on 28 September, reaching its maximum height about an hour before noon. The monsoon rainfall since 1 June had been less than normal. A day before the flood, of over 6 inches fall was registered. In a few hours there was a cloud burst over an extensive area. The tanks in the catchments of the river were full and the ground everywhere was super-saturated. The rain descended in sheets, flooded the small tanks, and overburdened their water weirs. Tanks burst one after the other. The largest of the intercepting tanks were the Palmakul tank and Parti reservoir, both located in the Yentair River Valley. The heaviest rainfall was recorded at Shamsabad in the vicinity of the two tanks. Both tanks were breached and they let large volumes of water into a river channel already running full. The combined waters rushed down the valley in an uncontrollable mass and hurled themselves against the bridges, buildings, and rampart wall of the flooded city. See, 'The Flood at Hyderabad: An Account of the Flood: Its Causes and Proposed Preventive Measures', Hyderabad, 1909, in M. Visvesvaraya Private Papers, Reel No. 2, NMML, Delhi, p. 3.
45. Ibid., pp. 4-5.
46. The Parti reservoir was situated 22 miles above Hyderabad. It was breached at about 5 a.m. and the flood in Hyderabad reach its peak at about 11 a.m. The flood crest must have travelled at an average rate of 4 miles per hour. Preliminary, op. cit., p. 10.

47. Ibid., p. 12.
48. Roscoe Allen, once the Chief Engineer for Irrigation, observed, 'The restoration of so many supply channels from the Musi River, the Lower Rajakalwa, the Shaligowrarum and others, brings out how very important a source for irrigation for Nalgonda District the Musi River is; and it further becomes very essential indeed that, in so far as possible, its waters should be conserved for irrigation purposes'. Preliminary, op. cit., p. 13.
49. It was believed that if the main object of the scheme was to make the river carry as large a quantity of water as possible, a question could be raised as to why one of the arms of the river had to be closed. This was done because the enlarged river channel above the anicut would be wide enough for the future requirements of the floods and the land lying unused could be reclaimed. Visvesvaraya felt that there was no reason to have a greater waterway for the river at Imli-Bun than was actually provided in the upper reaches. It was argued that on sanitary and other grounds, it was highly undesirable to have two river channels, one of which would be more or less always dry and in close proximity to a crowded quarter of the city. Ibid., p. 15.
50. 'Schemes for Flood Abatement and Drainage for Hyderabad City and Irrigation in the Musi Valley, September 6, 1909', in Sir M. Visvesvaraya Private Papers, Reel No. 2, NMML, Delhi, p. 7.
51. 'Mysore Drainage', in Sir M. Visvesvaraya Private Papers, Reel No. 2, NMML, Delhi, p. 9.
52. The Dewan's letter stated among other things: 'His Highness is confident that should you accept the offer now made to you, you will find in developing the vast irrigation resources of the land of your birth. His Highness is aware that you attach greater importance to opportunities for rendering public service than to mere official emoluments. Such opportunities will be open to you in works and projects which have to be carried out in Mysore.' See, Rangachar and Rao, 'A Brief Life Sketch', op. cit., p. 306.
53. In his address to the fourth session of the Mysore Civil Engineers Association, Bangalore in 1910, Visvesvaraya argued that in advanced countries like Italy, the canals were constructed and maintained by the government or a company and the control and distribution of water in the smaller canals were managed by societies and associations composed of cultivators themselves. In the matter of town planning, he argued that there had been a new awakening in the West. City, town, and village extensions were planned ahead with the same care as individual buildings, and there were associations in every important city of America for providing parks, widening roads, beautifying promenades and public places. All large cities—and many small ones too—were constantly undergoing renovation, alteration, and im-

provement. See, Speeches made before Assumption of Office of Dewan, 1910–12 (the fourth session of the Mysore Civil Engineers Association held in the Association's rooms, Bangalore, 14 November 1910), Speeches by Sir M. Visvesvaraya, Dewan of Mysore, 1910–11 to 1916–17, Bangalore, 1917, pp. 4 and 8.

54. M. Visvesvaraya, *Memoirs of My Working Life*, Bangalore, 1951, p. 46.

55. In 1910, the Mysore government sanctioned amounts for the first stage of the Cauvery Reservoir Scheme. The construction of a reservoir across the river for developing irrigation in the river valley within Mysore had been under consideration and discussion for over forty years. It came into prominence after 1902 and the erection of the hydro-electric power station at Sivasamudram. The summer flow was found inadequate for existing irrigation and power supply. With the increasing depths of mines, the mining companies required additional power. The Sivasamudram scheme was enlarged twice to meet the demand. Each enlargement added to the insecurity of the power supply. A storage reservoir was therefore found to be necessary. The reservoir at Kannambadi, about 8 miles above Seringapatam, was prepared in 1908 by Captain Dawes under the supervision of the Chief Engineer. The cost was estimated to be Rs. 4,40,000. The project was for irrigating 3,00,000 acres on both banks of the river, and generation of electric power to the extent of 70,000-HP for industrial and other purposes within and outside Mysore. The project was not carried out. The question was again taken up. In fact, fresh surveys started in 1910. Superintending Engineer Subbarao, carried on the operations assisted by his two Assistant Engineers. Simultaneously, estimates amounting to Rs. 8 lakh for the fourth installation on the Cauvery Power Scheme were prepared by the Chief Electrical Engineer. The combined project amounting to Rs. 253 lakh was submitted to the government in May 1911. *The Hindu*, 13 October 1911.

56. According to the agreement of 1892 with the Madras government, the latter's consent was necessary for constructing a reservoir across Cauvery. The negotiations were opened in October 1910. The magnitude of irrigation interests in Madras Presidency led to lengthy correspondence between the Madras government and the Government of India. The Madras government agreed to the construction of a reservoir of 11,030 m.cu.ft under certain conditions accepted by the Mysore government and confirmed by the India government. The mining companies agreed to purchase power at the rate of $ 12 per HP per annum for ten years from the date of power supply. The reservoir would not only generate power for the mines, but also provide sufficient storage for irrigating an additional 25,000 acres in

the valley. Ibid.; *The Daily Post*, 13 October 1911; C.G. Barber, *History of the Cauvery Mettur Project*, Madras, 1940, p. 13.

57. Note 2 October 1910, by the Secretary to Government in the Public Works Department, in Sir M. Visvesvaraya Papers, Reel No. 2, NMML, Delhi, p. 2.

58. Visvesvaraya stated that an additional power supply of 15,000 HP would be available whenever required by erecting an additional plant at Shimsha. The power could be used for railways, industries, and a variety of other purposes. He observed, 'If a forward policy is pursued and industries are encouraged to provide a market for the power, it will bring wealth to the country and additional revenue to the state. The scheme is one of great promise'. Ibid., pp. 2–3.

59. The Comptroller suggested that the balances of the principal funds other than government balance proper could be temporarily utilized for purposes like the one under consideration. Of the Rs. 86 lakh required, about Rs. 34 lakh could be found out of the surplus government balance. But for the remainder one had to temporarily utilize nearly half the Savings Bank deposit and State Insurance Fund and about one-third of the Cauvery Power Scheme Depreciation Fund. The figures were as follows:

Half of Savings Bank Deposits	Rs. 34 lakh
Half of State Insurance Fund Balance	Rs. 11 lakh
One-third of Cauvery Power Scheme Depreciation Fund	Rs. 7 lakh
Total	Rs. 52 lakh

'Cauvery Reservoir Scheme', in Sir M. Visvesvaraya Private Papers, Reel No. 2, NMML, Delhi, p. 65.

60. Assuming that the usual grant was fixed at Rs. 20 lakh, the government balance at the beginning of the several years would stand in the following way:

1 July 1912	Rs. 86 lakh
1 July 1913	Rs. 61 lakh
1 July 1914	Rs. 38 lakh

Thus after a number of years when the work had repaid Rs. 37 lakh of the capital expenditure, the government balance at the beginning of the year would be equal to the working balance required for the treasuries, viz., Rs. 75 lakh.

The above figures make it clear that, if the work for which the expenditure was required to repay (besides working charges and interest), capital expenditure to the extent of Rs. 52 lakh in a reasonable time, the expenditure perhaps would be managed with money in the hands of government without having to resort to external borrowing.

It was calculated that the cash balance of the treasuries (including Madras Bank and London Bankers) on 1 July 1911 would be about Rs. 93 lakh. Since a balance of Rs. 75 lakh was sufficient on that date, the commencement of the year would have an available sum of Rs. 18 lakh. The accretions to cash balance by increase of liabilities under debt heads, could be taken to be Rs. 10 lakh in a normal year. This would also be available for the purpose. Thus the actual sale of securities in the market during the year 1911–12 could therefore be possibly avoided. In the year 1912–13 against the requirements of Rs. 30 lakh, the accretion to debt head balances for the year could supply about Rs. 10 lakh and the sale of securities to the extent of about Rs. 20 lakh will therefore be necessary. For similar reasons, sale of securities to the extent of about Rs. 18 lakh would be necessary during 1913–14. If the ordering public works grant could be limited to Rs. 20 lakh during the three years in question, the actual sale of securities would be Rs. 23 lakh instead of Rs. 38 lakh. Ibid., p. 66.

61. The Comptroller argued that the real difference between the Rs. 86 lakh scheme and the less expensive scheme of Rs. 40 lakh devised by the Secretary was that only the former could be developed into a much larger project. The Rs. 86-lakh scheme contemplated the dam being constructed with foundations and width for a 39,000 m.cu.ft. reservoir. Thus it was necessary to have an unqualified assent of the Madras government to a 29,000 m.cu.ft reservoir before sanctioning the work on the Rs. 86 lakh basis. If the Madras government agreed to construct a smaller reservoir, then money would simply be wasted by having the foundation and width of the dam on the 39,000 m.cu.ft basis. Ibid., p. 68.

62. The Comptroller proposed that Mysore state should not build a reservoir larger than the one to which the Madras government had given its unqualified support. At the same time, to avoid waste, the Mysore government had to begin work with foundations suitable for only such a reservoir. Ibid., p. 68.

63. The Comptroller observed that '. . . what rough facts and figures I have been able to gather from the files sent to me from time to time regarding the project, I have been forced to the conclusion that taking into account the irrigation side only, neither the Rs. 86-lakh scheme taken by itself nor the complete scheme will be such a work as can reasonably be classified as productive. I know that strong views to the contrary are held by Mr. Visvesvaraya; but my conclusions are drawn from the public statistics of irrigation works in Mysore and in other parts of India. Indeed, a scrutiny of the statements appended to the latest irrigation review of the Government of India shows that the Government of India themselves have not attempted a storage

work contemplated in the scheme under consideration; and this is undoubtedly due to the fact—storage is so costly even in the most favourable circumstances that very few irrigation works which depend on it are remunerative; and under adverse financial conditions it becomes impossible.' (The Comptroller in this regard was citing the *Report of the Irrigation Commission*, *1901–1903*). Ibid., pp. 69–70.

64. Cauvery Reservoir Scheme, pp. 69–70.
65. The Comptroller felt that in the second stage of the completion scheme an additional outlay of Rs. 82 lakh on the irrigation side extended the irrigable area by 50,000 acres and after the completion of the third stage there would be a capital expenditure of about Rs. 160 lakh with a total irrigated area of 1,50,000 acres or more correctly only 1,25,000 acres. Thus, for the 25,000 acres of the first stage there was a need for only a less expensive electric reservoir. Possibly it was from these figures that the fear of extreme unproductiveness of the scheme on the irrigation side was entertained. Ibid., p. 70.
66. The Comptroller tried to base his arguments on the statistics pertaining to the Kurnool Canal Project of the Madras Presidency. The capital cost of the project was about Rs. 225 lakh, the average return was roughly half per cent per annum. The work proved to be a loss to the state exchequer to the tune of over Rs. 6 lakh. The total loss on this work up to the end of 1908–9 was calculated at nearly Rs. 2.5 crore. Thus irrigation works were viewed as a burden on the State's finances. The agricultural conditions of Mysore also did not hold out promise for the irrigation side of the project. Ibid., p. 71.
67. 'Cauvery Reservoir Scheme' p. 73.
68. Ibid., p. 74.
69. Visvesvaraya strongly asserted that the financial accounts issued by the Comptroller's office about irrigation channels of the Cauvery clearly showed that those works were paying about 5 per cent.
70. Visvesvaraya pointed out that the Comptroller on the basis of the opinions of the Madras Irrigation authorities had tried to establish the point that irrigation works in Mysore would be protective. The latter was apparently not aware that Irrigation Commission had held the view that a productive work could also be protective.
71. 'Cauvery Reservoir Scheme', p. 74.
72. The printed report of the Madras project provided several interesting details:

    (i) The average discharge for eight months of the irrigation season at the delta head in the 20 years ending 1909-10 was 425,555 m.cu.ft

(ii) The average quantity actually sent down for irrigation into the delta in this period was 294,084 m.cu.ft
(iii) The quantity surplussed or wasted at the Cauvery anikat was 131,471 m.cu.ft

See 'Note on the Effect of the Proposed Cauvery Reservoir on Existing Irrigation in British Territory', in Sir M. Visvesvaraya Private Papers, Reel No. 2, NMML, Delhi, p. 2.

73. Ibid.
74. Ibid.
75. Visvesvaraya agreed with Colonel Ellis, Special Superintending Engineer, Madras Presidency, that gauges be established above and below the Mysore Reservoir for the accurate measurement of the volumes impounded on daily basis. Ibid.
76. Ibid.
77.

Months	Average depths corresponding to average fortnightly discharge in feet	Average depths corresponding to maximum fortnightly discharge in feet	Average depths now proposed by Madras in connection with their new project in feet	Limit of depths above which impounding may be allowed as new proposal in feet
June	2.1	4.5	2.9	4.5
	4.5	5.5	2.9	4.5
July	4.8	6.5	6.1	6.1
	5.6	6.8	6.1	6.1
August	5.6	7.9	5.9	6.1
	5.0	6.2	5.9	6.1
September	4.5	5.2	5.8	5.5
	4.0	5.4	4.7	5.5
October	4.4	5.6	5.0	5.5
	4.4	5.2	4.2	5.5
November	4.0	5.4	3.8	5.5
	3.1	5.2	3.8	5.5
December	2.4	4.85	3.0	5.5
	2.0	4.0	1.7	4.0
January (1st half)	2.1	3.7	2.1	4.0

This table was based on the figures given on pp. 34 and 207–12, vol. III, Papers regarding the Madras Cauvery Reservoir Project. For more details, see 'Note on the Effect of the Proposed Cauvery Reservoir on Existing Irrigation in British Territory', op. cit., p. 3.

78. Visvesvaraya also referred to some statistics regarding the flow of water.

Year	m.cu.ft
1890–1	83,987
1894–5	47,605
1899–1900	46,545
1904–5	112,969
1905–6	64,708
1908–9	125,613

Ibid., p. 4.

79. Ibid., pp. 4–5.
80. Ibid.
81. Ibid.
82. Note dated 2 October 1910 by the Secretary to Government in the Public Works Department, in Sir M. Visvesvaraya Private Papers, Reel No. 2, NMML, Delhi, pp. 4–5.
83. Note on Irrigation Improvements in Mysore, in Sir M. Visvesvaraya Private Papers, Reel No. 2, NMML, Delhi, p. 3.
84. Ibid., pp. 7–8.
85. Aoda Dam in Gwalior State: Note by Sir M. Visvesvaraya, in Sir M. Visvesvaraya Private Papers, Reel No. 4, NMML, Delhi, pp. 1–2.
86. The Technical Adviser, at the request of M.Visvesvaraya, prepared a revised estimate for the reduced size of the tank. In this scheme he included the drainage area of the Panahar Nala. For more details see ibid.
87. *Aoda Dam in Gwalior State*, p. 3.
88. Ibid.
89. Ibid., p. 4.
90. Ibid.
91. Jayanta Bandopadhyay and Shama Parveen, 'Interlinking of Rivers in India: Assessing the Justifications', *Economic and Political Weekly*, 11–17 December 2004, vol. XXXIX, no. 50, p. 5308.
92. Peter P. Mollinga, *On the Waterfront: Water Distribution, Technology and Agrarian Change in a South Indian Canal Irrigation System*, Delhi, 2003, pp. 131–2.
93. To understand Visvesvaraya's ideas, see 'Economy of Water in Deccan Irrigation', in Sir M. Visvesvaraya Private Papers, Reel No. 1, NMML, Delhi, pp. 1–2.

94. For a deeper understanding on this issue, see Kathakali S. Bagchi, *Irrigation in India*, Delhi, 1995, p. 98.
95. Ibid., pp. 101–2; see also Ramaswamy R. Iyer, *Water: Perspective, Issues and Concerns*, Delhi, 2003, pp. 126–8.
96. Kathakali S. Bagchi, *Irrigation in India*, p. 104.
97. Ramaswamy R. Iyer, op. cit., p. 130.

CHAPTER 12

The Re-inheritance of Heritage: The Construction of the *Shubhankari* Tradition in Colonial and Post-Colonial Bengal

SANTANU CHACRAVERTI

This paper is about how the *Shubhankari* tradition came to be constructed and fashioned during colonial and post-colonial periods. First we clarify what the term *Shubhankari* meant. *Shubhankari* seems to have had different, albeit related, denotations. For most users the term denotes a corpus of computational verses composed by different, occasionally anonymous, authors, the most renowned of whom being one Shubhankar. Many of these verses (called *arjyas*) incorporated algorithms facilitating practical computations with indigenous units of measure and currency. Others posed simple arithmetic or algebraic problems. These verses were taught to children both in elementary schools and at home. The practical computational algorithms were supposed to be so ingenious that anyone memorizing them could solve complicated problems with ease and very often without resorting to pen and paper.

When we speak of the 'construction' of the *Shubhankari* tradition we are concerned purely with Bengalis, or more specifically the active agents of Bengali modernity. The first modern references in English to Shubhankar and *Shubhankari* were made by Europeans, e.g. by Francis Buchanan [1] and William Adam. The latter makes several references to *Shubhankari* in his celebrated Reports and the references are of distinct value to historians of the *Shubhankari* tradition.[2] Another, rather dramatic, reference to *Shubhankari* is by Reverend James Long.[3] But it is the present author's understanding that the European references are tuned to a different pitch and may not be analysed in the same breath as native pronouncements. I shall thus refer to European utterances on *Shubhankari* only in passing and not as integral elements in my narrative.

BANKIMCHANDRA CHATTOPADHYAY

One of the early explicit references to *Shubhankari* by a Bengali in the nineteenth century is the one by Bankimchandra Chattopadhyaya, whose powerful articulation of Bengali/Hindu/Indian nationalism is known. It will be useful to look at something Bankimchandra had to say regarding indigenous knowledge systems. In May 1873 he published an article entitled 'The Study of Hindu Philosophy'.[4] This article provides key insights into Bankim's understanding of the Indian heritage. He emphatically states that on account of certain characteristics of the Hindu culture, science, which showed such great promise in early India, could not really develop to its natural fruition. His words should be of some interest to the historian of modern Indian thought:

> . . . The intellectual history of Europe bears to that of India the same relation as *physiology* does to *pathology*; while the one presents the richer field for the investigation of the laws of the healthy and vigorous growth of civilization, the other furnishes greater facilities of studying it *under conditions of disease and death*.[5] [italics added]

Bankim nevertheless wishes that his readers understand and appreciate the achievements of the early Hindus and try to find out exactly how such achievements were laid to waste. In this context he asks a number of questions, one of which was, 'What is in fact the real place of Hindu philosophy in the history of Science?'[6]

The reference to 'history of Science' is intriguing and indeed Bankim goes on to show the possible areas of 'Hindu' thought that he considered important to the history of science and scientific method. Regarding Indian scientific achievements, he quite obviously refers to Aryabhata, and as an example of correct method he cites the Nyaya definition of 'cause'[7]. Such examples indicate the specific areas of past accomplishments that Bankim considers sufficiently commendable to be cited as 'achievements'.

Now for the reference to *Shubhankari*. In an article published in 1872, devoted to praising Dr. Mahendralal Sarkar's efforts to establish an Indian association for research in science, Bankim drew the readers' attention to both the excellence of ancient Indian scientific enterprise and what he considered to be its 'subsequent decline'.[8] As an example of the latter, Bankimchandra referred to *Shubhankari* mathematics, which he considered to be starkly inferior to the tradition of higher mathematics that constituted the pride of India. *Shubhankari* for him represented the crude, the gross, the vulgar.[9] The fact that it constituted a popular local tradition, catering to children learning mathematics and people concerned with practical problems of counting, does not seem to have held much importance for him.

BENGALI IDENTITY

Insofar as Bankimchandra's interest in Indian knowledge systems was concerned, it was the tradition of Sanskritic 'high science', embodied in ancient Indian astronomy, mathematics, and Nyaya epistemology that claimed his attention. Yet there was a problem with that particular tradition. It was 'Indian' rather than Bengali. None of the stalwarts of ancient Indian science could be claimed for Bengal.

But the nineteenth century was not merely the century of reconstruction of the ancient Indian heritage—a reconstruction couched in the model of ancient glory and subsequent 'medieval' decline. It was also the century of the refashioning of the vernacular, and of the creation of a vernacular literature and print culture. In Bengal this was the century when prose literature in regional languages was created and vernacular periodicals and newspapers went on to create a mainstream culture. This resulted in the creation of a Bengali nationalism that was at least as powerful in its emotional appeal as Indian or Hindu identity-awareness.

Bankim himself had clearly identified Bengalis as a 'nation' (*jati*). He had proclaimed that the Bengali nation needed a national history, and that national history was a useful *consequent to* and an important *constituent of* national pride.[10]

THE HUNT FOR HERITAGE

The important characteristic of a Bengali consciousness was that it could not really seek its glorious past in remote antiquity. Its historical roots were fairly recent and its character was more mass-based and popular than classical. Therefore the hunt for the past glories of Bengal could not really pin itself on the model of ancient glory and medieval decline. Insofar as Bengal was concerned, it was the post-ancient and the relatively recent which was the history and culture that mattered.

Yet, when the question of past scientific achievements came, Bengal's history seemed to have little to offer: one had to resort to the history of a larger identity, that of India. For Bengal had neither an Aryabhata nor a Bhaskara, neither a Charaka nor a Sushruta.

The seeds of a new recasting of tradition were being sown even while Bankim was lamenting the sad burial of India's past. His own journal, the *Bangadarshan*, in its Chaitra 1279 (March–April 1873) issue carried an anonymous article entitled 'Bangalar Bhagnansha' (Bengali Fractions).[11] This article attended to the indigenous method of fractional notation prevalent in Bengal and the author perforce had to pay attention to *Shubhankari* in this regard. It so happens, and as I have shown elsewhere, this much-neglected article is a crucial one in making sense of *Shubhankari* and seems to have been the first critical treatment of popular counting methods in Bengal.[12]

The formidable scientific superiority of, and the necessity of tutelage to, the West were of course readily accepted within the intellectual mainstream. Hence the glories of India were recounted in terms whose explicit or implicit referents belonged to Western modernity—Bhaskara's anticipation of the modern differential calculus, or how far Aryabhata anticipated the Copernican system.[13] But it was not exactly clear what traditional Bengal had to offer in terms of scientific excellence.

The closing years of the nineteenth century witnessed a striving to discover Bengal's cultural forms and norms, her own literary and cultural wealth. In order to do this one needed to pay attention to the popular and the folk. This new spirit was embodied in the launching of the Bangiya Sahitya Parishat in 1894. The Swadeshi ideal of course also strengthened the new mood. We find Ramendrasundar Trivedi, in an article of 1905, making a moving plea to his readers to help the Sahitya Parishat compile a history of Bengal. He wrote:

> Find us manuscripts; collect local words and usages, *bratakathas*, rural rhymes, rural songs, children's stories, accounts of villages, towns, sacred sites, deities, fragments of histories of sacred sites; fragments of chronicles, anything and everything that can be discovered or procured, and toss them in the direction of the Parishat's office. . . .[14]

But well before Ramendra made his plea, local traditions, local manuscripts, and folk tales were being collected by men such as Nagendranath Basu, Dineshchandra Sen, and Dakhkhinaranjan Mitra Majumdar. In fact, hunting for manuscripts took researchers to remote villages and this manuscript hunt came to be conceived as something of an adventure.[15] Mountain loads of manuscripts were collected during the last decade on the nineteenth and the early decades of the twentieth century and scholarly concern, peeping through nationalistic lenses, could gradually discern the contours of a pre-colonial Bengali culture. But what amidst this topography could be identified as a scientific heritage unique to Bengal? Astronomical and medical texts from medieval Bengal had been collected, but being in Sanskrit and considered mere commentaries on the ancient texts, they did not catch the imagination of the enthusiasts. There was of course the Navya Nyaya, but that was something that once again belonged to the elite Sanskritic tradition and moreover, Navya Nyaya's intense concern with terms, categories, and correct language use probably seemed more 'scholastic' and metaphysical than scientific in a view of science that was influenced by positivism.

But there was one field whose Bengaliness and vernacularity was undisputable and which could be included in the category of 'science' without any difficulty. This was the *Shubhankari* tradition.

THE *SHUBHANKARI* AS HERITAGE: PROSPECTS AND PROBLEMS

As a point to be scored in the game of self-assertion, *Shubhankari* had certain attractive features. First, it was compellingly vernacular, the verses exuding the heady smell of daily life and earthy common usage. Second, and what added to their mystique, was the fact that many of the verses were intriguingly cryptic, and only an expert in the tradition could make sense of all the verses. Third, the verses, as verses (meant to be memorized and transmitted through oral repetition), stood for the culture of orality, an easily recognizable feature of folk heritage.

But then, this tradition was in a certain sense quite alive in late nineteenth- and early twentieth-century Bengal and apparently in little need of 'rediscovery'. In fact, since the second decade of the nineteenth century, arithmetic books for the new schools that were coming up readily incorporated *Shubhankari* verses and methods. Arithmetic textbooks by Robert May and John Harley, both published in 1819 incorporated typical *Shubhankari* verses.[16] And students continued to be examined on their knowledge of *Shubhankari* methods throughout the nineteenth century. During the later part of the century we find school textbooks purely on *Shubhankari* methods.[17]

If we take a look at the contents of *Shubhankari* textbooks or at the *Shubhankari* material included in arithmetic textbooks of the nineteenth century, we find that it is almost wholly confined to what is practical computation—involving indigenous units of weight and measure. In fact, it is the persistence of these indigenous units through the colonial period that made *Shubhankari* verses and methods relevant. It is on account of the usefulness of this kind of teaching that William Adam finds the teaching in the *pathshalas* eminently useful,[18] and Reverend James Long applauds Shubhankar's verses in his *Descriptive Catalogue of Bengali Works* published in 1855.[19] Until 1894 there was a separate paper on *Shubhankari* methods in the primary and middle school scholarship examinations. Therefore, there were textbooks dedicated entirely to *Shubhankari*. One such book from the late nineteenth century is of course Panchanan Ghosh's famous *Shubhankari*.[20]

In the context of the utility of *Shubhankari* verses in practical computation, it must be borne in mind that quite a bit of *Shubhankari* pertained to the cowry currency of Bengal. In fact, it can be shown that almost the entire paradigm of its computation was organized along the lines of traditional cowry counting. During the early decades of the nineteenth century, the colonial government had been taking systematic steps to eradicate the cowry as a medium of currency, replacing it with coins of fixed value.[21] Insofar as the government was succeeding in its efforts (indeed, by the late nineteenth century one no longer hears of cowries as even unofficial currency anywhere in Bengal), the value of *Shubhankari* was being somewhat reduced. *Shubhankari* retained some value, for the other units of measure continued in existence, but, in an academic atmosphere increasingly exposed to the rich and rigorous methods of modern mathematical procedures, its value was under very real threat. Euclidean geometry, basic algebra, and the elements of astronomy had already been incorporated in the school mathematical curriculum. A candidate for the entrance examination to the University of Calcutta from 1858 was examined not only on arithmetic, algebra, and geometry, but also on elementary mechanics. And in the Bachelor of Arts curriculum a student was compulsorily exposed to arithmetic and algebra, geometry, plane trigonometry, mechanics, hydrostatics, hydraulics and pneumatics, optics, and astronomy.[22] Those taking mathematics as an Honours subject were of course subjected to a full course of still higher mathematics. Those who completed the Mathematics Honours curriculum of the University of Calcutta easily had more powerful mathematical tools at their disposal than a Brahmagupta or a Bhaskara. Moreover, while arithmetic textbooks continued to incorporate *Shubhankari* verses and to use the traditional Bengali notations for traditional units of currency, weight, and measure, the character of school arithmetic in colonial Bengal was itself undergoing rapid change during the nineteenth century. Increasingly, following changes in arithmetic instruction in England, school textbooks started giving more stress to 'basic' operations. For example, in Jadavchandra Chakraborti's *Patiganit* that came out in the 1890s, we find the author stressing in the Preface that the unitary method was a basic computational concept and was not only more fundamental than the 'rule of three' but all problems that could be solved by the latter rule could be solved by the former method (and in a more mathematically intuitive manner). In this context the author refers to the prevailing practice of teaching arithmetic in England.[23]

Actually by the second half of the nineteenth century, mathematical instruction in schools and colleges had easily outstripped the bounds of *Shubhankari* instruction. What then did *Shubhankari* have to offer in support of the claims of those who were intent on proving the excellence of indigenous heritage, apart from certain practical advantages it conferred in negotiating with indigenous units?

A TEXTBOOK WITH A DIFFERENCE

An answer to this question was given by an unknown author, one Mahendranath Mukhopadhyay of Bankura. His *Asthita Samadhan* was published in 1895.[24] This was a book written for schoolchildren, but it was not a usual textbook. But we will understand the nature of the book better if we attend to its title.

By the term 'Asthita Samadhan' the author means the 'Solution of *Asthita Panchak* Problems'. And *Asthita Panchak* in the Bengali language stands for mathematical problems of varying levels of difficulty that could call for algebraic methods. Some of these problems were what we would call simple equations and others were, interestingly, problems that would today be described as ones involving indeterminate equations (in the first degree).

The author clearly states the objective of this book: as one derives great pleasure from the mathematical problems posed by the great Bhaskaracharya in his *Lilavati*, similarly one may derive great pleasure from the Asthita type of problems present in Bengali prose and verse. Unfortunately, such problems, which were once staple among the *gurumahashayas* of Bengal, were gradually dying out. Yet, if one admits that the primary aim of mathematical instruction lay in cultivating the logical use of the intellect, then it is desirable that such problems be taught to children.[25]

At one stroke, with the reference to Bhaskara and *Lilavati*, the author places the tradition of mathematical instruction in the *pathshalas* on a level equivalent to the great tradition of classical Indian mathematics. Moreover, by insisting that by learning to solve problems widely discussed in the traditional Bengali mathematical lore students will learn to make logical use of their intellect, the author makes a strong claim in favour of traditional mathematical *vis-à-vis* the contemporary school instruction that was shaped on the European model.

The author tells us in his foreword that he had compiled *Shubhankari* problems from the *pathshala gurumahashayas* themselves. This is particularly interesting in light of the fact that in the book he has solved many problems that are clearly algebraical, in a manner that may be described as arithmetical. This method becomes particularly interesting when tackling problems that would today be described as ones involving indeterminate equations. In solving such problems the author proceeds in what would be described by historians of mathematics as the *regula falsi* method or the *rule of false position*[26]—what the layperson would describe as the intelligent trial and error method.

We know from extant *Shubhankari* manuscripts that problems involving indeterminate equations of the first degree do occur in the *Shubhankari* tradition. However, the manuscripts do not seem to offer any clue to how such problems were solved. Mahendranath Mukhopadhyay is important in that he possibly provides us with the only clue to how exactly such problems were solved within the Shubhankari tradition. However, we need not get into a discussion of such methods in this particular essay. We do not know how many people read Mahendranath Mukhopadhyay's work, or how far its contentions gained acceptance. But the very fact that *Shubhankari* manuscripts got collected during the early decades of the twentieth century shows that manuscript collectors had acquired an eye for them. *Shubhankari* manuscripts are to be found in the Asiatic Society vernacular manuscripts collection, the Visvabharati Bangla Punthisangraha, in Bishnupur, and in the University of Calcutta collection. All these manuscripts were collected during the early decades of the twentieth century.

LALITMOHAN BANDYOPADHYAY

Apart from the lure of the Asthita Panchak, the *Shubhankari* had other interesting features. Or that is how it was represented and interpreted by Lalitmohan Bandyopadhyay, whose crucial contribution to the understanding of the *Shubhankari* was published in *Bharatavarsha* in 1917 in two consecutive instalments.[27] These constitute an important moment in the modern assessment of *Shubhankari*. The first instalment is particularly fascinating as it tries to translate the arithmetic and logic of the indigenous approach in contemporary mathematical terms and is required reading for anyone wishing to make sense of the tradition.[28] The second instalment is also important as it provides important clues to *Shubhankari* methods of dividing mixed quantities by mixed quantities of the same type.[29] Lalitmohan's article presents a highly interesting engagement, the review of a cryptic algorithmic tradition in modern analytical terms. What is also important about the article is that its author provides an interesting bit of historical information. He tells us that by the second decade of the twentieth century few in the modern educated milieu had a firm grasp of *Shubhankari* computational principles.[30] (This corroborates what I learned from oral interviews of very elderly people who had studied *Shubhankari* methods at school in the 1910s. The interviewees declared that the common practice was to memorize *Shubhankari* verses without ever truly understanding their exact significance.)

ANOTHER TEXTBOOK

It was possibly this state of affairs that led to the publication of the famous textbook on *Shubhankari* by Madhusudan Deb, some time in the 1920s. The copy that I possess is the third edition published in 1936.[31] In this book, the author takes great effort to make the principles of *Shubhankari* clear to the student. The relative values of the traditional units of weight and measure are carefully enumerated, and how the *Shubhankari* verses aid computations involving these traditional units is shown step by step. Moreover, this particular book shares Mahendranath's concern with the *Asthita Panchaka* elements in the *Shubhankari* tradition—an entire chapter is devoted to *Asthita Panchaka* and the methods of solving such problems.[32]

DINESHCHANDRA SEN

But a rather forceful and very different kind of statement about the significance of the *Shubhankari* tradition was to come from someone who had little to do with mathematics. This was Dineshchandra Sen. His *Brihat Banga*, first published in 1935,[33] is devoted to tracing a cultural history of Bengal from the earliest times and exploring the Bengali cultural heritage in its fullness. A more than usual dose of Bengali chauvinism may be detected in the work. But there is an aspect of the work that places it squarely within the Sahitya Parishat tradition, with its deep emotional commitment to rediscovering Bengal at the roots of her local and vernacular existence. Thus Dineshchandra's work pays painstaking attention to local and popular cultural traditions. In this the second volume of the work devotes a considerable section to pre-colonial popular education and includes large chunks from *Shubhankari* manuscripts in his possession.[34]

As this work would have it, the harking back to the traditional is not merely a journey of emotive rediscovery of one's roots. It is also a matter of indicating the intrinsic value of certain elements in one's tradition against such Westernized elements that one has blindly imitated. It is with this in view that Dineshchandra offers an assessment of the *Shubhankari* heritage. He stresses on two related aspects of *Shubhankari.* The first is the organic naturalness that follows from its being a key element of the popular wisdom of rural Bengal, enshrined in traditional Bengali popular pedagogy. The latter, according to him, was able to communicate useful education in a natural way that Westernized school-college education could not. Thanks to this, *and this is the second aspect*, peasants of Bengal could undertake practical measurements of land areas, prices of goods, and so on, very often orally, and with a speed and facility that was ordinarily beyond the abilities of educated people taught in the modern manner and holding postgraduate degrees. Moreover, the estrangement from and neglect of traditional knowledge that 'modern' pedagogy entails had resulted in the manufacture of unnecessarily artificial Bengali terms to denote mathematical operations—when simple and elegant expressions existed in traditional computational lore. Dineshchandra insists that what has subsequently been much applauded and glorified as 'mental arithmetic' in erudite circles is none other than the age-old principles underlying the traditional expertise of country people in employing simple yet brilliant techniques to facilitate complex computation.[35] The nub of Dineshchandra's critique is that modern Europeanized pedagogy has so estranged us from the genius and pedagogy of our own people that we have failed to appreciate and imbibe its obvious virtues. Consequently, Dineshchandra calls upon modern teachers of mathematics to pay more attention to their indigenous traditions.[36] Apparently, it is not Dineshchandra's intention to discard the modern Westernized system of instruction, but to improve its efficacy by drawing upon (or 'appropriating') indigenous reserves.

HEMENDRANATH PALIT

Another person who took a great interest in *Shubhankari* in the pre-Independence era was Hemendranath Palit. He was a person from Bankura who was keenly interested in the local traditions of his district. So far as our concerns go, he was important for two reasons. The first was an abiding interest in the *Shubhankari* tradition itself, which led him to collect an impressive collection of *Shubhankari* verses. He did not live to see the publication of his very important work, *Shubhankari*, but more of that later. Second, it was primarily on account of his efforts that the identity of a historical Shubhankar, as a resident of Bankura, seems to have been built.[37]

AFTER 1947

The Government of India shifted to the metric system and decimal currency in 1957. Although land areas continued to be measured in the traditional manner, non-decimal subdivisions of the currency were banished and therewith the *Shubhankari* was rendered useless. Now the interest in *Shubhankari* was only academic. But the shift to the metric and the decimal was symptomatic of a much deeper commitment. Before we can get to understand the nature of the commitment, we need to digress a bit.

NUMBERS, UNITS, AND THE MODERN STATE

What immediately strikes the eye of a modern audience of *Shubhankari* is that the units of weight and measures it deals with are organized on blatant non-decimal lines. The dominating ratios are 1:4 and 1:16. There is also a modest spattering of 1:5 and 1:20. (The proportions 1/3, 1/9 and 1/27 occur at the level of fractions and in fact there is a well-developed system of signs to denote such fractions.[38]) Yet, at the level of pure whole numbers, the notation is the typical decimal place-value one, as one would expect from an Indian computational tradition. This mismatch between units of measure and the system of notation creates obvious problems of counting (just recall the problems one encounters when one uses the decimal notation to express square inches in terms of square feet). No decimal system of metrics had developed in Bengal or anywhere else in India. Had decimal metrics developed, the *Shubhankari* would have been rendered redundant or would have not emerged at all. One wonders why a culture that gave birth to a consistent decimal place-value notation did not bring its metrics in concordance with its system of numerical notation. The answer seems to be as follows:

Bringing about change in traditional metrics is an inordinately uphill task as it calls for radical change in peoples' habits and thinking. Such steps call for two things—*first, a centralized state that is endowed with a sufficiently radical vision of drastic change and second, has the power to impose on and inform its subjects on a massive scale.* The second criterion requires that such states have very powerful modes of communication at their disposal. The British colonial state in India fulfilled the second criterion but was understandably lacking in the first, which is why it continued the existing system of units with occasional modifications. Indeed, no state in the history of humankind prior to the Revolutionary Government in France fulfilled both the above criteria, and that is why it was the latter which was the first in the world to introduce the metric system, which brought its measurement units in tune with the decimal place-value notation (that had conquered Europe during the thirteenth to the sixteenth century).[39] Yet even that visionary and powerful government failed to inspire its people to accept the decimal-metrics in measuring time and as a result the world is still stuck with the Babylonian sexagesimal system of time (and incidentally also of angle) measurement.[40]

When India introduced the metric system and decimal currency in 1957, it was the result of a visionary programme to rationalize the system of metrics, on the assumption that modernization and development could only proceed on the basis of a totally rational and uniform system of measurement. It was the same zeal that led the Council of Scientific and Industrial Research (CSIR) to set-up a Calendar Reform Committee in 1952 to devise a uniform and truly scientific calendar.[41]

The Calendar Reform Committee's report was submitted to the CSIR in 1955 and the government, accepting the recommendations of the Committee, decided that a unified National Calendar be adopted for use with effect from 21 March 1956.[42] The new national calendar made a sharp break with indigenous calendrical traditions in introducing the tropical solar year in the place of the sidereal year that had been in use down millennia. How far the new national (*Saka*) calendar found acceptance outside governmental documents and pronouncements is of course a different question.[43]

The new calendar and the shift to the decimal and metric system and of course the importance attached to such an organization as the CSIR (and indeed the very existence of such an organization) indicated that the government of the independent nation had made a firm commitment to modern scientific research, value and practices. The commitment to modern science and technology came *in tandem* with a commitment to industrialization and planned growth. The new nation state was not officially blind to history and tradition, but nothing in the past could really serve as a model for the steps to be taken.

SHUBHANKARI IN THE INTELLECTUAL CLIMATE OF A NEW NATION STATE

In the new intellectual climate, *Shubhankari*, a tradition that stood for the rural, local and non-decimal, could hardly get much attention. It is understandable that in the 1950s and the 1960s we do not find much reference to *Shubhankari*. Notwithstanding the recommendations it had received in the Adam reports, it did not seem to offer a paradigm that needed emulation. However, notwithstanding a general decline in scientific or pedagogic interest in *Shubhankari*, new insight into *Shubhankari* came from another quarter—from Sukumar Sen, an acknowledged authority on the history of Bengali language, literature, and manuscripts. A discussion on *Shubhankari* appeared in the second part of the first volume of Sukumar Sen's *Bangala Sahityer Itihas*, first published in 1963, with the second edition coming out in 1965. Another interesting reference is to be found in the first part, which was first published in 1940 but went through several editions, the fifth edition being published in 1970. Combined together, the two references yield an exceedingly interesting insight into the nature of the *Shubhankari* tradition, and may be summed up as follows.

Computational verses, generally stamped or associated with the name of Shubhankar, are not only a feature of Bengal *but also of Assam*. The name of Shubhankar Das is by no means unknown in Assam. Manuscripts have been found in Assam with verses bearing the names of Shubhankar, Raghupati, Umapati Singha, Sri Narayan Das, Bakul 'Kayastha', and so on. The time of composition of one of the manuscripts has been dated as the early eighteenth century (the period conventionally associated with Shubhankar). Often the language and content of some of the verses found in Assamese manuscripts is similar to some found in Bengali manuscripts.[44] Now, Bengali and Assamese flowed along the same channel prior to the fifteenth century, from which time onwards these languages diverged into different streams. Prior to the fifteenth century, the ancestor of modern Assamese and modern Bengali belonged to the same stock of *abahattha*—a common name for all the various regional variations of the corrupted *Prakrit*.

Now it so happens that the most famous verse associated with the name of Shubhankar begins with the famous couplet, *Kuroba Kuroba Kur,oba Lijje/Kathai Kuroba Kathai Lijje.* It is Sukumar Sen's contention that these first two lines bear the footprints of *abahattha*—the particular variety that was the ancestor of modern Bengali and Assamese (and also incidentally of modern Oriya).[45] Further, a strong tradition in Bengal identifies Shubhankar as a Kayastha and indeed accounting and related mathematics was the traditional professional specialization of the Kayastha caste. It is interesting that the computational rhymes that have been collected from Assam went by the name of *Kayathali Arjyas* (Kayastha verses). It is Sen's contention that the computational tradition associated with Shubhankar hails back to a period to the fifteenth century or even earlier, rather than to the eighteenth century and it is from the common stock of *abahattha* computational lore that both Assam and Bengal derived their local computational algorithms.[46]

Interestingly, Sen's fascinating pointers have not been followed up by later researchers, for reasons we shall explore later.

The late 1960s and the early 1970s witnessed dissatisfaction and disappointment with the performances of the Indian nation state at various levels. Notwithstanding certain important achievements in industrialization, modernization and welfare, it had been rendered abundantly clear that in this world's largest democracy, 'development' had contributed little to the improvement of the actual conditions of the bulk of the *demos*. Suddenly far-left critiques on the one hand, and the traditional Gandhian fascination with the rural and the local on the other, seemed to have valid points in their favour.

In this changing intellectual climate the *Shubhankari* again started gaining an audience. Another point that acted in favour of *Shubhankari* was the growing global interest in the ethnosciences from the early 1970s.[47] The first attention to *Shubhankari* in this new intellectual atmosphere is one by Pradip Majumdar, one of the key investigators in the Indian mathematical tradition. His article, published in 1982, is a brief and informative discussion on the *Shubhankari* tradition.[48] Paramesh Acharya's *Banglar Deshaja Shikkhadhara*, published in 1989, which is the first comprehensive treatment of pre-colonial pedagogy in Bengal, examines in detail the *pathshala* tradition and devotes some space to *Shubhankari.*[49] The next important mention is the one by Haradhan Datta, which appeared in the early 1990s. Datta's article is not an analytical or mathematical treatment of *Shubhankari* but is a rather interesting treatment of the historical questions surrounding Shubhankar and *Shubhankari.*[50]

The Indian interest in ethnomathematics or non-elite mathematical traditions was epitomized in Ashok Jhunjhunwala's *Indian Mathematics: An Introduction*, which was published in 1993.[51] By highlighting the computational methods used by carpenters and accountants trained in the indigenous tradition, Jhunjhunwala opened up a totally unexplored area of enquiry. As a mathematician rather than a historian or anthropologist, Jhunjhunwala was more interested in analytical treatment of the methods that he enumerated. The book immediately caught the attention of mathematicians but unfortunately remained unnoticed by historians of folk or popular culture. Jhunjhunwala's work of course contains nothing on *Shubhankari* and there is no indication that the author was aware of the existence of such a tradition.

Analytical treatment of *Shubhankari* re-emerged in Bangladesh. An article entitled *Shubhankari Arjya*, written by Shahidullah Mridha, was published in 1995 in a special 'Mathematics Issue' of the magazine, *Mizanur Rahamaner Traimasik Patrika* published from Dhaka.[52] Shahidullah's article is one that, after a long time, offers a mathematical treatment of *Shubhankari* computational methods and is recommended reading for anyone desiring to gain a preliminary understanding of Shubhankari mathematics. Its sources are primarily arithmetic textbooks of the pre-Independence period, some of which we have already mentioned. Yet the article has a number of shortcomings, not the least of which is that it has been written solely from a mathematician's standpoint, and tells us nothing of the social, cultural, and economic milieu in which *Shubhankari* developed. As I have shown in detail elsewhere, such a discussion is not only crucial in its own right but is indispensable for an understanding of *Shubhankari* mathematics itself.[53] The very same number of the magazine carries another interesting article by one Araj Ali Matubbar, who also had considerable experience of working as an *amin*.[54] This is on traditional mensuration techniques and incorporates the *Shubhankari* and gives an indication of the evolution of a tradition, of how in the practical day-to-day working of an *amin*, measurements with traditional units could get wedded to measurement with more modern units.[55] The article deserves extended analysis, but that must wait for another occasion.

In 1997 was published an anthology of articles on history of Indian science entitled *Bharater Itihase Bijnan O Prajukti*. This anthology contained an article by Nandalal Maity.[56] This is probably the first article in Bengali whose sole and declared purpose is to explore the Indian tradition of Lokaganit or folk-mathematics as a whole, as against the traditional obsession with India's classical mathematical tradition. Maity's article deals with *Shubhankari* and the historicity of Shubhankar, though as that treatment is rather brief, it whets the appetite rather than satisfying it.[57]

A recent addition to the literature on *Shubhankari* is the publication of the long-awaited work by Hemendranath Palit. It has been re-edited by Ashok Kumar Palit and published by Burdwan University Press in 2001.[58] This slim but rich volume is a highly useful annotated anthology of traditional *Shubhankari* verses, followed by painstaking explanations by the author and editor. It is possibly the most important contribution to the treatment of *Shubhankari* that has been published since Independence.

POST-INDEPENDENCE APPRAISAL OF *SHUBHANKARI*

A few things immediately catch one's eye on perusing the post-Independence writings on *Shubhankari*. Unlike the colonial investigations on *Shubhankari*, the post-colonial explorations are generally less passionate and less emotionally charged—proving the superiority of an indigenous tradition seems less of an issue than what it was prior to 1947. But as compared to the pre-Independence investigations the post-colonial explorations are less stimulating and illuminating. Nothing that has been written on *Shubhankari* in the post-colonial period, except for Palit's annotated anthology of *Shubhankari* verses, matches the depth of Lalitmohan's article, and often seems redundant when one remembers the detailed explanations of *Shubhankari* mathematics in Madhusudan Deb's textbook. Let us try to see what is at issue here.

Shubhankari embodies two quite distinct elements. One is the element of practical computation and the art of negotiating with traditional units of weight and measure. At the very basis of this practical element is a certain paradigm of counting that follows from the traditional counting of cowries. The other is not practical at all, but might be described as non-practical mathematics. Belonging to this second element are playful explorations of highly interesting numbers and interesting arithmetic and algebraic problems.

It is important to understand both these elements. The first involves an understanding of how cowries were counted and also an understanding of the economic role of cowries in pre-colonial Bengal and their fluctuating relative value *vis-à-vis* the silver *taka* or rupee. An appreciation of the second point involves the understanding of an aspect of popular culture of pre-colonial Bengal, where arithmetic and algebraic problems and exploring strange and interesting numbers are part of the play element in popular culture—pastime and riddle-lore.[59] What the examination of these elements together entails, is the founding of an 'anthropology of traditional Bengali counting and number use'. Such an anthropology is yet to emerge in published form.[60] The question is why such a narrative did not emerge. There seem to be several factors that are implicated in the answer.

While our academia has woken up to the ethnosciences, thanks to international interest in the same, the interest is still somewhat superficial. During the nationalist period, the nationalist zeal to discover indigenous treasures led scholars to undertake serious and often pathbreaking investigations, but the same passion was absent in a post-colonial scenario. Besides, mathematicians who take an interest in *Shubhankari* prefer to base their interpretations on published work. Neither their training nor their proclivities incline them towards dusty and illegible manuscripts. They tend to lack interest in exploring cultural and economic contexts, whereas historians are more comfortable with manuscripts and contexts but tend to lack interest in traditions of computation. All these factors finally added up to the fact that in evaluating and understanding the *Shubhankari* tradition, the mathematician's and the historian's approaches have not converged.

Another quite different kind of limitation in the investigation of the *Shubhankari* tradition has resulted from its stemming originally from a Bengali Swadeshi spirit and identity-consciousness. The main issue being to prove the excellence of traditional Bengali wisdom, the result has been to ignore the above-mentioned clue provided by Sukumar Sen. While Sen clearly indicated the existence of *Shubhankari* type verses in Assam and indeed gave explicit references to manuscripts, no subsequent scholar on *Shubhankari* took it on himself to investigate such manuscripts. And Sen's reference to '*Kayathali arjyas*' in Assam does in fact open up an entire world of possible exploration. The existence of a traditionally literate Kayastha community associated with the scribes' and accountants' profession has been an important social reality not only in Bengal and Assam but also in the Bihar region. One wonders whether *Shubhankari* type computational verses were available in the Bihar region also and whether the *Shubhankari* tradition was a Bengal variant of a much wider computational tradition prevalent in pre-colonial eastern India. If such was indeed the case, then the question arises about possible links to popular computational traditions elsewhere in India. These are questions that have remained outside the ambit of the scholarship on *Shubhankari*. Actually, all this comes also as a self-criticism, for the present author, in his own investigations into the *Shubhankari* tradition, has failed to follow up on the valuable clue provided by Sukumar Sen four decades ago.

NOTES

1. Cited in Kalikinkar Datta, *Survey of India's Social Life and Economic Condition in the Eighteenth Century, 1707-1813*, Delhi, 1978, p. 45.
2. William Adam, *Report on the State of Education in Bengal (1835 & 1838) including Some Account of the State of Education in Bihar and*

a Consideration of the Means Adapted to the Improvement and Extension of Public Instruction in Both Provinces, ed. Anathnath Basu, Calcutta, 1941, pp. 143, 232–3, 238.

3. Cited in Dineshchandra Sen, *Bangabhasha O Sahitya*, Calcutta, 1991, rpt., vol. II, p. 737.
4. Bankimchandra Chattopadhyaya, *The Study of Hindu Philosophy*, compiled in Jogesh Chandra Bagal, ed., *Bankim Rachanavali, English Works,* Calcutta, 1969 (rpt. 1990), pp. 142–8.
5. Bagal, ed., *Bankim Rachanavali*, p. 142.
6. Ibid., p. 147.
7. Ibid., pp. 147–8. Bankim writes 'Aryabhatta' instead of the correct 'Aryabhata', thereby making a mistake that has had a long and distinguished history.
8. Bankimchandra Chattopadhyaya, 'Bharatavarshiya Vijnanashabha', *Bangadarshan*, Bhadra, 1279 BS, Calcutta, vol. 1, 1974 (1988), pp. 205–12.
9. Ibid., p. 209.
10. Bankimchandra Chattopadhyaya, *Bangalar Itihas Sambandhe Kayekti Katha*, Calcutta, vol. 7, 1982, pp. 277–83.
11. Anonymous, *Bangalar Bhagnangsha*, Calcutta, vol. 1, 1974 (1988), pp. 473-9.
12. Discussed in some detail in Santanu Chacraverti, *Subhankari: An Indigenous Tradition of Elementary Mathematical Instruction*, Kolkata, 2007.
13. An article in the *Dawn*, October, 1903, is entitled 'Was Bhaskaracharya Acquainted With the Principle of the Differential Calculus?'. See Haridas Mukherjee and Uma Mukherjee, *The Origins of the National Education Movement*, Calcutta, 1957, p. 325. For an early reference to the Copernican theory in connection with Aryabhata's proposition of diurnal rotation of the earth see Bankim's view in Bagal, ed., *Bankim Rachanavali*, p. 147.
14. Brajendranath Bandyopadhyay and Sajanikanta Das, ed., *Ramendra Rachanabali*, vol. 6, Kolkata, 1363 BS, pp. 141–2.
15. At least that is how it has been portrayed by Dineshchandra. Sen, *Bangabhasha O Sahitya*, vol. 1, Calcutta, 1986 (1991), pp. 17–18.
16. See discussion in Buddhadev Bhattacharya, *Bangasahitye Vijnan*, Calcutta, 1980, pp. 6–9.
17. The most famous and the one still available is of course the one by Panchannan Ghosh, *Shubhankari*, Calcutta, 1893.
18. William Adam, *Report on the State of Education in Bengal*, ed. Anathnath Basu, p. 146.
19. Dineshchandra Sen, *Bangabhasha O Sahitya*, vol. II, p. 737.
20. Panchanan Ghosh, *Shubhankari*.

21. D. Chakravarty, *Nineteenth and Twentieth Century Coins of India*, Calcutta, 1979, pp. 9–27. Also, Dharma Kumar, ed., *The Cambridge Economic History of India*, New Delhi, 1991, pp. 95, 276.
22. Pramathanath Banerjee et al., eds., *Hundred Years of the History of Calcutta*, Calcutta, 1957, pp. 63–4.
23. Preface to the First Edition of the book published in 1299 BS (i.e. 1892–3) that has been included in Jadabchandra Chakraborty, *Patiganit*, New Edition, Calcutta, 1339 BS.
24. Mahendranath Mukhopadhyay, *Asthita Samadahan*, Bankura, 1895.
25. Ibid., preface, pp. i-ii.
26. Ibid., pp. 107–70.
27. Lalitmohan Bandyopadhyay, 'Shubhankar', *Bharatavarsha*, Fifth Year, vol. 1, no. 5, Kartik, 1324 BS, pp. 670–3, and Lalitmohan Bandyopadhyay, 'Shubhankar', *Bharatavarsha*, vol. 1, no. 6, Agrahayan, 1324 BS, pp. 860–3.
28. Ibid., vol. 1, no. 5, Kartik, 1324 BS, pp. 670–3.
29. Ibid., vol. 1, no. 6, Agrahayan, 1324 BS, pp. 860–3.
30. Ibid., vol. 1, no. 5, Kartik, 1324 BS, p. 670.
31. Madhusudan Deb, *Shubhankari*, Calcutta, 3rd edn., 1936.
32. Ibid., pp. 183–202.
33. Dineshchandra Sen, *Brihat Banga*, vols. 1 & 2, first published in 1935. The edition used in this essay is Dineshchandra Sen, *Brihat Banga*, vols. 1 & 2 (rpt., Calcutta, 1993).
34. Ibid., vol. 2, pp. 896–903.
35. Ibid., pp. 896, 902–3.
36. Ibid.
37. For Hemendranath Palit's views on Shubhankar as the inhabitant of Bankura see Hemendranath Palit, 'Bishnupurer Itihaser Kayekti Natun Katha', *Prabasi*, Agrahayan, 1341 BS, pp. 242–5. An excellent summary of Hemendranath Palit's views is Haradhan Datta, *Sekaler Shikkhaguru*, Calcutta, 1384, pp. 263–4.
38. Anonymous, *Banglar Bhagnangsha*. A more contemporary discussion is to be found in Chacraverti, *Subhankari: An Indigenous Tradition of Elementary Mathematical Instruction*.
39. For a delightful, albeit brief, account of how the decimal place-value notation gradually took hold of European culture, John D. Barrow, *Pi in the Sky: Counting, Thinking and Being*, New York, 1992, pp. 99–101.
40. It should be mentioned at this point that there is no advantage intrinsic to a decimal system of notation. The number ten *possesses no special charm nor does it enjoy any special privilege vis-à-vis* any other number. The great achievement of Indian scholars lay not in hitting upon ten as a base number, but in developing a consistent place-value notation. A consistent place-value notation can have any

number as base, e.g. our modern binary system uses a place-value notation that has the number two as its base. Eight, twelve or sixteen, or any other number would do as well as base in a place-value notation. But once a base is chosen, it is convenient if one uses units of weight and measure that build upon the base number, for only then can the numerical notation yield the best advantage in the writing of units and in the conversion from a smaller to a higher unit, or vice versa. The great achievement of the French Revolutionary Government lay in the appreciation of this principle.

41. M.N. Saha and N.C. Lahiri, *History of the Calendar in Different Countries Through the Ages*, Delhi, 1992, Foreword.
42. Ibid.
43. For a discussion of the degree of acceptance or non-acceptance of the *Saka* tropical calendar see S.K. Chatterjee, *Indian Calendric System*, Delhi, 1998, pp. 18–30.
44. Sukumar Sen, *Bangala Sahityer Itihas*, vol. 1, part 2, 2nd edn., Calcutta, 1965, pp. 524–7.
45. Sukumar Sen, *Bangala Sahityer Itihas*, vol. 1, part 1, 5th edn., Calcutta, 1970, pp. 10, 56.
46. Sukumar Sen, *Bangala Sahityer Itihas*, vol. 1, part 2, pp. 524–7.
47. The sensational event which made ethnomathematics a charismatic discipline overnight, was the publication of Claudia Zaslavsky, *Africa Counts*, Boston, 1973.
48. Pradip Kumar Majumdar, 'Shubhankara: A Great Mathematician of Bengal', *Mathematics Education*, vol. XVI, no. 2, pp. 21–6.
49. Paramesh Acharya, *Banglar Deshaja Shikshadhara*, Calcutta, 1989, pp. 117–22.
50. Datta, *Sekaler Shikshaguru*.
51. Ashok Jhunjhunwala, *Indian Mathematics: An Introduction*, Delhi, 1993.
52. Shahidullah Mridha, 'Shubhankari Arjya', *Mizanur Rahamaner Troimasik Patrika*, vol. 12, no. 4, Issue 39, 1995, pp. 454–72.
53. Chacraverti, *Subhankari: An Indigenous Tradition of Elementary Mathematical Instruction*.
54. Ayub Hosain, 'Araj Ali Matubbar O Khetraphal Prasange', *Mizanur Rahamaner Troimasik Patrika*, vol. 12, no. 4, Issue 39, 1995, pp. 452–3.
55. Araj Ali Matubbar, 'Saral Khetraphal Ba Kalikasha', *Mizanur Rahamaner Troimasik Patrika*, vol. 12, no. 4, Issue 39, 1995, pp. 404–51.
56. Nandalal Maity, 'Lokaganit: Ganiter Bikash O Samriddhite Ek Phalgudhara', in Aparajita Basu, ed., *Bharater Itihase Bijnan O Prajukti*, Calcutta, 1997, pp. 45–66.
57. Ibid., pp. 52–6, 61.

58. Hemendranath Palit, ed., Ashok Kumar Palit (re-ed.), *Shubhankari: Rarh Banger Ganit Padabali*, Burdwan: Burdwan University, 2001.
59. Chacraverti, *Subhankari: An Indigenous Tradition of Elementary Mathematical Instruction*.
60. The concept of 'an anthropology of numbers' is taken from the title and contents of an enthralling work, Thomas Crump, *The Anthropology of Numbers*, Cambridge, 1991.

CHAPTER 13

Finding a Home for the History of Science in Post-Colonial India (1950–1960)[1]: The Influence of Joseph Needham and the Role of UNESCO

DHRUV RAINA

This essay is a prelude to a larger study on the influence of Joseph Needham on the first generation of historians of science working in independent India and their attempt to inaugurate an equivalent of the Needhamian project on the science and civilization of India, within a few years of the commencement of Needham's *Science and Civilization in China* (hereafter *SCC*) project. The focus is not as much to identify specific historiographical influences on their work,[2] though their importance is not being ruled out, as it is to examine Needham's role in the Indian project(s) in the decade 1950 to 1960. The plausible non-emergence of a Science and Civilization in India has been explored elsewhere.[3] This essay concerns itself with the network of Needham's Indian collaborators, and the attempt to forge a community of historians of science working on India. In doing so I venture to understand what Needham meant to his Indian colleagues. In particular, the larger study draws heavily upon the correspondence between Needham, Debiprasad Chattopadhyaya, Abdur Rahman, and S.N. Sen, between 1945 and Needham's death in 1995.[4] This correspondence throws light on the attempts to institutionalize the history of science in India in the first two decades following the end of British rule in India. It also reveals how international networks of scholars from the newly independent nations and the developed world were established during the era of the Cold War; and this was a task that Needham, in his incarnation within the world of science policy, had devoutly committed himself to.[5]

By the early 1950s Needham had more or less dedicated himself entirely to the study of the science and civilization in China,[6] carving for himself a niche in the Western academic sphere as a historian of Chinese science. By this time he was probably quite marginal to the department of history and philosophy of science at Cambridge, which in the post-war years had decisively revised its research and teaching priorities.[7] This was also the decade when movements for decolonization were raging through much of Africa and Asia; and decolonization was not just about the shift of power from European colonizers to Asian or African elites; it was as much a cognitive movement that sought to reconstruct a new identity within academic institutions that had or would be established.[8] It is this conjuncture in Needham's research priorities and the ones that were a consequence of decolonization in the Third World that is particularly interesting, for it opens a window to the reception of Needham in the Third World.

Even a decade before India declared Independence of British rule, scientists here had forged ties with members of the Cambridge Left.[9] With the processes of decolonization well under way, scientists from Third-World countries were looking for avenues to forge collegial ties across distances with their counterparts in the West.[10] For the Indian Needhamians, Needham was a figure endowed both with intellectual and institutional authority who could forward their claims and work to Western audiences still unfamiliar with the subcontinent. For his Indian counterparts, his importance in science and the history of science in the West was often disproportionately blown out of context, but in those early years he was perhaps the only soul in empathy with their dilemmas. On a less serious note, what Hardy was to Ramanujan, Needham was to this generation of historians of science from India. The analogy holds only so far as it suggests the distribution of power and influence or patronage in the world of science[11] even after the formal end of colonialism.

This was an issue very central to Needham's own concerns within UNESCO.

A founding thesis of the cross-cultural studies of science may be rephrased for our present purposes: while there may be a similarity in the controversies about science and society across countries and nations, there will be differences in the representations and images of science, the patterns of debate, and the rhetorical devices used to mobilize support for a set of programmes.[12] The similarities may possibly reside in the inauguration of a discipline across countries, but the pattern of institutionalization will be conditioned by the place of science in a given cultural context. In that sense the establishment of disciplines is embedded within the sociology of positionality.[13] It would be interesting to discursively approach the issue of Needham's Indian network with a brief digression into the literature on scientific internationalism, for that provides the disciplinary frame for discussing the collegial relationships and civility that brings together scientists from different worlds.

A distinction is drawn between two modes of scientific internationalism: non-governmental and inter-governmental. In the former case communities of scientists organize themselves to serve a common transnational aim. This mode serves the ends of science and science alone. Inter-governmental internationalism on the other hand is a vehicle for the promotion of cultural, political, and other interests—heterotelic internationalism was promoted by governments for extra-scientific purposes, which made of science a vehicle of politics by other means.[14]

Cambridge was the fount of influence for the early generation of physicists in India, several of them having worked with Fowler, Eddington, and later Dirac; the crystallization of disciplinary history of science and the theoretical debates in Cambridge from the 1930s had their impact on practising scientists in the Indian world.[15] Between the 1930s and the years of the Cold War the discourse of scientific internationalism would undergo change. The purely autotelic voices of the 1920s were joined by a heterotelic chorus that became particularly loud in the 1930s. This chorus was orchestrated in the form of movements of scientists in popular fronts and the unions of scientific workers created at the time, which was to survive in to the decades well after the Cold War had commenced.[16] During the inter-war years, the changing social and cognitive environment of science and the complex evolution of international regimes witnessed a trifurcation of positions. These Elzinga has called the *traditionalist*, the *leftists* sedimented in Bernal's social function of science, and the *liberal pragmatists* in a sort of middle position adopted by those working towards the enlargement of the agenda of UNESCO.[17] The period from the end of the Second World War to the 1970s is one where the social responsibilist ethic of science was in decline, and where there was constant trading between the autotelic and heterotelic approaches. The context of the historiography of the internal and external is provided by the ideology of the Cold War that separates the ideational contents of science from the politics of science.[18] The internal-external historiography, then, possibly served as a crucial cultural narrative of the Cold War that portrayed an image of science as impartial and neutral.[19]

DECOLONIZATION AND GLOBALIZATION

The colonization of the South Asian region was completed by the end of the nineteenth century, but the process of decolonization was underway by the middle decades of the twentieth century. The distinct historical and cultural locations of the nations under colonial rule shaped the institutionalization of modern science in these nations along different lines. In other words, as these nation states emerged from decades of anti-colonial struggle, their paths to modernization revealed a variety of political and cultural complexions.[20] The process of decolonization was premised on a sovereign commitment to societal and economic development. The programmes of development were often conceived, debated, refined, and revised during the period of anti-colonial struggle, and found expression in projects and plans that were put in place once independence from colonial rule had been achieved. Modernization through science and industrialization were the crucial ingredients of these projects.[21] Intellectual decolonization was reflected in a change in historical perspective and emphasis, a change whose impact was evident in the social sciences three decades later.

Speaking of a historiographic revolution in the last decades of the twentieth century, Sandra Harding pointed out that post-colonial science studies had affected important changes in the historiography of science. For one, it had helped chart out fresh perspectives on the integrity of the European sciences. Second, it had clarified the nature of the relationship between European and non-European cultures. As a corollary to the above, one may add that these studies revised the understanding of the nature of interaction between European and non-European sciences, and went on to provide a different perspective on how modern science itself was reconstituted as a result of the encounter with the knowledge forms of the non-West.[22] The process of decolonization was not a purely cultural or academic endeavour. It involved the refiguration of the institutions of colonial science, dovetailed to serve the politico-economic policies and programmes of the new nation states. This task was rarely conceived as one of the demolition of British legacies but one of pragmatically assimilating elements that were suited to the post-colonial developmental agenda.[23] In South Asia, however, the imperatives of hard earned sovereignty and the pressures of maintaining economic self-reliance created a pressure to leap-frog into the regime of international science.[24] If on the one hand there was an interest in the past of science in India, this was indeed subservient to the more demanding pressures of the fast globalizing scientific research system.

Independence movements had planned for the rapid installation of educational and scientific infrastructure after colonialism. The main actors in the political formations that spearheaded these movements were involved in the planning process and in drawing up programmes for the post-Independence institutions of scientific, technological, and industrial development.[25] Since scientific leadership was closely involved in the anti-colonial struggle, it acquired a great deal of latitude with the post-Independence political leadership in negotiating and fashioning the destinies of the scientific and technological institutions of the independent nations.[26] The political legitimacy conferred by the state on science and vice versa facilitated the rapid building of scientific institutions. In the case of India and Pakistan, a number of strategic areas of scientific research were protected from bureaucratic and political intervention and this created the illusion of autonomy of science from the social institutions that legitimated it.[27]

TWO MARXISTS AND A PHYSICIST: THE INDIAN NEEDHAMIANS

Debiprasad Chattopadhyaya, Abdur Rahman, and S.N. Sen were associated with these movements in Indian science and politics from the beginning of their careers, that is, from the 1950s. Chattopadhyaya was a student of the philosopher Surendranath Dasgupta and remained a communist up to his death. His work from the beginning was influenced by Needham's oeuvre. He travelled to England in the early 1980s and met Needham, who later wrote a foreword to his three-volume history of science of India.[28] Rahman was trained as a chemist and probably met Needham while he was a student in England in the late 1940s; when he returned to India in the 1950s he arrived with letters of recommendation from both Bernal and Needham addressed to Jawaharlal Nehru.[29] S.N. Sen was a physicist, and worked for some time with Needham at UNESCO before he returned to India and commenced his researches on the history of science under the umbrella of the Council of Scientific and Industrial Research (hereafter CSIR) and later the *Asiatic Society*. He shared Needham's ecumenical picture of science, but was not comfortable with dialectical materialism or Marxist historiography.[30]

Needham was acquainted with Rahman and Sen from the late 1940s; his friendship with Chattopadhyaya commenced in the 1970s. Historically, nevertheless, we encounter a deeper impression of Needham's work in Chattopadhyaya than in the researches of the other two. Chattopadhyaya's study of *Lokayata* and ancient Indian materialism, first published in 1959, was already conversing with the still emerging Needhamian corpus. He encountered similarities between the Tantric tradition and Taoism and suggested that by drawing comparisons from

Needham's analysis of Taoism it might be possible to throw light on what appeared obscure in Tantrism.[31] The arguments from Needham's *SCC* that Chattopadhyaya draws upon are a matter for separate study. Despite their ideological and professional differences, Needham remained an important source of inspiration for all of them, and forwarded their research papers and books to Western audiences, long before post-colonial theory and post-modernism assaulted the bastions of Eurocentrism.

The subsequent discussion will take up certain aspects of Needham's relations with one of these three very important figures in the history of science. Whether Needham was successfully mobilized in enabling their vision of the history of science in India, it is still too soon to tell. What is certain is that each one of them had very different motivations for pursuing the history of science that shaped the genre of history within which they were located.

Sen, unlike Rahman and Chattopadhyaya, was located within a liberal pragmatist historiographical framework, which respected very clearly the boundaries between the external and the internal. As mentioned earlier, while sharing Needham's ecumenical picture he was not necessarily empathetic towards historical materialism or the Marxist history of science. He nevertheless anticipated the 'social turn' with some reluctance, but his own researches never veered around in that direction.[32] He co-authored the National Institute of Science (hereafter NIS) sponsored *A Concise History of Science in India*[33]—that has just now been revised after three decades—and co-authored bibliographies on Indian astronomy, published several papers on ancient Indian mathematics and astronomy spread over several publications, and produced an authoritative history of the field sciences, and technical education in colonial India.[34]

Chattopadhyaya and Rahman belonged to the Left, wherein the history of science was not a chronicle about a disembodied entity but interlocked with very contemporary concerns such as those of development, modernization, and the transformation of attitudes. This was the source of solidarity between these two and their comrades of the Cambridge Left. Chattopadhyaya was a philosopher and authored several books on the history of

Indian philosophy and science that were largely preoccupied with explaining the absence of a scientific tradition in India along Western lines.[35] Chattopadhyaya remained a member of the Communist Party till his death in the early 1990s. He fought a long lasting polemic against the votaries of the caste system. During a heated discussion with my colleague Irfan Habib and I, a few years before his death, he confessed to having strived all his life in fighting an ideological battle against the recidivist elements of Indian society who he felt had impeded the advance of science. He published several volumes that were attempts towards an Indian Science and Civilization project. A set of edited volumes appeared in 1982[36] which comprised a collection of already published papers by leading scholars on the subject. In the mid-1980s, he published two gigantic volumes comprising a majority of chapters authored by him and a small number written by scholars who had collaborated with him on the project. His historiography was located in classical Marxism, and the history of science was a forum for him to resolve disputes repeatedly thrown up by Hindu chauvinist interpretations of Indian history and science in particular. The sciences in Indian antiquity, in this interpretation, were erased by the forces of idealism and the history of science was thus a morality tale about the scientific method and development.

Rahman pursued a double career like Bernal, primarily as policy maker and planner, and also as a historian of science. Until the early 1950s, he appears to have been associated with the undivided Communist Party of India. His interest in history was anchored in very contemporary concerns. He reckoned that the history of science had an important role to play in destabilizing standard conceptions of history, entertained by the West about science in India, and by Indians about their own history. This project thus closely engaged with the contemporaneous challenges posed for Eurocentrism as well as the efforts towards intellectual decolonization.[37] As pointed out above, he was primarily involved in the formulation, development and implementation of science policy in the country;[38] and in fact was one of the chief architects of science policy in the 1960s in the CSIR;[39] from which he proceeded to carve out an institute of science studies, that had a unit working on the history and philosophy of science.

Unlike Chattopadhyaya and Sen, Rahman pursued the option of establishing an institution or organization, since he possibly felt that the history of science could not be the pursuit of amateur historians or retired scientists pursuing their disciplines in their spare time. And he had commenced searching for an institutional option from the beginning of his career, an option that would not be realized till almost its end. And so the correspondence between Needham and his Indian colleagues may be read as a history of the attempts to find an institutional home for the history of science in post-colonial India. This is paradoxical, because the 1950s saw a rapid growth in the number and size of universities and research institutes; and yet it was not till the 1980s that the history of science found an institutional basis in the Indian academic world, despite the fact that Indian National Science Academy (hereafter INSA) was supporting a journal and a number of projects; the Asiatic Society was also doing the same.

TROUBLED BEGINNINGS

I have elsewhere discussed the role of INSA, earlier NIS, in officially inaugurating research into the history of science in India.[40] An academy dedicated solely to supporting research in the pure science, INSA decided to organize the first meeting on history of science in South Asia in New Delhi in 1950. In other words, it was to be a meeting blessed by the sovereign Indian state, sponsored in part by UNESCO; it would announce the need to produce a sort of Science and Civilization in India (hereafter SCI). Over the next few decades, many such beginnings would be seen and many projects aborted. Historians in the 1980s would begin trying to understand the absence of a Needhamian SCI. Although Needham did not attend the conference, he was supposed to be the star speaker, and much later was drawn into a controversy that did not so much affect his reputation as his perception of scholarship in India.

In February, 1950 Alexander Wolsky[41] from the UNESCO

Science Cooperation Office for South Asia in New Delhi informed Needham that a meeting was being organized on the History of Science in South Asia, and requested Needham to attend. He also apprised him of the scholars likely to attend the meeting.[42] Subsequent letters from Wolsky to Needham indicate that UNESCO was an important player in the organization of the meeting, for it was suggested that the meeting could be postponed to mid-July 1951 or September 1951 if Needham were certain to attend.[43] Further, it suggests that UNESCO was putting up its star historian of science and wished to play a central role in setting the agenda for the history of science in India. But as time went by, Needham's attendance seemed unlikely. His wife Dorothy was recovering from tuberculosis at the time.[44] Interestingly, Needham requested Wolsky to invite Rahman to the conference. Rahman had not been slated as one of speakers at the meeting—he had probably not yet been initiated into the ranks of historians of science in India and was thus not on the list prepared by NIS.[45] Wolsky did not give up hope till August 1950 that Needham would attend: '. . . we do not give up the idea of having you here as soon as possible. As you know, the symposium is supposed to discuss . . . '.[46] The themes of the conference and the notice were prepared by the physicist D.S. Kothari, Professor of Physics at University of Delhi. While Kothari was a physicist in the renaissance mould, with the mind of a polymath, he was no historian of science. He nevertheless visualized the importance of the discipline; and in fact was responsible for the introduction of a course on the history of science as part of the undergraduate physics curriculum at University of Delhi. The world of science was intervening in institutionalizing the history of science. Wolsky circulated the cyclostyled notice of the conference in September 1950 among leading historians of science of the time such as Needham, Sarton, Pelseneer, Cortesao, and Singer.[47]

The meeting was held, and none of them attended despite all the fanfare that preceded it. Nevertheless, Needham's paper on the 'unity of sciences' was circulated there[48] and the proceedings of the Symposium were published in *Nature* in April 1951. Needham however wrote a very scrupulous anonymous review of the papers and sent it to *Nature.* On 17 May 1951, *Nature* wrote back to Needham concerning his review and remarked that they could not print such an article 'without having a signature appended to it' and they sought his permission to include his name as author of the review.[49] In the meanwhile, Wolsky had published his own review of the proceedings in the *Archives Internationales d'Histoire des sciences* (1951).

In India, all hell broke loose, when the Needham review appeared. It must have been particularly embarrassing for Needham and UNESCO. Here was Needham being staged as the opponent of Eurocentrism, committed to the study of science and culture, taking on the claims of these Indian scientists in the process of rewriting their history after two centuries of colonial rule. I shall not discuss the details of the controversy, for I shall bring it up in a forthcoming paper. The question was one of timing; it was also about the clash of two cultures. The mood of Indian scientists at that time was quite upbeat. The legitimation of modern science within Indian society was initiated by Indian scientists suggesting that India had an ancient scientific tradition and that modern science was not foreign to them. Two years after the formal end of colonial rule they were setting out to reinvent their identities.[50] Needham, on the other hand, was anti-racist, anticolonial, and anti-Eurocentric: that should have endeared him to his Indian colleagues. In fact, Needham's break with scientism was reflected in his opposition to Eurocentrism and his desire to restore ethics into the discussion on science.[51]

So what was Needham's objection to the papers presented at this inaugural meeting? Scientists such as Needham, Pagel, and Bernal had worked towards the creation of an interdisciplinary network that involved non-scientists and in their historical method were critical of presentism and uneasy with anachronism.[52] Needham in particular did not even strictly adhere to positivistic distinctions between science and non-science. And thus while committed to the notion of history of science as important for science, they were simultaneously making a statement about the historiography of science. Thus the advancement of science must be accompanied by a development in the historiography of science in order for the relation to be a synergetic one. Properly executed historiography was of the essence that entailed reflections on the rules of historical contexts and historical explanations.[53] When he encountered these historical papers by his Indian scientist colleagues, Needham was taken aback by the ahistorical conception of science and the patent presentism that underpinned some of the papers. There were questions about the chronology of the Vedas that made Needham uneasy as well. At Needham's end, the papers betrayed an absence of historical method, which he had considered of prime importance in order to make history count for science. For his Indian counterparts, it was a question of recreating a glorious image of the past of Indian science in order to legitimate their own activity and ensure government support of science.

Although the controversy left several parties quite disturbed at the time, it left no trace in the Indian literature five years or even two years down the line. Rahman kept Needham informed of the controversy, but there was nothing to report after 1952.[54] It is rather difficult to explore what the reception to Needham's remarks were after 1953—it was something that was completely erased from the institutional memory of the INSA or the National Commission of the History of Science. However, the promise of Needham's commitment to non-Western science continued to inspire the programme, and Needham's work continued to be rhetorically invoked as paradigmatic for future programmes. But I think it left a very deep scar on Needham's mind; and made him vulnerable to the suggestions of the ground realities of the history of science in India as portrayed by his comrades Rahman and Chattopadhyaya. This probably explained his remark on his trip to India in 1958: 'I met two historians and several nationalists.'[55] I suppose the historians he was referring to were Irfan Habib from Aligarh Muslim University, and D.D. Kosambi of Tata Institute of Fundamental Research. Both of them came from a Marxist background, and had revolutionized the study of Indian history. The Marxist paradigm of Indian history took birth just as the freedom struggle was at its heights[56] and by the 1950s some absolutely startling work had commenced on the nature of colonialism, agrarian relations, etc.

But other than Irfan Habib, who had made contributions to

the history of medieval technology, Needham was not acquainted with this network of Indian Marxist historians. Furthermore, in terms of influence, the world of Indian Science was dominated by the so-called 'Bernalists' almost till the early 1970s. These Bernalists had studied in England in the 1940s and early 1950s or collaborated with physicists such as Bernal, Dorothy Hodgkin, P.M.S. Blackett, and several others.[57] Those of them such as Amulya Reddy who had studied at Imperial College were closer to Hyman Levy. The influence of these Bernalists began to decline as the Nehruvian age came to an end. The picture Rahman and subsequently Chattopadhyaya painted for Needham was reflective of this changing reality in the world of science. Interestingly, in the late 1990s, when a right-wing Hindu government came to power for the first time, the first task they undertook was to take control over the institutions and ministries of educations and research. They then proceeded to revise school and college curricula, and textbooks and to demolish research networks that they felt had been contaminated by Marxist historians and their liberal historian colleagues.

THE FIRST DECADE

As pointed out earlier, the history of science in India figured marginally in Needham's work; but as the years went by he learned about India and what was happening through his correspondence with these three historians of science. He was regularly informed of the progress of their careers and their attempts to find an institutional home for the history of science. While all three of them were affiliated to various scientific societies such as the CSIR, the Asiatic Society, and the INSA, it was not till the 1980s that two of them could move out of the project mode into an institute dedicated solely to the history of science. The Indian Science Congress had been held annually since 1914, and Bernal was the invited guest of honour in 1949. Rahman attempted to use this opportunity to introduce a new history of science section at the Science Congress. He was to inform Needham that despite Bernal's presence and his brilliant lecture, whose significance went unobserved, the attempt proved futile. What was needed was a book on the subject and

Needham's comment on a possible schema was solicited. The chapters were to deal with the nature of science, the multiple temporalities of the scientific disciplines, the possibility of extrapolating the scientific method to other disciplines, the historicity of the scientific method, and the status of technology as science.[58] As mentor or protégéur, Needham was constantly kept abreast of every attempt undertaken by these Needhamians on the history of science: to establish a department, to procure a position, to register a society, to attend a conference; and in this capacity he was constantly called upon to further letters of recommendations for jobs, and positions, fellowships, travel grants that make the business of academics a form of life. Thus in May 1952, we learn that Rahman was offered a position at the University of Delhi, but that the salary he was offered wasn't appropriate and Needham's intervention was sought to resolve the matter.[59] The very next year, Rahman put forward a proposal to the National Institute of Science to develop facilities for research on the history of science in India. The letter was written on the occasion of the 41st Indian Science Congress, this time blessed by the presence of Julian Huxley, who according to Rahman, delivered a very 'disappointing' lecture.

Rahman needed a letter of recommendation from Needham to impress upon the council of the NIS the need for facilities to pursue his research.[60] The problem seems to have been that Rahman was just beginning work in the area. In fact, most historians of science at the time were practising scientists embarking on a different trajectory. Rahman was effectively using his proximity to Needham and Bernal to certify his credentials to deliver on the history of sciences of India. Needham's recommendation reached the council whose meeting was held in May that year; and Rahman was assured by Kothari of the facilities necessary for pursuing his research.

The Indian journal *Science and Culture* had sent Sarton's *The History of Science* for review to Rahman.[61] In his review Rahman objected to Sarton's periodization of Indian history on the ground that Sarton had inherited James Mill's imperial chronology of Indian history.[62] This became a sort of *cause celebre* in Rahman's historiography. By November of the next year, Rahman had succeeded in founding the Indian Society for the History of

Science.[63] The Society was formally inaugurated at the session of the Indian Science Congress held in Calcutta.[64] The Indian Society for the History of Science was patronized by eminent scientists and this remained the pattern within the official history of science.[65] All was not fine with the Society, and Rahman complained to Needham that the society was frittering away money in re-issuing a 'book of hardly any utility', while evoking poor excuses for denying him the same.[66] However, by early 1959 the responsibility for setting up a library in the history of science for the NIS had fallen upon Rahman; and he had arrived, since he was now on the board of the NIS. Rahman informed Needham that the compilation of primary material was of the greatest urgency,[67] and he received a grant from the NIS to plan and prepare an annotated bibliography of source material. The second stage of the project involved period studies and the detailed studies of specialized books. The NIS had by now appointed him as Director of the project on science in medieval India.[68] This was one of the foci for the history of science that emerged in the 1950s, the other was a project that resulted in the publication of *A Concise History of Science in India*. But that had little to do with UNESCO and the network of historians around Needham.

In the period after Independence, as part of the larger project of decolonization, the scientific community sought to reconstruct its identity and aligned itself closely with the project of the developmental state. Research into the history of science was seen as an effort towards the legitimization of science as cultural activity. In these efforts towards institutionalization, historians of science in India worked through collegial circles and networks of progressive colleagues in Europe, who were themselves attempting to recover a humanist legacy for science during the post-World War II period. In this scheme, international organizations played a central role in supporting networks of collaboration between scholars in the developed and developing world. What was different from the programme that emerged in Cambridge or France was that the history of science was patronized by a scientific society that did not enlist specialist historians or experts from the humanities: it was an interdisciplinary formation inaugurated and developed by scientists within the world of science. From the very beginning this remained the historiographic orientation of the officially supported history of science in India.

NOTES

1. This paper began to take form over a discussion with Gregory Blue and S. Irfan Habib at a café in Liège several years ago. Without John Moffet's rapid recovery of the letters that form the core of the discussion, it would not have reached its present form. An Exchange Fellowship of the Indian National Science Academy, New Delhi and the Royal Society, London, enabled me to work at the British Library, the Needham Research Institute, and the University of Cambridge Library. Institutional affiliation was provided by the School of Oriental and African Studies, London, the Needham Research Institute, and the Department of History and Philosophy of Science at Cambridge University.
2. A discussion on the subject is taken up in D. Raina and S.I. Habib, eds., *Situating the History of Science*, Delhi, 1999.
3. Raina, *Images and Contexts*, 2003.
4. Chattopadhyaya, Needham and Sen died within a few years of each other.
5. There are several axes along which the Needhamian project could be configured. One set has to do with his dedication to science and socialism, the institutional commitment to UNESCO, and the project on the scientific and cultural history of mankind. P. Petitjean, 'Needham, Anglo-French Civilities and Ecumenical Science', in S. Irfan Habib and D. Raina, eds., *Situating the History of Science*, Delhi, 1999, p. 152. Also A. Elzinga, 'Revisiting the Needham's Paradox', in the same volume.
6. Needham, *Science and Civilization in China*, 1977.
7. A.K. Mayer, 'Setting up a Discipline, II: British History of Science and "The End of Ideology" 1931–1948', *Studies in History and Philosophy of Science*, vol. XX, 2003, pp. 1–33.
8. Raina, *Images and Contexts*, 2003.
9. We have a list of the books Nehru had read while he was imprisoned by the British authorities between 12 February 1934 and 3 September 1935. This includes 183 books, and among the science books are those by H.G. Wells, Aldous Huxley, Bertrand Russel, J.B.S. Haldane, A. Pinkevich, Gerald Heard, and Patrick Geddes. B. Singh, *Jawaharlal Nehru on Science and Society: A Collection of his Writings and Speeches*, Delhi, 1988, p. xiii. His letters written after 1933, following his return from Europe and the Soviet Union bear titles like 'Darwin and the Triumph of Science', 'The Good and Bad Applications of Science', 'Science goes ahead' (ibid., p. xii).

10. T. Schott, 'World Science: Globalization of Institutions and Participation', *Science, Technology and Human Values*, vol. XVIII, no. 2, 1993.
11. S. Turner, 'Forms of Patronage', in S.E. Cozzens and T.F. Gieryn, eds., *Theories of Science and Society*, Bloomington, 1990.
12. Elzinga, 'Introduction: Modes of Internationalism', p. 1.
13. R. Collins, *Conflict Sociology*, New York, 1975.
14. Elzinga, 'Introduction: Modes of Internationalism', 1996, pp. 3–4.
15. In a paper presented at the Colloquium on engaged biologists in 2003, on the 1931 Congress, Chilvers mentions that an Indian scientist from Calcutta and one from Allahabad were also present—this may have to be checked out, but it is most likely that the scientist from Allahabad was the theoretical astrophysicist, M.N. Saha.
16. Elzinga, 'Introduction: Modes of Internationalism', p. 9; Petitjean, 'Needham, Anglo-French Civilities and Ecumenical Science', 1999; G. Werskey, *The Visible College*, London, 1978.
17. Elzinga, 'Introduction: Modes of Internationalism', 1996, pp. 9–10.
18. Ibid., p. 12; H.F. Cohen, *The Scientific Revolution: A Historiographical Inquiry*, Chicago, 1994.
19. S. Shapin, 'Discipline and Bounding: The History and Sociology of Science as seen Through the Externalism-Internalism Debate', *History of Science*, vol. 30, 1992, pp. 334–69; M.A. Dennis, 'Historiography of Science: An American Perspective', in J. Krige and D. Pestre, eds., *Science in the Twentieth Century*, Harwood, pp. 1–26.
20. Raina and Habib, 'Introduction' to the Chapters on 'Science in Twentieth Century South and South East Asia', 2005.
21. S. Visvanathan, *Organizing for Science*, London, 1985; Z. Baber, *The Science of Empire: Scientific Knowledge, Civilization and Colonial Rule in India*, New York, 1994; D. Raina and S.I. Habib, 'Introduction' to the Chapters on 'Science in Twentieth Century South and South East Asia', *History of Mankind*, vol. 7, UNESCO, forthcoming.
22. S. Harding, *Is Science Multicultural? Postcolonialisms, Feminisms, and Epistemologies*, Bloomington, 1998.
23. D. Raina and S.I. Habib, *Domesticating Modern Science: A Social History of Science and Culture in Colonial India*, Delhi, 2004.
24. D. Raina and A. Jain, 'Big Science and the University in India', J. Krige and D. Pestre, eds., *Science in the Twentieth Century*, 1997, pp. 859–77.
25. Visvanathan, *Organizing for Science*, 1985; D. Abrol, 'Colonized Minds or Nationalist Scientists: The "Science and Culture Group" ', in R. Macleod and D. Kumar, eds., *Technology and the Raj*, London, 1995, pp. 112–33.
26. Raina and Jain, 'Big Science and the University in India', 1997.
27. Raina and Habib, 'Introduction' to the Chapters on 'Science in Twentieth Century'.

28. D.P. Chattopadhyaya, *History of Science and Technology in Ancient India: The Beginnings*, Calcutta, 1986.
29. Correspondence at the *Needham Research Institute*, Cambridge.
30. Sen, 'Changing Perspectives in the History of Sciences', 1966.
31. Chattopadhyaya, *Lokayata: A Study in Ancient Indian Materialism*, Delhi, 1959, p. 64.
32. Sen, 'Changing Perspectives in the History of Sciences', 1966.
33. D.M. Bose, B.V. Subbarayyapa, and S.N. Sen, *A Concise History of Science in India*, Delhi, 1970.
34. S.N. Sen, 'The Character of the Introduction of Western Science in India during the Eighteenth and Nineteenth Centuries', *Indian Jour. of Hist. of Science*, vols. 1–2, 1988, pp. 112–22.
35. Chattopadhyaya, *Science and Society in Ancient India*, Calcutta, 1979.
36. Chattopadhyaya, *History of Science in India*, 1982.
37. Rahman, 'A Conceptual Framework of History of Science in India', 1984.
38. Teich and Young, eds., *Changing Perspective in the History of Science*, Rahman with Alvi had contributed a paper on Fatahallah Shriazi.
39. A contemporary of Rahman, who later went on to become a leading electrochemist in India, once remarked to me that in the 1950s and 1960s, on any issue of technology transfer or science policy, Rahman was always the first to 'blow the whistle'.
40. D. Raina, *Images and Contexts, The Historiography of Science and Modernity in India*, Delhi, 2003.
41. Alexander Wolsky was a Hungarian biologist, principal scientist at UNESCO between 1948 and 1954. He later went on to teach experimental embryology and genetics in Biology at Fordham University and retired from Marymount College in the United States [Tarateta, 2005]. This further explains his association with the Needham, the former embryologist.
42. Needham papers at Needham Research Institute (hereafter NRI), Cambridge, Correspondence: Wolsky, 16/2/1950.
43. Ibid.: Wolsky to Needham, 6/3/1950.
44. Ibid.: Needham to Wolsky, 26/7/1950.
45. Ibid.: Rahman to Needham, 13/6/1950.
46. Ibid.: Wolsky to Needham, 16/6/1950.
47. Ibid.: Wolsky to Needham, 15/9/1950.
48. Ibid.: Wolsky to Needham, 15/1/1951.
49. Ibid.: Letter from *Nature*, 17/5/1951
50. Raina and Habib, 'Introduction' to the Chapters on 'Science in Twentieth Century South and South East Asia', 2005.
51. Petitjean, 'Needham, Anglo-French Civilities and Ecumenical Science', 1999, pp. 196–7.

52. Mayer, 'Setting Up a Discipline, II: British History of Science and "the End of Ideology", 1931–1948', 2003, pp. 8–12.
53. Ibid., pp. 7–8.
54. Needham papers at NRI, Correspondence: Rahman to Needham, 25/2/1952.
55. This was recounted to me by Shigeru Nakayama at Liège, and is found scattered through his correspondence.
56. R. Thapar, 'The Contribution of D.D. Kosambi to Indology', *Interpreting Early India*, Delhi, 1993.
57. R.S. Anderson, 'Peter Blackett in India: Military Consultant and Scientific Intervener, 1947–1972—Part One', in *Notes Rec. R. Soc. Lond.*, 53 (2), pp. 253–73.
58. Needham papers at NRI, Correspondence: Rahman to Needham, 12/3/1950.
59. Ibid.: Rahman to Needham, 20/5/1952.
60. Ibid.: Rahman to Needham, 23/4/1953; cf. [Needham's reply 16/5/1953].
61. G. Sarton, *The Study of the History of Science*, rpt., 1957.
62. Needham papers at NRI: Correspondence: Needham to Rahman, 30/3/1954.
63. Ibid.: Rahman to Needham, 12/11/1955.
64. Ibid.: Rahman to Needham, 27/8/1956.
65. Ibid.: Rahman to Needham, 12/3/1957.
66. Ibid.: Rahman to Needham, 4/7/1957.
67. Ibid.: Rahman to Needham, 12/3/1959.
68. Ibid.: Rahman to Needham, written some time in May 1959.

BIBLIOGRAPHY

Abrol, Dinesh, 'Colonized Minds or Nationalist Scientists: The "Science and Culture Group" ', in R. Macleod and D. Kumar, eds., *Technology and the Raj: Technical Transfers to India 1700-1947,* London, 1995, pp. 112-33.

Anderson, Robert S., 'Peter Blackett in India: Military Consultant and Scientific Intervener, 1947-1972—Part One', *Notes Rec. R. Soc. Lond.*, 53 (2), 1999, 253-73.

Baber, Zaheer, *The Science of Empire: Scientific Knowledge, Civilization and Colonial Rule in India*, New York, 1996.

Bose, D.M., B.V. Subbarayappa, and S.N. Sen, *A Concise History of Science in India*, Delhi, 1970.

Chattopadhyaya, Debiprasad, *Lokayata: A Study in Ancient Indian Materialism*, Delhi, 1959.

———, *Science and Society in Ancient India*, Calcutta, 1979.

———, ed., *History of Science in India*, 2 vols., Delhi, 1982.

———, *History of Science and Technology in Ancient India: The Beginnings*, 2 vols., Calcutta, 1986.

Collins, Randall, *Conflict Sociology*, New York, 1975.

Chilvers, Christopher, Paper presented at the colloquium on engaged biologists, Paris, May 2003.

Cohen, H. Floris, *The Scientific Revolution: A Historiographical Inquiry*, Chicago, 1994.

Dennis, Michael Aaron, 'Historiography of Science: An American Perspective', in John Krige and Dominique Pestre eds., *Science in the Twentieth Century*, 1997, pp. 1–26.

Elzinga, Aant, 'Introduction: Modes of Internationalism', in Aant Elzinga and Catharina Landström, eds., *Internationalism and Science*, 1996, pp. 3–20.

———, 'Revisiting the Needham's Paradox', in S. Irfan Habib and Dhruv Raina, eds., *Situating the History of Science: Dialogues with Joseph Needham*, Delhi, 1999.

Harding, Sandra, *Is Science Multicultural? Postcolonialisms, Feminisms, and Epistemologies*, Bloomington, 1998.

Mayer, Anna-K., 'Setting up a Discipline, II: British History of Science and "the end of ideology", 1931–1948', *Studies in History and Philosophy of Science*, XX, 2003, pp. 1–33.

Needham, Joseph, *Box Containing his Correspondence with his Indian Colleagues*, Cambridge.

———, *Science and Civilization in China*, vols. 1 to 5, 1977, Cambridge.

Petitjean, Patrick, 'Needham, Anglo-French Civilities and Ecumenical Science', in S. Irfan Habib and Dhruv Raina, eds., *Situating the History of Science: Dialogues with Joseph Needham*, Delhi, 1999, pp. 152–97.

Rahman, Abdur, 'A Conceptual Framework of History of Science in India', in A. Rahman, ed., *Science and Technology in Indian Culture: A Historical Perspective*, Delhi, 1984, pp. 21–43.

Raina, Dhruv and Ashok Jain, 'Big Science and the University in India', John Krige and Dominique Pestre, eds., *Science in the Twentieth Century*, 1997, pp. 859–77.

Raina, Dhruv, *Images and Contexts: The Historiography of Science and Modernity in India*, Delhi, 2003.

Raina, Dhruv and S. Irfan Habib, *Domesticating Modern Science: A Social History of Science and Culture in Colonial India*, Delhi.

———, 'Introduction' to the Chapters on 'Science in Twentieth Century South and South East Asia', in *History of Mankind*, vol. 7, UNESCO, 2005.

Sarton, George, *The Study of the History of Science*, 1957, New York.

Sen, S.N., 'Changing Perspectives in the History of Sciences', *Science and Culture*, 31, 5, 1966, pp. 214–19.

———, 'The Character of the Introduction of Western Science in India during the Eighteenth and Nineteenth Centuries', *Indian Journal of History of Science*, 1 and 2, 1988, pp. 112–22.

Shapin, Steven, 'Discipline and Bounding: The History and Sociology of Science as Seen Through the Externalism-Internalism Debate', *History of Science*, 30, 1992, pp. 334–69.

Schott, Thomas, 'World Science: Globalization of Institutions and Participation', *Science, Technology and Human Values*, XVIII, 2, 1993.

Singh, Baldev, ed., *Jawaharlal Nehru on Science and Society: A Collection of his Writings and Speeches*, Delhi, 1988.

Tarateta, Maria, 2005, 'Longtime Professor at Fordham and Marymount Dies', http://www.fordham.edu/Campus_Resources/Public_Affairs/Inside_Fordham/Inside_Fordham_Archi/February_2005/In_Brief/

Teich, M. and R.M. Young, eds., *Changing Perspectives in the History of Science*, Heinemann, 1973.

Thapar, Romila, 'The Contribution of D.D. Kosambi to Indology', in *Interpreting Early India*, 1993, Delhi, pp. 89–113.

Turner, Stephen, 'Forms of Patronage', in Susan E. Cozzens, and Thomas F. Gieryn, eds., *Theories of Science and Society*, Bloomington, 1990.

Visvanathan, Shiv, *Organizing for Science*, Oxford, 1985.

Werskey, Gary, *The Visible College*, London, 1978.

CHAPTER 14

Successes and Failures in the Organization of Research for Industrial Development in the CSIR System of Laboratories

DINESH ABROL

The Council of Scientific and Industrial Research (CSIR) system of laboratories was established in 1942 against the backdrop of the Second World War.[1] It was set up as an Autonomous Society, though the funds were to come fully from the government, with the expectation that its leadership would not only meet the research needs perceived then by the colonial government to be of strategic importance for the country but would also serve industry.[2] Even before India became independent, work had got underway on the establishment of nine national level laboratories. The National Chemical Laboratory (NCL) and the National Physical Laboratory (NPL) were established by the Board of Scientific and Industrial Research (BSIR—the precursor of the CSIR) and functioned out of the University of Delhi until their new buildings were constructed in Pune and Delhi respectively. The first CSIR laboratory to be established under the system of CSIR laboratories, of which the foundation stone of its present building was laid in 1945, was the Central Glass and Ceramics Research Institute (CGCRI). Out of the proposed nine national level laboratories at least in the case of seven laboratories relationships had to be established with the existing industries to serve immediate needs. Except for the NCL and the NPL where the industries targeted were yet to come up, in the early years research in the CSIR system was in the nature of applied research.[3]

The CSIR leadership had to build arrangements to connect its research work with the demands arising out of government and industry. These varied in their level of flexibility in the selection of R&D programmes but there did exist sufficient space for the leadership to ultimately cultivate a balance in the research portfolios over the period in respect of each and every laboratory.[4] In the mandates of these laboratories the scope and functions included undertaking basic and applied research for meeting the immediate and future needs of industries on the widest possible basis. Most CSIR laboratories are now over forty-year old research-performing organizations and have gone through the process of change being made in their programme priorities regularly by its leadership. Presently there exist in the CSIR system of laboratories in total 38 laboratories and 50 field centres. Over 5,300 scientists are employed, among whom 3,000 are PhDs. Constituent establishments work in areas as diverse as aerospace and oceanography, genomics and geophysics, and, food and drugs.

Till the mid 1980s, all the changes affected in the role and functions of the CSIR system of laboratories were essentially of incremental nature. Notwithstanding their significance for the organization and management of R&D in respect of meeting the immediate priorities the leadership considered important, most of these changes fitted reasonably well with their original aim of contributing in the widest possible manner to the achievement of technological self-reliance for industrial development. Perhaps, therefore, the forces that were even getting somewhat affected adversely in respect of their strategic interests by these incremental changes could afford to remain a mute witness to the history of change-making in respect of the mandates and programmes of the CSIR system of laboratories.

However, during the past two decades the Indian S&T policy makers have directed the CSIR system to restructure its research and development orientation with the aim of shifting the laboratories to the paradigm of self-financing R&D and making them a global R&D platform providing competitive R&D and high quality science-based technical services world over. Continuing with the process of integrating the CSIR into a global division of scientific and technological labour, national policy makers are now talking of reinventing the CSIR to direct its laboratories to venture into new uncharted areas for acquiring the position of a most preferred partner globally, in at least one niche domain or to play a leading and catalytic role in networking teams from universities, industries, and elsewhere to exploit emerging opportunities. The aim is to make breakthroughs at the leading edges of such research and technology or to identify opportunities for gain through strategic partnerships for large intra-organizational impact by involving and associating the Indian research community to do world-class research in underpinning technologies.

This shift in the role of CSIR represents a paradigmatic change in respect of the mandate, priorities, and governance of research. The changes currently underway are radical and the process of change is itself also expected to have a significant impact on the CSIR system. Due to the significance of these changes, there arise a number of challenging research questions for those who are in the profession of science, technology, and society (STS). The question being addressed here is whether or not the CSIR system can actually be transformed in accordance with the vision and strategy proposed by the reforming leaders, and to what extent; and who will benefit, who will be adversely affected in what way by the successful implementation of these changes in the mandate of the CSIR system of laboratories.

In this study I therefore trace the history of institutionalization of organization and management of R&D and related S&T activities with a spotlight on the nature of inertia shown and changes made readily over the period of more than sixty years. The study analyses the reforms undertaken from time to time over the period of last sixty years in respect of the changes sought in mandates, priorities, and governance of research and the experience of these developments in terms of the impacts realized on the performance of CSIR system as a whole. I briefly delve into the history of patterns of establishment of knowledge interests in the process of earlier reforms, with the aim to delineate the dynamics of development of new research functions, mandates, and programmes. In the process, this study aims to help us to develop an understanding of how and why the constituencies are responding to the changes that the Council is experiencing, this time under globalization, in respect of the redefinition of its knowledge interests. It traces the actual pattern of linkages established by the CSIR with global and local actors during and before the implementation of recently contemplated changes from the standpoint of how the present and future interests of the Indian people may be affected by the changes. The study also aims to provide some guidance on what kind of practice of R&D management is under development under the reforms currently underway and how it would be shaping the future patterns of successes and failures in the CSIR system. Finally, I hope to enable the public to understand how, in which direction, and with what effect different forces have been responsible for influencing the changes that the CSIR system earlier implemented and is now in the process of implementing.

CONTRADICTORY INFLUENCES

In the evolution of the practice of science, technology, and industry (STI) planning, the period 1950–80 can be characterized as a phase during which government still cared for the paradigm of STI planning for 'national development'. After gaining Independence, India chose to declare its faith in the paradigm of planning, and its political leadership believed in the suitability of the path of development of mixed economy and regional balance. The state sector was identified as a leading player in the process of industrialization. The paradigm of planning that India chose after Independence was to a large extent shaped by the Nehruvian perceptions of 'self-reliance' and 'national development'. During this period the national system of STI developed *in tandem* with publicly funded R&D institutions with the responsibility of supporting the selected path of development. Development entailed the growth of domestic private capital, consumer goods, and domestic markets and the state sector had been given the responsibility to provide to the domestic private capital highly subsidized inputs in the form of basic raw materials, energy, transport services, R&D services, and capital goods.

This basic Nehruvian perception of how India should be pursuing the goals of national development and self-reliance continued to inform also the priorities of the CSIR laboratories. The Nehruvian signature to the contradictory Industrial Policy Statement of 1948 welcoming foreign capital and foreign collaboration had a considerable influence on the patterns of successes and failures that the country experienced very early in respect of the utilization of CSIR.[5] When the country was faced with a balance of payments crisis due to unplanned expenditure on procuring obsolete, inappropriate 'technology' from abroad, the CSIR laboratories did their best to contribute to the process of substitution of imports through the development of indigenous know-how. It is important to note that during the final years of the period of 1950 to 1980, the country saw the rise of forces that have proved to be essentially antithetical to Nehruvian aspirations of self-reliance and social justice. During those years the CSIR came under much criticism from the political leadership as well as from business groups. Its budgetary support began to dwindle by the mid-1980s to the level that it had just enough money to pay salaries to its staff. But the irony is that the CSIR had to officially face the accusation from the Abid Hussein Committee in 1986 that its laboratories reinvented the technologies available with the foreign companies at a high cost.

Analytically speaking, the period of 1950 to 1980 can be divided into two distinct sub-phases. During the first phase, 1950 to 1966, state intervention did not have explicit characteristics of channelling the productive system into organization of the activities and structures (e.g. innovations made in the course of the production and investment process itself and the establishment of organized in-house R&D and design engineering for the generation of new technology) beyond the development of domestic capabilities for the purposes of operations and maintenance of production in the facilities undertaking manufacturing of capital and consumer goods. During this phase, the CSIR system expanded the capabilities of its laboratories for a variety of industries covering several fields, including agriculture and defence. During this whole period, Pandit Nehru was himself around as the prime minister. His government undertook planning for the creation of scientific and industrial knowledge and capital equipment on the one hand, while foreign and private monopolies were allowed to exist with the expectation that would help the country to obtain the required technology on the other.[6] During this period a total of 26 research laboratories were established under CSIR.

BUILDING ON THE STRENGTH OF NEHRU–BHATNAGAR ALLIANCE

During this period the S&T programmes chosen by the CSIR laboratories for implementation were innovative and largely compatible with the priorities of self-reliant development. Programmes on wasteland recovery using biological methods at CBRI, the development of milk powder from buffalo milk at CFTRI, beneficiation of high ash coking coals for coke making, low temperature carbonization, coal gasification, etc., at CFRI; a programme for low grade raw materials utilization for the benefit of import substitution; utilization of economic and medicinal plants at CIMAP and other regional laboratories of CSIR; and the development of capabilities for scientific instrumentation are some examples to illustrate the point that in the early years the conception of planning for S&T in the CSIR was linked to the use of S&T in planning.

In the initial years, the CSIR system of laboratories worked to provide S&T support in metrology, standardization, testing, and database on indigenous raw materials/resources to the indigenous industry which was then in its nascent stages of growth. Let me cite an illustrative example of the kind of extremely useful applied work that the CSIR did during this period, which the policy makers have characterized these days as an infrastructure creation phase. CFRI, Dhanbad, a laboratory set up with the mandate of working on coal and oil, showed that with the washing circuits developed at the laboratory, it was now going to be possible for industry to beneficiate inferior coking coals available in India. This had a major impact on the fortunes of the Indian steel industry. Now the Indian steel industry was clearly a viable proposition for India, it was not required to import any more coking coal. During this period a large number of CSIR laboratories (CFRI, NML, CMRS, CLRI, and CGCRI) had their work devoted substantively to the immediate needs of industry and their contribution was quite significant. Linkages were established for the development of capabilities required for technology choice, assessment of available raw materials, utilization of raw materials through beneficiation, development of detailed engineering reports, assistance in the erection of imported standard equipment, and support in the form of technical services. In case of the existing industries having big business in place or where for its benefit the state was to establish infrastructure related industries on an immediate basis, the linkage regime was developed in favour of strong industry service orientation. In the case of those laboratories where the corresponding industries were non-existing autonomy existed for the work of longer-term nature. Even for these laboratories the situation changed during the next phase of policy development.

An example of a very early resistance put up by Indian scientists for the expansion of space in the civilian R&D areas and for a larger, longer-term S&T effort for local needs, can illustrate the point that an important trend within the process of S&T development was the struggle between the aspirations of scientists and domestic private capital. The struggle took place as a result of a difference in perceptions over the problem of whether the domestic private capital can be allowed to treat the S&T system as a system of in-house R&D capabilities where shorter time horizons guide the resource allocation. The tussle was between the local scientific elite, which had an interest in the establishment of higher science and technology investments, and the domestic private capital, which had almost uniformly cultivated the outlook of short-term gains to local technological development efforts. To give a concrete example, it recommended through the First Reviewing Committee that the government should avoid opening new laboratories and the CSIR should be directed to concentrate on the efforts needed to promote the processes and products already developed during the period of 1942–8. This committee was set up in 1948 by the government to review the functioning of an infant CSIR. In the wake of Independence, this was an opportunity for the country to determine the future priorities of CSIR. It had the representatives of big business as members in large number. At the time when this recommendation was made, the CSIR laboratories were in the process of just being established. It had only nine laboratories under its wing.

This recommendation was fortunately overturned when the Indian scientific community protested.

PLACE OF THE CSIR SYSTEM IN THE THEORY AND PRACTICE OF S&T PLANNING

The theory and practice of S&T planning was impacted by Soviet experiments in the organization of scientific and technological research. Right from the beginning, the scientists and technologists were willing to cultivate S&T for 'development'. In the choice of programmes and projects, the thinking of S&T planners was in no way a product of colonized minds.[7] However, in terms of the evaluation of the practice of planning and policy formulation, it needs to be recalled here that in the early years from 1947 to 1955 decisions on the setting up of scientific institutions and their funding were made by a relatively unstructured policy process. Thanks to the commitment of the political leadership, in the First Five Year Plan, even before the S&T planning process really got underway, eleven national laboratories under the CSIR were set up. During the Second Plan period, the object was to develop the existing facilities and to bring the work of scientists in the CSIR laboratories to bear as closely as possible upon important problems in different fields of national development. During this plan, the Planning Commission was expanded and the responsibility for integrating science into development fell to it. The responsibility of the Planning Commission was spelled out in 1959 as the setting up of independent committees and panels of scientists as needed, and taking their views and recommendations into consideration in the planning of economic development and the attainment of the aims. During this important formative period, the CSIR system was the only scientific agency to be included in the preparation of the plans. A panel was set up to devise schemes for the research activities of the CSIR. It is important that even when the Planning Commission included none of the other scientific agencies of the government into the framework of a national plan for science, the CSIR had a place in the Indian system of development planning.

The role of Scientific Policy Resolution (SPR) of 1958 in

orienting the subsequent events for the CSIR system of laboratories can be seen clearly from the recommendations made at the first and second conferences of scientists, technologists and educationists. Based on the perceptions of the SPR, the first and second conferences focused largely on the accomplishments and gaps in relation to the creation of adequate S&T manpower and infrastructure, funds for scientific research, and facilities for quality education. The second conference did recognize the problem of relevance and utilization, and recommended steps for improving linkages and directing the R&D work into relevant channels through projection. However, it is important to underline that the concern was still essentially in terms of working for avenues for the utilization of infrastructure already created and not in terms of analysing the pattern of demand for science and technology.

Yet even the paradigm that dominated the prevailing practice of advanced capitalist countries with the linear model of innovation had made a substantial impact, at least at the level of evaluation of results coming out of Indian S&T institutions. As far as the norms and environment to be guaranteed to the scientific research institutions is concerned, these were embedded in the understanding which emphasized that science is an endless frontier and should be allowed to operate under a social contract in which the government provides funds for scientific research in the general expectation that, sooner or later, it would lead to technological development. Therefore, during the latter part of this period there were voices arguing within the CSIR scientific community that the government had planned too tightly and needed to provide funds for scientific research mainly in the expectation that eventually it would help industry with technological development. Together with a view of STI processes as an 'autonomous' neutral development, their perspective was not conducive to the shaping of ethos of the emerging mass of S&T professionals towards a systemic planning of S&T. In some parts of the CSIR, its contradictory influence was evident in the R&D orientation to the extent that their priorities were more in line with the perspective of giving the country basically 'a wide base of science capacity' not necessarily geared to any particular strategy of integration into socio-economic development but capable of responding to contingencies and opportunities arising out of development plans, to be utilized by anyone who wants to make use of it. The result was that during this period the practice of planning of S&T in the CSIR system could not be connected in a systematic way to the interventions being made by the government for the use of S&T in planning. There was not much attention paid to the views and recommendations of scientists and technologists in the planning of economic development and the attainment of national aims. Thus, though there is enough evidence to show that the dominant policy culture of CSIR scientists still cared for and did not ignore the use of S&T in planning during this period, their preference for a passive response to S&T planning also meant shorter-term priorities and ad hoc-ism often gaining the upper hand.

Of course, as far as the CSIR priorities of technology development are concerned, the dominant conception of 'development' was to catch up with the world. Not that there were no conflicting conceptions of development within the Indian scientific community. For example, D.D. Kosambi of TIFR had a very different conception of the priorities with his emphasis on solar energy, which the CSIR also pursued as a priority in the beginning. Yet it is true that the dominant conception of 'development' in CSIR did not turnout to be very different from the development priorities of the other policy cultures—political-bureaucratic, academic, industry-market, and civic. The CSIR remained mainly focused on the contributions that science and technology could make to that part of the industrialization process that was to transform the modern sectors of the economy. And everywhere in all the sectors, agriculture, industry, defence, science and education, to which the use of S&T was to be devoted, the perspective was developed in such a way as to catch up an import substitution.

CONTRADICTORY INFLUENCES, 1967 TO 1980

In the second phase, state intervention was designed to organize quite explicitly the linkages of the S&T system with the productive system for the establishment of activities and structures leading to the development of scientific and technological capabilities for investment and innovation. My analysis of the experience of the CSIR system during this phase indicates that the leadership had come to a greater appreciation of the commercial and political realities of industrial technology (Alam and Langrish 1984).

The source of technology transfer changed in the second phase. Technology transfer from the Soviet bloc began to decline in favour of increasing collaboration with transnational corporations. The policy was changed to introduce regulations and measures that sought to increase the bargaining power of domestic private capital *vis-à-vis* the foreign companies. It was the Round Table of Scientists and Technologists organized by the government in 1967 which recommended that in cases where know-how of a high standard could be developed within the country, foreign collaboration was undesirable. Change was considered necessary because during the first policy period, despite the open-door policy of foreign investments and technology imports, the extent of such investment had remained low. And there had developed a serious balance of payments crisis. Import of technology came under modified regulations through the Foreign Investment Board (1969), the Industrial Licensing Policy Instruments and Foreign Exchange Regulation Act (1973), and Technology Development Fund Scheme (1976). Industries were classified into categories where foreign collaboration was permitted or not permitted (Nayar 1983).

Research institutions were made to direct their efforts to areas where foreign technology was not permitted. The CSIR held 'Industry R&D get-togethers' to identify the problems of research with the aim to affect import substitution. National Chemical Laboratory, which had enjoyed a balance of 50:50 between basic and applied research, was asked to alter the balance to 20:80. Efforts were made to introduce in-house engineering design facilities by setting up a full scale organization for the design, erection, and commissioning of pilot plants and industrial plants in order to assist the industry to bridge the gap between research and application. The direction of the Third Reviewing Committee of CSIR was that 'in the circumstances (as it is placed) there is every reason for the government, mainly through CSIR, to take a large range of activities, while continuing to press industry to do its duty'. The CSIR was pressurized by the government to submit itself to the discipline of the 'user departments' and industrial sponsors. Government directives even included involving user departments in the supervision of research and the sponsorship for each and every pilot plant. CSIR scientists had to struggle to obtain the recommendation that the Council should continue to be able to take up any type of activity in the field of science lacking a sponsor. During this period, the CSIR system took the responsibility of devising the supply of technological know-how packages in five important industrial areas—electronics, drugs and pharmaceuticals, pesticides, food, and energy. Policy support in the form of a new patent legislation (1970), commitment of not importing technology so as to give a boost to indigenous technological know-how was provided to the national laboratories. In the areas of drugs and pharmaceuticals, pesticides, and food, considerable success was achieved. The laboratories succeeded in transferring technological know-how for more than 100 drugs and 35 technical pesticides. In the area of food processing, know-how was transferred in respect of milling processes, processing of spices, by-product utilization, equipment design, and packaging of foods.

Of course, the irony is that this phase is being criticized today as the period in which the country was reinventing the wheel. But what we need to appreciate is that even the decision of the government to select import substitution as the new paradigm of development was dictated by the imperatives of the path of development of mixed economy in which the big business had a significant place. Domestic private capital preferred to exploit domestic markets for established products. As these markets were quite large and not yet fully exploited, import substitution was easier. Import substitution was a necessity because domestic private capital was going to incur otherwise serious losses. Its bargaining power *vis-à-vis* foreign capital and people was under serious threat. The system selection exercise, which essentially meant what kind of innovation system the country could build, was also dictated by the way domestic capital had developed or was willing to develop itself in the immediate future, in respect of the new technological capabilities. The same is true of the choices made in respect of technological systems selected for meeting the objectives of industrial, agricultural, and social development. For example, during this policy period when the government took the decision to implement the technology system of high input responsive and high yielding varieties to bring about the Green Revolution in agriculture, the CSIR system had to respond by taking up the development of indigenous know-how for agro-chemicals.

Inconsistency in implementation, lack of co-ordination, and perusal of contradictory policies were of course an accompanying characteristic of the period. The idea of planning was introduced into the S&T system to remedy the shortcomings. The first S&T Plan (1974–9) adopted the sectoral approach, identifying 24 major sectors in order to take a 'sectoral systems' perspective of science and technology. But the elements of innovation system, whether in the government apparatus, or in industry, were not ready for a sectoral system approach. In the words of S.K. Subramanian, the Secretary NCST, a number of scientists, technologists, and administrators had participated in drawing the plan, but in their individual capacities. A number of them failed to exhibit the same enthusiasm when they began to view the programmes from their official stand. This was particularly in case of several central ministries. Further, the plan was not uniformly strong in all sectors, but was skewed towards the so-called elitist programmes.

However, through this experience too ministries ended up with different types of projects, which led to some changes at the level of the individual departments and ministries. The Science and Engineering Research Council, creating a National Information System for S&T, and establishing a scheme for registration of R&D by private sector, was also a part of these efforts. Annual plan discussions by the Planning Commission with the relevant ministries began to devote attention to the S&T component, and several of the ministries started making special provision for S&T and setting up special committees for implementation of S&T programmes. But the failure was monumental when it came to bringing about partnerships between the different players, agencies and industry, for the implementation of identified programmes. The NCST and its poor implementation are analytically quite instructive.

In the CSIR, even after the exercise of NCST, the linkage regime remained oriented to the requirements of technology absorption, where industry was merely a user of available capacities and did not encourage collaborative programmes with the state S&T sector for indigenous technology development. For the domestic private capital, the CSIR institutions were mere problem-solving mechanisms. In all partnerships that came about, private capital was only a user and not a partner willing to invest in the development of technology absorption or innovation. They were ready to use the fruits of S&T if the proposed use of indigenous S&T was profitable and available in a readymade form.

During the period 1977 to 1985 the country saw several new tussles at the level of ideas for S&T planning. There was a tussle over what should be the nature of modifications in the conception of 'development'. In 1977, with the advent of the first non-Congress government at the centre, the debate on the place for production by the masses in the process of S&T planning was given a big boost in official circles. In the beginning of 1981 India took its first IMF loan. In 1983 the country brought out its first technology policy statement where the contradictory influence of market and plan was evident. From a reading of the Technology Policy Implementation Committee (TPIC), it is quite clear that policy makers were divided over the extent of the mix of market and plan to be used. Finally, the country moved to the phase of internal liberalization where market-friendly S&T reforms were given emphasis.

MARKET-FRIENDLY REFORMS AND THE CONSEQUENCES FOR CSIR

After 1985 the advocates of liberalization argued for policy reforms on the basis that the Indian corporate sector was likely to have a better incentive under liberalization, than to link itself directly with the structures of CSIR institutions. They believed that the Indian corporate sector would be able to involve the publicly funded Research and Technology Development (RTD) structures in the execution of these tasks in a better and cost effective way. They maintained that the tasks of technology acquisition, absorption, improvement, and development would be better tackled in an environment of strong competition and removal of controls. They argued for the implementation of stronger patent protection, freedom for collaboration between foreign companies and Indian business for the acquisition of know-how and joint R&D, immediate lifting of controls on internal competition, increased emphasis on export-led growth to foster major innovations and public R&D laboratories to be made autonomous, and market-friendly knowledge-intensive business services selling organizations (Desai 1988).

In a way, almost for the past two decades now, they have had increasing sway over S&T policy making. But the important fact is that during this period, under the influence of their neo-liberal policies, the proportion of national resources spent on R&D by India declined steadily from a peak of 0.91 per cent in 1986–7 to 0.71 per cent in 1995–6. This decline was acutely felt in the CSIR at all levels. Although under the ratio of national resources spent on R&D came back to 0.81 per cent in 1998–9 (Government of India 2002), what is equally important to note is that much of the enhanced allocations went to the departments of defence, nuclear energy, and space. From the standpoint of protection of the long-term interests of CSIR system, increase in industry R&D expenditure is a welcome trend. But even during the 1990s, when the government implemented mainly the liberalization policy, R&D expenditure fell in 12 out of 28 broad industries and even where it got raised, the R&D to sales ratios stagnated or declined.

SHAPING THE CSIR SYSTEM OF LABORATORIES UNDER LIBERALIZATION

In the case of the CSIR—the largest civilian R&D agency of the country—goals set for the year of 2001 were to move towards the path of self-financing by generating over Rs. 7 billion from external sources, as against Rs. 1.35 billion in 1994–5, of which at least 50 per cent would be from industrial customers (up from 15 per cent in 1994–5); developing at least ten exclusive

TABLE 14.1: INDUSTRIAL RESEARCH AND DEVELOPMENT EXPENDITURE IN (Rs. 10 MILLION) AND ITS PERCENTAGE OF SALES TURNOVER IN BRACKETS FOR LEADING INDUSTRY GROUPS

Industry Group	1994–5	1995–6	1996–7	1997–8	1998–9
Electricals & Electronics	128.86 (0.6)	136.10 (0.6)	185.60 (0.28)	204.06 (0.30)	200.03 (0.29)
Defence Industries	79.48 (1.9)	79.85 (1.7)	144.79 (4.44)	180.23 (5.22)	218 (6.11)
Metallurgical Industries	51.79 (0.2)	51.75 (0.2)	123.95 (0.48)	123.16 (0.57)	142.56 (0.78)
Drugs & Pharmaceuticals	165.81 (0.4)	198.73 (0.4)	266.34 (0.63)	287.49 (0.63)	377.50 (0.76)
Transportation	170.42 (0.9)	226.88 (1.0)	133.11 (0.47)	201.15 (0.79)	152.83 (0.60)
Fuels	81.33 (0.1)	106.02 (0.1)	110.66 (0.11)	99.20 (0.09)	156.68 (0.13)
Chemicals (other than fertilizer)	243.46 (0.8)	309.16 (0.8)	180.49 (0.51)	174.67 (0.47)	195.14 (0.51)

Source: Department of Science & Technology.

and globally competitive technologies in niche areas; holding a patent bank of 500 foreign patents (up from 50); realizing 10 per cent of operational expenditure from intellectual property licensing (up from < 1 per cent); and deriving annual earnings of $ 40 million from overseas R&D work and services (up from < $ 2 million).

Over two-thirds of the CSIR's income continued to come from budgetary support. External cash flows continued to account for only about one-fifth, and this has remained at the same level over the period of last one and half decade. Government (budgetary support plus contract revenue from government agencies as grants-in-aid) continued to be the major source of finance, accounting for over 80 per cent. The amount of cash generated from industry continued to about 25 per cent of the external cash flows (or about 6 per cent of total receipts) and that too has not shown much variation across the years. Foreign earnings do not work out to be more than 1.7 per cent.

The vision 2001 envisaged that CSIR would hold a valuable portfolio of at least 1000 Indian patents and 500 foreign patents. It is well known that the propensity to patent also differs considerably among R&D organizations and inventors. However, the policy makers were insistent to get the CSIR as a whole to be measured by patenting activity. The Council has been

TABLE 14.2: THE 'BIG FIVE' CSIR LABORATORIES BASED ON PERFORMANCE DURING 1999–2002 FOR PATENTS FILED AND GRANTED ABROAD

CSIR Laboratories	Filed 1999–2000	Granted 1999–2000	Filed 2000–01	Granted 2001–01	Filed 2001–02	Granted 2001–02
CIMAP	68	1	38	7	31	25
IICT	32	6	72	4	92	17
NCL	18	12	63	16	43	16
CDRI	16	0	8	6	27	6
IICB	3	1	18	0	9	5
Total	199	35	452	56	580	86

Source: IPMD, CSIR

TABLE 14.3: POOR CSIR PERFORMERS FROM 1999–2002, FOR PATENTS GRANTED ABROAD

Name of the Laboratory	Granted 1999–2000	Granted 2000–1	Granted 2001–2
CBRI	0	0	0
CCMB	0	1	0
CEERI	0	0	1
CMRI	0	1	0
CSIO	0	0	0
CSMCRI	0	0	0
IMT	0	1	0
NAL	0	0	0
NBRI	0	0	1
NGRI	0	0	0
NIO	0	0	0
NML	0	0	0
RRL (BHU)	0	0	0
RRL (BHO)	0	0	0
RRL (JM)	0	0	0

successful in securing 591 Indian patents and 101 foreign ones cumulatively (CSIR 2001). Available evidence also tells us that much of it is accounted for by a select set of five laboratories.

Gupta (2005) has analysed the whole count data of Indian patent output of 2880 (1978–2000) inventors who produced 6877 patents in India. Of these, 62 (2 per cent) inventors produced 1225 (18 per cent) patents, 691 (24 per cent) inventors produced 4166 (61 per cent) patents and 2189 (76 per cent) inventors produced only 2711 (39 per cent). The patenting in US by inventors from CSIR also tells a similar pattern. There are 461 inventors who produced 7555 patents in US. Analysis indicates that again 19 inventors (4 per cent) produced 155 patents (20 per cent), a highly skewed pattern of inventor propensity. There are 48 (10 per cent) inventors who produced 227 (30 per cent) of the patents, while 413 (90 per cent) inventors produced 528 of the (70 per cent) patents.

Provisional figures from CSIR indicate that in the last five years, only 4 per cent of patents in force have been licensed. Compare this achievement with the target of 10 per cent of operational expenditure of CSIR to be realized from intellectual property. This record of the failure of the CSIR 2001 Vision and Strategy is yet another facet of the liberalization period.

The strategy contained a whole range of misapplied targets, to which the CSIR system could not be expected to respond favourably. The vision and strategy talked of globalizing the CSIR system, making it a global platform of R&D. It talked of 'patent or perish', when we know that for several parts of the system this was a misapplied target. It had put too much faith in market forces. It went on to withdraw from the societal missions, where the strengths of CSIR system had been applicable.

The primary function of the CSIR is to undertake R&D directed towards continuous improvement of indigenous technology and adaptation and development of imported technology. But during this period, the primary function failed to receive the support of CSIR leadership. Take the available evidence in this regard from the report of the Comptroller and Auditor General of India (CAG, Report No. 5 of 2003). This CAG Report tells us that of the 984 technologies developed by 23 laboratories/institutes, 607 technologies, including 247 developed before 1996–7, were yet to be transferred. It also informs that these laboratories were unable to furnish specific information on the actual expenditure on the development of technologies. Reasons for non-transfer of 246 developed technologies showed that 77 technologies were not found fit for transfer, while 87 required further improvements/developments. Besides, in 82 cases (including 34 developed prior to 1999–2000), the negotiations for transfer were underway. CSIR sustained a loss of Rs. 99.31 lakh due to violation of its guidelines on technology transfer. Royalty/Premia of Rs. 134.58 lakh remained unrealized in 17 cases of technology transfer. During 1980s there was at least an attempt to reposition the CSIR for the objective of development of technologies for societal missions and rural industrialization. Efforts were initiated in respect of several technology missions. During the 1990s structures created for this purpose were dismantled. In the year 2000, the management even closed down the Division of Societal and Technology Missions (STMD), a nodal point set up for the co-ordination of this work. But today this important CSIR–society linkage is virtually dead. Due to the withdrawal of support to societal programmes of technological research, CSIR–society linkages are completely absent; no more mission based work on drinking water, leather, utilization of flyash bricks; even the linkages sought to be established with S&T-oriented NGOs are viewed with suspicion.

Publicly funded institutions have a very important role to play in technology transfer to marginalized people. But they continue to lack in the contacts with the bridging organizations. To establish an effective connection with the resources and opportunities available to the poor, the CSIR organizations would need the help of bridging organizations. In the absence of this connection, technologies that these organizations have created are having very little success among the rural and urban poor (Abrol et al. 1998). A complete failure of the policy of liberalization is evident in the field of technology transfer to the weaker sections.

OPENING UP OF THE CSIR SYSTEM TO MNCs

The national S&T policy was radically modified to open up the publicly funded structures of R&D and technology assimilation to the help of foreign capital for a wide range of policy objectives.

TABLE 14.4: TRANSFER AND DEVELOPMENT OF TECHNOLOGIES IN CSIR LABORATORIES

Name of Laboratory/ Institute	Technologies		
	Developed	Transferred	Lying untransferred
1. Central Building Research Institute (CBRI)	92	81	11
2. Centre for Biochemical Technology (CBT)	23	10	13
3. Centre for Cellular and Molecular Biology (CCMB)	05	02	03
4. Central Food Technology Research Institute (CFTRI)	133	27	106
5. Central Glass & Ceramics Research Institute (CGCRI)	31	11	20
6. Central Institute of Medicinal & Aromatic Plants (CIMAP)	27	17	10
7. Central Leather Research Institute (CLRI)	65	04	61
8. Central Road Research Institute (CRRI)	49	25	24
9. Central Salt & Marine Chemicals Research Institute (CSMCRI)	28	09	19
10. Indian Institute of Chemical Biology (IICB)	33	02	31
11. Indian Institute of Chemical Technology (IICT)	150	130	20
12. Industrial Toxicological Research Centre (ITRC)	08	04	04
13. National Aerospace Laboratories (NAL)	05	–	05
14. National Botanical Research Institute (NBRI)	50	03	47

15. National Chemical Laboratory (NCL)	NA	NA	NA
16. National Environmental Engineering Research Institute (NEERI)	37	–	37
17. National Geophysical Research Institute (NGRI)	04	01	03
18. National Metallurgical Laboratory (NML)	35	20	15
19. National Physical Laboratory (NPL)	49	11	38
20. Regional Research Laboratory Bhopal (RRL-Bho.)	15	07	08
21. Regional Research Laboratory Bhuvaneshwar (RRL-Bhu)	112	–	112
22. Regional Research Laboratory Jammu (RRL-J)	14	05	09
23. Regional Research Laboratory Trivandrum (RRL-T)	19	08	11
Total	984	377	607

Foreign companies were invited to freely access the CSIR institutions through the mechanism of contract work. The current policy is to freely permit the foreign companies to set up R&D facilities and venture capital funds in the country for the benefit of development of R&D and technology development. All of these have a totally new set of strategic implications for the processes of integration of CSIR with the emerging global system of S&T. See Box 14.1 for the highlights of this important development. Today the foreign companies appear to be better placed than the national companies in respect of the access to CSIR laboratories (Box 14.2). The country has been induced to open up the national laboratories to foreign companies for contract research in those fields where the Indian companies'

BOX 14.1

SOME CSIR-MNC TIE-UPS

- Duponts for identification of molecules and drugs.
- Parke Davis for supply of medicinal plants.
- Abbot Labs for synthesis of organic molecules.
- General Electric Company for intermediate for polycarbonates.

SOME IMPORTANT QUESTIONS

- Are emerging links with foreign companies helping to upgrade Indigenous R&D?
- Are the needs of national institutions met through these collaborations? While IIT tie-up with IBM is strong on resource generation, it is weak in respect of collaboration.
- Will local industry be adversely affected if R&D is used by MNCs to enhance their own competitive advantage?
- Are forward looking linkages generated through collaborations between industry and publicly funded institutions or would they lock out local industry under the new IPR regime?
- Would this policy take us away from the problems of modernization of the existing labs and lock them into a linkage regime which has as narrow a focus as it had before?

BOX 14.2

S&T INSTITUTIONS AND LOCAL INDUSTRY PARTNERSHIP

- Income from royalty and premia for the CSIR laboratories remains a small component.
- For the CSIR laboratories consultancy continues to be still the main source of income from private sector units.
- Pressure for raising the external cash flow exists: CSIR needed more than 2000 projects to meet its external cash flow needs.
- This put tremendous pressure on the system in which only 5000 scientists were working, with negative consequences for CSIR systems and coherence.

interests are the highest. It is significant that the CSIR institutions have been induced to get into the collaborative arrangements with foreign companies in those sectors where the strategic interests of Indian industry are high.

This option has been preferred by the CSIR partly because it has for the last two decades been under heavy criticism for not being able to strike it big in the area of indigenous technology development. But this is hardly a good way of proving itself, as the process of putting these institutions into foreign contracts can also hurt the national system of innovation. Even in this latest phase of the post-liberalization regime, the collaborative R&D programmes in which publicly-funded laboratories would be partnering local industry are still a far cry (see Table 14.5).

TABLE 14.5: TARGETS AND ACHIEVEMENTS OF R&D PROGRAMMES

Item	Targets for 2001	Achievement in 2001
Financing	Rs. 7 billion from external sources, of which at least 50 per cent will be from industrial customers	Secured only Rs. 2.45 billion from external sources in 2000–1, of which only about 25 per cent is provided by industrial customers. The amount of cash generated from industry is only about 7 per cent of total CSIR receipts
Patenting	Hold a patent bank of 500 foreign and 1000 Indian patents; realize 10 per cent of operational expenditure from intellectual property licensing	Secured only 591 Indian and 101 foreign patents; only 4 per cent of patents in force licensed; Council's patent holding accounted for by a small set of five laboratories, see attached tables for labs with strong and weak patent performance
Foreign earnings	Derive annual earnings of $40 million from overseas R&D work and services	Only about $ 5 million from foreign sources for R&D work and services

Technology development	Development of at least ten exclusive and globally competitive technologies in niche areas	Not even in reckoning; technology development as a function has acutely suffered in the recent period
Technologies for human welfare	A vital source of S&T for societal missions to offer technology with a human face	Only 1 per cent of expenditure of 1000 crore budget spent on rural technologies, mission division dismantled

POST-LIBERALIZATION CSIR R&D MANAGERIAL PROCESSES

R&D managerial processes are under pressure to focus on globally fashionable areas that have more relevance to the priorities of the developed world. There is a trend to weaken support from CSIR headquarters to those laboratories that are not covered by the globally fashionable areas. There is evidence of the growing fragmentation of long-term programmes due to the unplanned introduction of contract research. Work has been abandoned with the retirement of scientists in a number of important research areas (superconductivity, low temperature physics, coal gasification, silicon for photovoltaic applications, etc.). Research work has suffered acutely during this period in respect of the programmes of both commodity and regional laboratories. Even the weak project monitoring culture that the CSIR tried building in the previous period has been thrown into the dustbin; when asked by the CAG, CSIR laboratories were unable to furnish specific information on actual expenditures on the development of technologies. Foresight exercises were not systematically undertaken for the laboratories during this period. Even the structures of Technology Advisory Boards (TABs) that the CSIR was asked to introduce by the Abid Hussain Committee have not been kept alive to fill the foresight gap.

Although since 1996, there has been a lot of talk of the laboratory directors being turned into CEOs of the CSIR laboratories, and the DGCSIR being the CEO of CSIR, during this very period the CSIR system also chose to abandon the approach of application and technology missions. In actual practice, there has been no major change in the culture of CSIR R&D management. Too much media hype regarding achievement of a new CSIR culture (e.g. DGCSIR as CEO of CSIR, 'CSIR Incorporated', 'Team India achieved') is beginning to create cynicism. In actual practice, the scientific community of CSIR has been disregarded as a stakeholder in the running of CSIR. During this period, till very recently, there was even a trend towards the appointment of university-bred persons as directors. Increased reliance on the external performance evaluation and externalized training in R&D management has not helped the CSIR to improve in any significant way its planning, monitoring, and evaluation practice. For example, there has been no serious attempt to undertake systematic project costing but even then to camouflage and obfuscate there is talk of measuring the outcomes of CSIR on the basis of economic valuation (CSIR SWA 2004).

There is a reduction in the transparency of decision-making. The leadership is blocking information that used to be available to the CSIR scientists on the laboratory/programme/project performance and R&D managerial processes. There is virtually a complete information blackout: information is being kept away from the CSIR scientists and parliament. Even for the in-house policy researchers, access to information available with the MIS (IMPACT) on human resources, ECF mobilization, and allocation of resources, R&D priorities and technology development is not easy. Similarly, the CSIR scientific community does not have free access to reports of Advisory Body (AB), Society and Governing Body (GB) and CSIR audit reports (CSIR SWA 2004).

The order of day in CSIR is to strengthen the bureaucratic culture. The current CSIR Joint Secretary determines international visits, and shapes the rights of scientists. Management decisions are being largely made by directors without the participation of scientists. In most CSIR laboratories, the practice of collegiate management is largely a slogan; the CSIR authority structure does not relish in actual practice a delegation of powers to scientists. One-man rule is getting strengthened at the laboratory level. The CSIR SWA is accusing the management of victimizations galore, the promotion of unnecessary litigation, the emasculation of dissenters, and the de-recognition of unions (CSIR SWA 2004).

KELKAR COMMITTEE AND ITS PROPOSALS FOR CSIR RESTRUCTURING

Going by the broad thrust of the Kelkar Committee report, it appears that the recommendations are 'tuned' to the view that seeks an untrammelled globalization of everything—including R&D. The recommendations of the Kelkar Committee were prepared to feed into the 'India Shining' campaign, with a complete 'disconnect' to the realities of the user base of the technologies sought to be developed by CSIR system. Kelkar Committee recommends: enhance effectiveness by doing the right things such as exploring uncharted areas, availing emerging opportunities, developing world class research, and fostering research-university orientation and network of reinforcing partnerships. It is clear that the proposals are not rooted in ground realities. These proposals are also likely to meet the fate of CSIR 2001Vision (CSIR SWA 2004).

The CSIR was set up to undertake mission-oriented scientific and industrial research, as a result of which its institutional framework is not conducive to foster a research university orientation; conditions prevailing in respect of human resources in quite a large number of CSIR laboratories are in fact far more favourable for the provision of R&D and S&T services for the purpose of technology adaptation, technology absorption, dissemination of generic technology, environmental monitoring and protection, safety, regulation, etc. Several CSIR laboratories are even today capable of contributing in a big way to the task of industrial upgrading in many sectors, but no proposals have been made by the Kelkar Committee. Quite a few sectors of CSIR are amenable to undertaking societal concerns-linked mission and application-oriented research, but no agenda has been set by the Committee for this objective. Although existing network programmes are largely patchwork, the Kelkar Committee chose not to review these programmes explicitly. There was not even an attempt to analyse the reasons for the failure of technology development (CSIR SWA 2005).

My own reading of the past indicates that technology development has been failing due to poor systemic connections in terms of links with local resource utilization and local markets and the lack of bridging organizations (Abrol et al. 1998). Even from the Kelkar Committee side there are no recommendations on the development of bridging organizations, technology incubation, or on the arrangements for demand creation for local resources and local capabilities. The same applies to the promotion of collaboration with S&T based NGOs for technology application and incentives for diffusion through appropriate legislation and policies with respect to waste recycling, wasteland recovery, energy conservation, technology absorption, and the saving of materials.

Certainly the Kelkar Committee has not assessed the track record of past implementations of autonomous business units and activity portfolio review; the recommendation of utilization of economic valuation-based indicators is completely inappropriate. When there is no culture for evidence-based policy-making, no environment for free and frank discussion, and no freedom of expression, it is dangerous to suggest to the CSIR system it should be practising economic valuation.

ALTERNATE VISION OF THE ORGANIZATION AND MANAGEMENT OF R&D

To bring out the ensuing tension in respect of the organization and management of R&D I will cite here the view of CSIR Scientific Workers Association (SWA) regarding how the CSIR system should be guided today by its policy makers. In its view, the proposed CSIR Jewels Scheme, incentives including monetary rewards, external performance audit based on economic valuation, and human assets accounting and staff alignment, are all misplaced solutions. It points out that the CSIR authorities have a predisposition towards the misuse of 'autonomy'. Autonomy has been used to strangulate accountability. In the case of CSIR, it has led to the strengthening of one-man rule. It argues that the master–servant relationship inherent in CCS (RSA) rules is not at all conducive for a research organization like the CSIR. In a publicly funded R&D organization, accountability must be ensured from the top to the bottom. It argues that the delegation of powers, project costing and participative management must be legislated by the government on a statutory basis. It argues that CSIR bye-laws need to be amended wholesale to reflect appropriate R&D managerial decision-making processes to prevent the leadership from practising a bureaucratic culture and one-man rule. Therefore, in contrast to the Kelkar Committee's recommendation the scientists themselves suggest that it is more appropriate to protect the CSIR as a statutory organization.

The CSIR SWA also differs on how and what directions the CSIR has to take. Incentives based on economic valuation of laboratory performance indicators are misplaced strategy. There is no doubt that the state-of-art in evaluation research advises quite clearly against the use of economic valuation in research evaluation in the case of publicly funded strategic civilian research. What is needed is to institutionalize programme controls and assessments based on process indicators and online monitoring. Even today the CSIR system, as it has significant capabilities in these areas, can be made to develop technology application and development missions for concerns such as wasteland recovery, upgradation of the non-farm sector (artisan technologies included), agricultural implements and agro-processing, the development of plant based health systems, processing of economic and medicinal plants for the promotion of local uses, the upgradation of food vending, fruit and vegetable processing, the management of drinking water and sanitation, environmental remediation, all-weather rural roads, slum upgradation, and waste recycling. Better results in respect of reputation and economic value will accrue to the CSIR system if it undertakes such missions seriously.

The CSIR SWA has been asking the CSIR to strengthen its interaction with SIDBI, CAPART and other user departments and agencies to improve the effectiveness of CSIR laboratories, particularly with respect to the supply of R&D based services to small-scale industries. It has sought from the CSIR to strengthen technological programme-based approach to the dissemination of emerging and new generic technologies to society as a whole. It has suggested that the CSIR needs to encourage consortia formation for the formulation of sectoral and regional area-based technological programmes across and within laboratories. It has asked the CSIR to institutionalize a transparent and participative, flexible, programme-based project culture. It has also recommended to the CSIR authorities to reinforce partnerships with consulting and engineering design organizations. It is in favour of encouraging a diverse set of users, including S&T based NGOs, to set things right in respect of the improvements needed in the utilization of capabilities. CSIR must undertake transparent and participative area/sector/technology field based exercises to select uncharted areas and emerging opportunities in the relevant fields. It has recommended that the CSIR needs to undertake an immediate assessment of human resource potential using a combination of expert assessment and self-assessment, so as to optimize resource utilization (CSIR SWA 2004).

HUMAN RESOURCE MANAGEMENT IN CSIR

From the late 1970s onwards in the formation of promotion policies, the staff and scientist representatives used to be officially involved as members of the committees. Long and concerted deliberations between CSIR management and the scientific community helped the promotion policy of CSIR to evolve in a way that it largely met the aspirations of the scientists and at the same time satisfied the human resource planning of the management. However, as the CSIR SWA has reported, there was no participation of the CSIR scientists in the formulation of the latest promotion policy, introduced in January 2001. The CSIR SWA has asked the authorities to recognize the difference between government organizations run on bureaucratic patterns and a research organization which must nurture creativity and innovation. In the new policy, the most contentious issue is the confidential report on the evaluation of scientists. This (CR) system was replaced in 1990 by Annual Performance Appraisal Report (APAR), which was 75 per cent transparent. Promotion was based on one's total score in three components, annual performance, peer review, and interview. It was fairly pragmatic and transparent and generated faith in the CSIR system.

However, the CSIR SWA points out that the Apex Committee's recommendations were thoroughly retrograde as it gave extraordinary importance to the confidential report. In its view this policy is in violation of the law of the land and against natural justice. CSIR has unilaterally introduced a personnel policy since January 2001, deviating from the APAR, by putting very high premium on Confidential Report (CR) in a scientist's career. And by deciding a system of extremely high thresholds (of 80 to 90 per cent) for eligibility to face the interview board for promotion to the next higher post, the SWA claims that this system denies scientists an opportunity to face the interview board. This is not only against natural justice, but promotes sycophancy and conformism and kills creativity and innovation. The CSIR SWA points out that science will thrive better in an environment that is open and trustworthy. It states that the CR system is a legacy of colonial bureaucracy and has bred distrust. The earlier MANAS had elements of open and transparent mechanisms that guided the performance evaluation of scientists. CR has become, or has come to be seen as, an instrument of oppression or punishment, rather than human resource 'development' (in any sense of the word).

The CSIR Scientific Workers' Association is of the view that the present R&D structure of the labs is highly centralized with near dictatorial powers for the directors, and no accountability. They may or may not be benevolent dictators or feudal landlords. It is left to the whims and fancies of the directors to restructure activities or remove divisions/departments within the labs. As a result, there is no design for acquiring leadership or managerial skills within the system nor are these skills evaluated at the assessment time. The SWA is of the view that though the director-general and directors of institutes have assumed the mantle of CEOs but continue to practise values that are alien to the culture of an R&D organization. The holders of top management positions, it says, have wielded unlimited powers without institutional mechanisms of accountability in the CSIR system. Institutions like Research Councils and the Governing Body, which used to perform some overseeing functions, have been reduced to legitimizing devices by the appointment of cronies and loyalists to the ruling party. The SWA points out that during the last decade, various participative mechanisms within the institutes, such as committees, were either scrapped or reduced to sycophantic clubs (CSIR SWA 2004).

In January 1999, the Associations of CSIR Scientists (CSIR-SWA) and the Federation of Employees Unions were unilaterally derecognized by the authorities. A democratic and three-decades-old time-tested process of dialogue was discontinued on the pretext of forming Service Associations as per the rules of 1993. The SWA leaders were harassed, the members were prohibited to hold meetings in institute premises, and the notice board facility was withdrawn. All sorts of communication channels between the management and associations were abruptly stopped.[8]

To date, tension remains between the scientists association and the CSIR management on the issue of its right to function as a trade union and to continue interventions in the organization and management of R&D. It is not clear whether the CSIR SWA will be able to rise to the occasion and succeed in making an impact on the processes of formation of future strategy and vision of CSIR.

TOWARDS A RECAPTURE OF THE HISTORY OF THE CSIR SYSTEM IN THE RADICALLY EVOLVING MOMENTS OF TENSION

The policy practice of planning for the domestic production of capital and basic goods, which transformed the face of Indian industry during the post-independent period, was an important influence on the priorities of the CSIR laboratories. In a number of industries, parts of these laboratories were virtually made to serve as in-house research and development laboratories. These parts acted for these industries as their main domestic suppliers of inputs. Of course, there were other parts in which S&T personnel worked without any user for collaboration in sight. Parts of CSIR laboratories were innovative and they do not fit the description of those analysts who have assumed that the CSIR system of laboratories was mainly geared to pursuing research either to obtain international scientific recognition or to develop facilities and programmes for the production of otherwise cheaply obtainable technological knowledge off the shelf from foreign companies.

However, since the early 1980s the New Economic Policy (NEP) advocates have been continuing to state in their pronunciations on the CSIR system that this system of laboratories failed in linking itself to the production system. As a result, the CSIR manpower should be subjected to competition within national and global markets to make them more effective. In my view, the NEP advocates are quite off the mark. One cannot generalize that in the pre-liberalization era as a whole these institutions were bereft of links with the productive system. Second, even in the pre-liberalization era, the autonomy of research institutions was limited in character. Evidence is in favour of the position that the autonomy had to be constantly negotiated by scientists with industry. In terms of role and direction, their mandates changed with changes in external demand. The use of these institutions by industry was often myopic, the networking practice followed was more akin to the supplier-based mode of innovation. Consequently, a large number of CSIR laboratories have actually been in need of an urgent renewal of their strategic capabilities.

The paradigms of R&D are changing in respect of further development of non-renewable energy, development and utilization of biomass and solar energy, regulation of use of materials and energy for the sake of environmental and health problems, development of systems for transport and information, and many other related aspects. As a result, the assumptions are also changing concerning what kind of technologies would be needed to transform the Indian economy in the direction of sustainable development. Due to these developments several kinds of deficits exist in the STI. The ongoing structural adjustment process will not be able to correct the systemic deficiencies of the Indian system of STI. The processes of economic liberalization are unlikely to take care of the deficits that have appeared on account of earlier policies.

In the absence of appropriate interventions, the post-liberalization processes will only accelerate the pace of subordination of CSIR laboratories to the private capital. The linkages that CSIR laboratories are today experiencing under the NEP capacity are forcing scientists to work much more for the transfer of knowledge than for the creation of technology. Today again parts of the CSIR laboratories are working far more intensely for the problems that are amenable to the consulting mode of links. Systemic mismatches are going to grow due to the fragmentation of the longer-term programmes that these institutions were pursuing. With the growing trend of contract research, the ethos of publicly funded institutions is changing fast. In the CSIR laboratories, the space for public interest research is on the decline. It seems that under the conditions of the NEP type of readjustment, the scientific and technological goals of people's oriented development will no longer work as guideposts. The deficits are getting aggravated; their repair is essential if the country is to make a transition to the path featuring both global competitiveness and sustainable development. The new and emerging requirements of technology detouring are unlikely to be taken care of by the business groups.

The nature of failure that the country is therefore witnessing is systemic; it cannot be corrected through the development of private sector driven public-private partnerships. The ongoing process of adjustment is imitative; it lacks local correspondence. It has been launched by the government without a correct assessment of the growing deficits and mismatches experienced by the publicly funded S&T component of the system. It is quite clear that there is no consensus as such on the vision and strategy of how the CSIR has to reinvent itself today in the new environment. The balance of forces is still tilted in favour of the school that wants the CSIR to integrate itself with the global S&T system. But there exists a considerable tension in the system over this paradigmatic shift. The way things are, integration has serious consequences for the nation as a whole. Its systemic impacts need to be studied carefully.

CONCLUSION

We have provided a historical account of the origins and institutionalization of the organization and management of research and development for industrial development in the CSIR system. The degree of success achieved is determined by the ability of its leadership to establish proper relationships with users for the generation and transfer of knowledge and technology. This suggests that the kind of relationships which the constituent laboratories were made to forge with the different types of users have had a contradictory impact on the capability building processes and the building of a participatory culture of planning, evaluation, and monitoring of R&D.

It has been argued that contrary to the widespread belief of the CSIR being largely bereft of linkages with industry, its laboratories are in many cases in-house R&D laboratories for state industry and big business groups. The suggestion is that combined with the determining influence over the evolution of research organization of the state sector and big business with whom the poorly managed relationships seem to have been responsible for undermining the implementation of a long-term outlook right from the formative moments of CSIR laboratories. Appropriate routines of planning, monitoring, and evaluation of research in and about the CSIR system were never devised.

Even the failure to establish open, transparent, accountable, collegiate, and democratic structures seem to have its roots in the bureaucratic culture of laboratory management that the leadership chose in the early years. When the CSIR was a small establishment, the damage this culture of bureaucratic management could do was not serious. It may be even argued that it had a legitimate function when in the early years the CSIR laboratories were being mobilized to contribute to national development. But it seems that the way the constituent laboratories of CSIR went on to recruit personnel and organize them to deliver has resulted in a culture of research organization that continues to encourage the laboratory management and scientist included to reproduce a bureaucratic order of 'authority formation'. CSIR management has developed a vested interest in the same management practice to such an extent that it is today rigid in sticking to the position of not allowing power to devolve; decision making remains largely in the hands of Director-General and Laboratory Directors, with whose sanction the rest of bureaucracy is functioning on a day-to-day basis. These practices have certainly had an adverse impact. It seems that even the minimal changes that the scientific community has been seeking in the organization and management of R&D are being today resisted by the management.

It is thus quite clear that the CSIR system has failed to develop not only the desired R&D management routines appropriate to the development stage of the country, but also the culture of democratic participation. This is significantly responsible for the lack of learning and correction. This factor makes the CSIR system vulnerable to the changes that are imposed on it without involving the scientific community at large.

Finally, as a way of providing an advance warning, as the impact of recent changes will be large and significant in terms of interests and orientation of knowledge development from the standpoint of society, the policy makers should not take this process further without due consideration of systemic impacts that the proposed paradigmatic change entails.

NOTES

1. See the observations made on the origins of CSIR in the saga of Indian science since Independence in a nutshell by Pushpa Mittra Bhargava and Chandana Chakravarti, 2003, pp. 29–30. The CSIR originated against the backdrop of the Second World War, but it was the efforts of the National Planning Committee to formulate the application of science and technology in the economy and the success of the Board of Scientific and Industrial Research (in whose research results the Indian industrialists showed a lot of interest) that speeded up the establishment. The Council was set up through the untiring efforts of A. Ramaswami Mudaliar and Shanti Swarup Bhatnagar. The latter was Director-General, Scientific and Industrial Research, on the Board of Scientific and Industrial Research, which had been set up through the efforts of A. Ramaswami Mudaliar in April 1940 to co-ordinate the work of the existing organizations, and to receive research proposals from various research institutions, universities and industries. The Board's effective functioning in the very first year led not only to a number of war-related problems being solved, but also prepared the background for the utilization of the laboratory results by industry. Consequently, in 1941, an Industrial Research Utilization Committee was set up, whose responsibility was to advise the government on the selection of industries that could utilize the results of research emanating from the implementation of the schemes of the BSIR. Against this backdrop, there

grew a demand for a concrete agency that would establish and maintain scientific and industrial research laboratories in the country, leading gradually to the establishment of the CSIR as a system of laboratories, which would cater to the needs considered to be of importance in the economy.

2. The CSIR system of laboratories was established to organize scientific and industrial research with funds from the colonial government. The policy makers had also chosen to set this organization up as an autonomous society to allow its constituent establishments the autonomy that they were required to have for operational reasons (in respect to co-ordination with industry and with government agencies). In August 1947, ten days after Independence, through the initiative of Prime Minister Jawaharlal Nehru, the constitution of CSIR underwent a small but significant development. The prime minister became the President of the CSIR; this is a privilege the CSIR continues to enjoy even today. In the early years, this arrangement, it was also called by some the Nehru–Bhatnagar effect, gave a leeway to the leadership to negotiate its space in the landscape of the emerging relationship between government and industry. However, it may be noted that the CSIR had to be ever ready to take up the immediate needs of the government and big business whenever called upon, as a part of the arrangements established by individual laboratories with the system as a whole.
3. Immediately after Independence, the policy makers had formed the First Review Committee of the CSIR system of laboratories, with the key representatives of big business as members, to organize the facilities of scientific and industrial research for national development. This committee laid stress on immediate efforts for transfer of research results to production organization. The committee was against the planning and establishment of more laboratories. Further expansion of the CSIR was put on hold for a short period. See Report of the First Reviewing Committee, CSIR, 1948.
4. As stated, the Council was set up as an autonomous body. It had the prime minister providing it an umbrella of security, with which it could bring changes in the research programmes and established new laboratories as and when necessary. No doubt it had a set of detractors, but its problems were compounded only when it showed serious weaknesses in organization and the management of R&D.
5. Prime Minister Jawaharlal Nehru confirms this in a statement made in the parliament in 1949 when he explains why the government had accorded national treatment to foreign companies. The government's expectation was that 'scientific and industrial knowledge and capital equipment can best be secured along with foreign capital' (Nehru in the parliament, April 1949). Given this expectation, even industries such as

oil and fertilizers, which were to be in the main, the preserve of the public sector, were kept open to multinationals. Foreign companies controlled 85 per cent of the trade in tea, 60 per cent in shoes, 70 per cent in soaps and 100 per cent in oil distribution, when we started in earnest our planning process in the 1950s. The framework of the efforts to utilize science and technology for the national 'development' was marked by this contradictory start.

6. During this sub-phase, India undertook financial and technical collaboration with both the West and the Soviet bloc. The signing of financial-cum-technical collaborations was tilted heavily in favour of turnkey technology transfers. All the early mining, power, steel, and chemical plants in this country which came up during the 1950s and 1960s were set up in this way. During this period, the first three Five-Year Plans concentrated on building domestic manufacturing capacity in capital goods.
7. Dinesh Abrol, 'Science and Culture Groups', in Deepak Kumar and R. Macleod, eds., *Technology and Raj*, Delhi, 1995.
8. It may be noted that CSIR-SWA has been the only association continuously representing the scientists, engineers, and some technical personnel for the past three decades and has been given due recognition and facilities by the CSIR and its constituent laboratories and institutes. Its office-bearers have been members of the official committees that framed the personnel policies such as NRAS (1981–8) MANAS (1988–92) Revised MANAS (1992–2000). It is also registered as a Trade Union since 1971.

BIBLIOGRAPHY

Abrol, Dinesh, *SPR-A Different Perspective*, Delhi, November 1985.

———, '*Recommendations for Consideration of the CSIR Review Committee*', Submitted to the Abid Hussein Committee from NISTADS, 1986.

———, '*Science and Culture Group: National Scientists or Colonized Minds*', in Deepak Kumar and R. Macleod, *Technology and Raj*, Delhi, November 1995.

Abrol Dinesh, Usha Menon, L. Pulamte and P.V.S. Kumar, *Evaluation of CSIR Rural Technologies*, Delhi, 1998.

Alam, Ghayur and John Langrish, *Government Research and its Utilization by Industry: The Case of Civilian Industrial Research in India, Research Policy*, vol. 13, no. 1, 1984.

Atma Ram, 'A Few Thoughts on Applied Science in India', Twenty-eight Acharya Jagdish Chandra Bose Memorial Lecture, Calcutta, 30 November 1966.

Bhargava, P.M and Chandana Chakrabarti, *The Saga of Indian Science Since Independence: In a Nutshell*, Hyderabad, 2003.

Bhatnagar, D.K and M. Rahman, *Linkages in CSIR Projects: A Study*, CSIR, 1985.

Business India, *CEO of CSIR INC*, 28 June to 11 July 1999, pp. 50–7.

Comptroller and Auditor General of India, *CAG, Report No. 5*, Delhi, 2003.

CSIR SWA, *Save CSIR: An Appeal from CSIR Scientific Workers Association*, Delhi, February 1988.

———, *Sordid Affairs*, Delhi, August 2004.

———, *Performance of CSIR under NDA*, Delhi, September 2004.

———, *Comments on the Kelkar Committee*, Delhi, February 2005.

CSIR, *Industrial Research Planning Committee Report*, Delhi, 1945.

———, *Report of the First Reviewing Committee (Chairman, Ardershir Dalal)*, Delhi, 1948.

———, *Report of the Second Reviewing Committee*, Delhi, 1954.

———, *Report of the Third Reviewing Committee (Chairman, A. Ramaswami Mudaliar)*, Delhi, 1964.

———, *Proposals for the Fourth Five Year Plan (vol. 1)*, Delhi, May 1964.

———, *Development of Policies and Perspectives of Science in India*, Delhi, April 1964.

———, *Report of the Committee of Enquiry (Vol. 1)*, Delhi, February 1970.

———, *Towards a New Administrative Culture in CSIR*, Delhi, June 1991.

———, *Strategy Perspectives for CSIR*, New Delhi, February 1992.

Dedijer, Steven, 'Industry and Research in India', Talk at the Shri Ram Institute, Delhi, 13 June 1964, p. 10.

Desai, Ashok, ed., *Technology Absorption in Indian Industry*, Delhi, 1988.

Government of India, *Scientific Policy Resolution*, No. 131/CF/57, March 1958.

———, *R&D Statistics*, Department of Science and Technology (DST), 2002.

Gupta, V.K., *How Indian Scientists Collaborate in Patenting*, Annals of Library and Information Studies, 2005.

Hussain, Abid et al., *Report of the CSIR Review Committee: Towards a New Perspective*, Delhi, December 1986.

Jolly, Vijay, 'CSIR: Profiting from R&D', in Sumanta Ghosh et al., *World Class in India: A Case Book of Companies in Transformation*, 2000, pp. 291–317.

Nayar, Baldev Raj, *India's Quest for Technological Self-Reliance*, Delhi, 1983.

NCST, *National S&T Plan; & Various Reports of the 24 Selected Sectors*, Department of Science and Technology, 1974.

Planning Commission, *Report of the Working Group on Scientific Research for the Third Five Year Plan*, Delhi, December 1959.

Rahman, A., *Anatomy of Science*, Delhi, 1972.

Contributors

DINESH ABROL is Scientist at the National Institute of Science, Technology and Development Studies, New Delhi. He is deeply involved with the science movement in India.

ARUN BANDOPADHYAY is Nurul Hasan Professor of History, University of Calcutta, and formerly Dean, Postgraduate Faculty of Arts, University of Calcutta. His interest of research varies from economic history, business history and history of science and environment.

RAJ SEKHAR BASU is Reader at the Department of History, University of Calcutta. His book entitled *Nandanar's Children: The Pariyar's Tryst with Destiny, Tamilnadu, 1850-1956* is forthcoming from Sage, New Delhi.

SANTANU CHACRAVERTI has been working on various themes in the history of science since 1991. He was admitted to Ph.D. from Jadavpur University in 1997 for his research on 'Ramendrasundar Trivedi and Bengal's Response to Modern Western Science.' His postdoctoral work includes *Subhankari—An Indigenous Tradition of Elementary Mathematical Instruction*, published from the Asiatic Society. His articles and monographs have been published in various journals and anthologies. His other interests include environment and human rights. He is presently working with DISHA, a Kolkata based civil society group engaged with environmental issues.

PRATIK CHAKRABARTI received his Ph.D. from the Centre for Historical Studies, JNU, New Delhi. He teaches History of Medicine at the University of Kent, UK. He is the author of *Western Science in Modern India: Metropolitan Methods, Colonial Practices*. His second book, *Materials and Medicine: Trade, Conquest, and Therapeutics in the Eighteenth Century* is forthcoming. He is now finishing his third book, on medical research in colonial India, 'Laboratories in the Empire: Medical Research in British India, 1890-1940'.

ROHAN D'SOUZA teaches at the centre for Science Policy, Jawaharlal Nehru University, New Delhi. He is the author of *Drowned and Dammed: Colonial Capitalism and Flood Control in Eastern India* (2006).

S. IRFAN HABIB has been with NISTADS, working in the area of history of science. He has co-authored a book with Dhruv Raina called *Domesticating Modern Science: A Social History of Science in Colonial India* (2004). He has edited two books titled *Situating the History of Science: Dialogues with Joseph Needham* (1999) and *A Social History of Science in Colonial India* (2007). He has also published a book *To Make the Deaf Hear: Ideology and Programme of Bhagat Singh and His Comrades* (2007). Its Hindi edition came out in September 2008 and the Tamil version was published in March 2009. S. Irfan Habib has recently joined the newly instituted Maulana Abul Kalam Azad Chair at the National University of Educational Planning and Administration, New Delhi.

DEEPAK KUMAR is Professor at Zakir Hussain Centre for Educational Studies, Jawaharlal Nehru University, New Delhi. He is the celebrated author of *Science and the Raj*.

MADHUMITA MAZUMDAR is Assistant Professor at the Dhirubhai Ambani Institute of Information and Communication Technology in Gandhinagar, Gujarat. Her research interests straddle the domains of the social history of science in Bengal and the cultural politics of nationalism in colonial and post-colonial India. She wrote her Ph.D. dissertation on the subject of 'Science and the Nationalist Agenda in Colonial Bengal, 1904-1947'. She is currently working on a cultural and political history of science communication in India.

SUJATA MUKHERJEE is Professor of History and Director, Centre for Gandhian Studies, Rabindra Bharati University. She works on history of science, technology and medicine, and on Gandhian studies.

DHRUV RAINA is Professor at the Zakir Hussain Centre for Educational Studies, Jawaharlal Nehru University, New Delhi. He is one of the most reputed writers on the history and Philosophy of Science.

MRIDULA RAMANNA retired as Reader and Head, Department of History, S.I.E.S. College, Mumbai. She has authored *Western Medicine and Public Health in Colonial Bombay, 1845-1895* (2002), contributed chapters to edited volumes, *The Encyclopedia of Pestilence, Pandemics and Plagues* (2008), *The Dictionary of Medical Biography* (2007), *History of Medicine in India: The Medical Encounter* (2005), *History and Gender: Some Explorations* (2005), *Colonialism as a Civilizing Mission: Cultural Ideology in British India* (2004), *The Spanish Influenza Pandemic of 1918-19, New Perspectives* (2003), *Disease and Medicine in India* (2001) besides research papers in *Economic & Political Weekly, Social Scientist, Indian Historical Review, Journal of the K.R. Cama Oriental Institute,* and *Indica.*

ARABINDA SAMANTA is Professor of History, Burdwan University. He has worked on malaria epidemic and is interested in health and environmental history.

UNIVERSITY OF CALCUTTA
Department of History, Monographs

JAYANTA KUMAR RAY, *Administrators in a Mixed Polity*

HARAPRASAD CHATTOPADHYAYA, *Internal Migration in India: A Case Study of Bengal*

HASI BANERJEE, *Political Activity of the Liberal Party of India*

BINAY BHUSAN CHOUDHURY, *The Tribal Protest Movements in Eastern India 1760-1922: Origin, Ideology and Organisation* (in Press)

ARUN BANDOPADHYAY, *The Agrarian Economy of Tamilnadu: 1820-1855*

SEKHAR BANDYOPADHYAY, *Caste, Politics and the Raj: Bengal 1872-1937*

HARAPRASAD CHATTOPADHYAYA, *From Marsh to Township East of Calcutta: A Tale of Salt Water Lake Township (Bidhan Nagar)*

JAHAR SEN, *Essays in Indo-Nepal Trade: A Nineteenth Century Study*

SURANJAN DAS AND SEKHAR BANDYOPADHYAYA (eds.), *Caste and Communal Politics in South Asia*

NIRBAN BASU, *The Working Class Movement: A Study of Jute Mills of Bengal 1937-47*

MUNTASSIR MAMOON AND JAYANTA KUMAR RAY, *Civil Society in Bangladesh: Resilience and Retreat*

HARAPRASAD CHATTOPADHYAYA, *Migration Between India and Nepal: A Sociocultural Study*

SURANJAN DAS AND JAYANTA KUMAR RAY, *The Goondas: Towards a Reconstruction of the Calcutta Underworld*

JAYANTA KUMAR RAY (ed.), *India-Nepal Cooperation Broadening Measures*

B.K. JAHANGIR AND JAYANTA KUMAR RAY (eds.), *India-Bangladesh Cooperation Broadening Measures*

HARI VASUDEVAN (ed.), *Commercialization and Agriculture in late Imperial Russia: Essays on Russian Economic History*

HASI BANERJEE, *Sarojini Naidu: The Traditional Feminist*

ARJAN DE HAAN AND SAMITA SEN (eds.), *A Case for Labour History: The Jute Industry in Eastern India*

SUBHO BASU AND SURANJAN DAS (eds.), *Electoral Politics in South Asia*

BHARATI RAY (ed.), *Women and Politics: France, India and Russia*

BHARATI RAY AND DAVID TAYLOR (eds.), *Politics and Identity in South Asia*

BASUDEB CHATTOPADHYAY, *Crime and Control in Early Colonial Bengal 1770-1860*

SURANJAN DAS, *Kashmir and Sindh: Nation-Building Ethnicity and Regional Politics in South Asia*

BHASKAR CHAKRABARTY (ed.), *Exploring Regional Security: South and Central Asia*

SURANJAN DAS AND P.K. BANDYOPADHYAY, *Food Movement of 1959: Documenting A Turning Point in the History of West Bengal*

SHANTANU CHAKRABARTY, *Cooperation in South Asia: The Indian Perspective*

RAJ SEKHAR BASU AND SANJUKTA DAS GUPTA (eds.), *Narratives of the Excluded: Caste Issues in Colonial India*

S. MASWOOD, A. DEY AND R. BISWAS (eds.), *Between Tradition and Modernity Aspects of Islam in South Asia*